Impact Studies

Series Editor: Christian Koeberl

Springer
Berlin
Heidelberg
New York
Barcelona
Hong Kong
London
Milan
Paris
Tokyo

Jüri Plado Lauri J. Pesonen (Eds.)

Impacts in Precambrian Shields

with 98 Figures and 43 Tables

Springer

DR. JÜRI PLADO
Institute of Geology
University of Tartu
Vanemuise 46
51014 Tartu
Estonia

PROFESSOR LAURI J. PESONEN
Division of Geophysics
University of Helsinki
Gustaf Hällströmin katu 2
00014 Helsinki
Finland

ISBN 3-540-43517-4 Springer-Verlag Berlin Heidelberg New York

Library of Congress Cataloging-in-Publication Data
Impacts in Precambrian shields / Jüri Plado ; Lauri J. Pesonen (eds.).
p.cm. -- (Impact studies)
Includes bibliographical references.
ISBN 3-540-435174 (alk. paper)
1. Geology, Stratigraphic—Precambrian. 2. Cratering. 3. Meteorites. I. Plado, Jüri, 1969-
II. Pesonen, Lauri J. III. Series.

Springer-Verlag Berlin Heidelberg New York
a member of BertelsmannSpringer Science+Business Media GmbH
http://www.springer.de

Printed in Germany

Camera ready by editors
Cover design: E. Kirchner, Heidelberg

Printed on acid-free paper SPIN 10868654 32/3130/as 5 4 3 2 1 0

Preface

The present volume is the result of activities within the scientific programme "Response of the Earth System to Impact Processes" (IMPACT) of the European Science Foundation (ESF). The ESF is an association of 67 national member organisations devoted to scientific research in 24 European countries. The IMPACT programme is aimed at understanding meteorite impact processes and their effects on the Earth System. Launched in 1998 for a duration of 5 years, the programme is now supported by 15 ESF membership countries. The programme deals with all aspects of meteorite impact research and operates through workshops, exchange programs, and short courses.

The 4th IMPACT programme workshop *"Meteorite Impacts in Precambrian Shields"* took place on May 24-28, 2000, in Lappajärvi, western Finland. A total of 84 scientists from 19 countries from Europe, North America, and Africa participated in the workshop. During the workshop, 43 oral, 31 poster, and several video presentations were made. An exhibition of impactite rocks from Finnish craters and two excursions were organised. The excursion to impact melt rock outcrops of the Lappajärvi structure took place during the workshop. The Karikkoselkä and Sääksjärvi impact structures in south-central Finland were visited during the post-meeting excursion.

Scientifically, the workshop focussed on the characterization of meteorite impact structures of all ages and sizes occurring in the Precambrian shields, stable parts of the continents composed of Precambrian rocks with little or no sediment cover. Precambrian shields, together with platforms (collectively referred to as cratons), represent areas where most of the impact structures on Earth have been found (Fig. 1). Especially rich in impact structures are the mid-latitude shields, and those where intensive research programmes have been carried out. The Precambrian shields provide opportunities for a wide range of scientific studies of impact structures, because of good exposure and a generally homogeneous target.

The workshop consisted of nine scientific sessions on various topics of impact cratering, including the global perspective, structural aspects, geophysical and geochemical signatures, dating, shock effects, and impacts into the sea. A special session was devoted to the Popigai impact structure in Siberia. This volume, "*Impacts in Precambrian Shields*", contains fourteen papers, most of which result from the workshop. All the manuscripts were reviewed by at least two referees and were selected on the basis of originality and scientific value.

The volume begins with two introductory contributions, which present an overview of meteorite impact structures in two prominent Precambrian shields: the Fennoscandian Shield and the Canadian Shield. Abels et al. give a detailed overview of thirty confirmed impact structures in Fennoscandia, and discuss possible features that await demonstration of their meteoritic origin. They compare the crater ages of the Canadian and Fennoscandian shields with the global record and show that the peak of early Paleozoic craters is restricted to these regions due to similarities in their geological history. However, an enhanced accretion rate of meteorite impacts for these times cannot be excluded.

Dence lists data of eleven Canadian impact craters in Precambrian rocks, focusing on the Brent structure, Ordovician of age. This structure is well understood due to several drillcores and extensive studies. Therefore, it provides an excellent example for the study of the main structural elements of simple impact craters. Particular attention is given to mineralogical signatures that reflect the shock pressure attenuation with depth. Also, crater morphometry, structure, geophysical signature, and the distribution of pseudotachylites and allochthonous rocks are discussed in terms of varying crater diameter.

A group of papers on the 100 km wide Popigai impact structure in the Anabar Shield of Siberia follows. Masaitis introduces the geological aspects of this Late Eocene (35.7 Ma) crater. Pilkington et al show that geophysically Popigai is characterized by a large negative gravity anomaly (35 mGal) and an extensive magnetic low. Two-dimensional gravity and magnetic models are created by geophysical fields and petrophysical data, collected partly during the International Popigai Expedition in 1999. Armstrong et al. describe their first results on dating of four zircon grains from Popigai impact melt rocks. They conclude that "group B" zircons (isometric grains without any traces of melting along the grain boundaries) have neither crystallized from the impact melt, nor suffered any shock-induced Pb-loss.

Naumov describes the general scheme, features and mineral assemblages of impact-generated hydrothermal systems, based mainly on studies of three Russian craters, Popigai, Kara, and Puchezh-Katunki. He concludes that hydrothermal solutions are weakly alkaline and have higher activities of silica creating favourable conditions for formation of Fe-smectites and zeolites.

Gurov et al. present geological and petrographical features of the 165 Ma old and extensively eroded Zapadnaya impact structure on the Ukrainian Shield. As many others, this structure hosts high-pressure mineral phases in impactites, such as coesite and impact diamonds. The diamonds are composed of cryptocrystalline aggregates of hexagonal and cubic phases.

Recently intensified research activities have resulted in three papers about the Bosumtwi crater in Ghana. This 10.5 km wide, well-preserved structure formed in the Precambrian shield of the West African Craton about 1.07 Ma ago. Wagner et al. analyse the Landsat Thematic Mapper and ERS radar data in order to constrain the surface and subsurface geology. These remote sensing data reveal a detailed picture of the crater rim, including arrays of radial and concentric fracture zones. Two concentric structural features at distances of about 10 - 11 and 17 -19 km from the lake centre are established. Boamah and Koeberl report geochemical features from around Bosumtwi to find signatures of impact in the lateritic soils of the tropical rainforest environment. They compare the geochemical data of soil samples with the geochemical signals (such as the K-content) obtained by airborne radiometric mapping of the structure. Compared to average upper continental crustal composition, the Bosumtwi soils are more siliceous and ferruginous and dominated by quartz, Fe- and Al- oxides, and kaolinite. Concentric areas with a high radiometric signal of potassium reflect possible K mobilization due to the impact event. Artemieva has mathematically modelled a Bosumtwi-sized crater to simulate the formation of impact melt and tektites. Various impact angles and

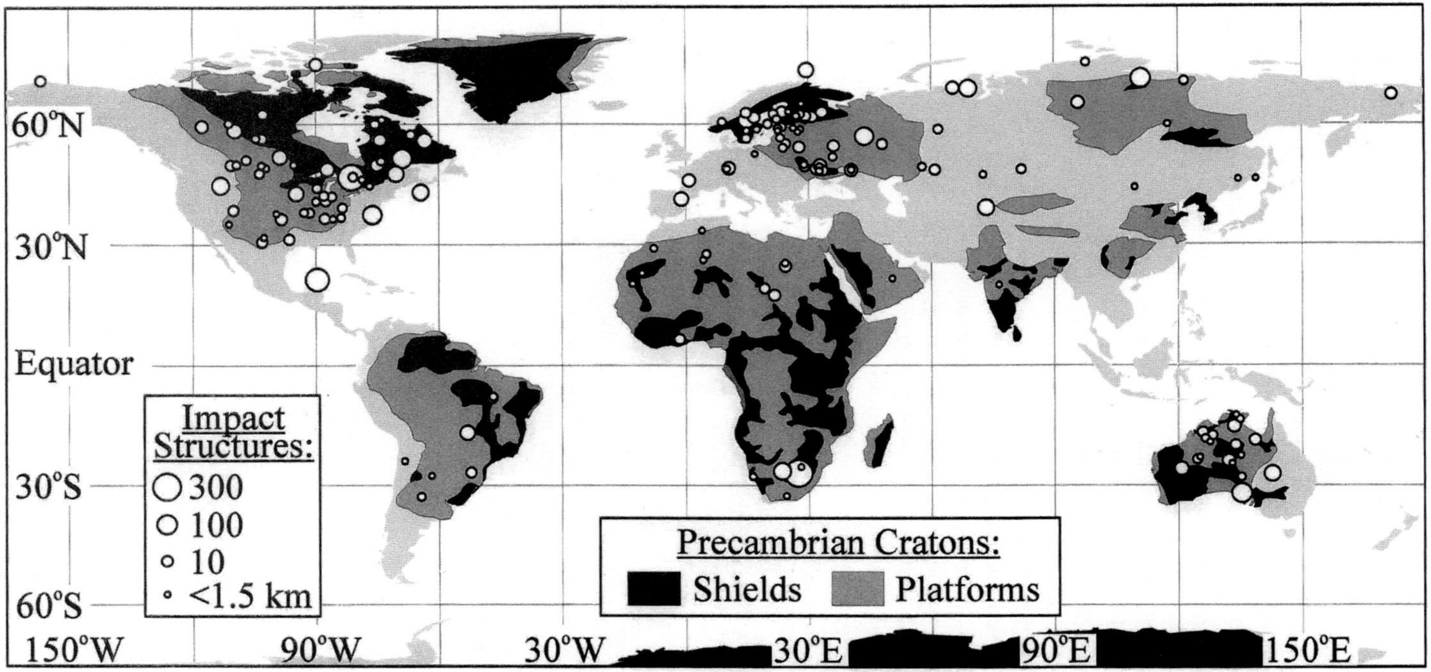

Fig. 1. The global map of proven meteorite impact structures in Precambrian shields (dark grey) and platforms (light grey).

projectile velocities are included in the models. It is found that the amount of possible tektite melt increases with low impact angle and increasing projectile velocity.

The three following papers describe impact structures found close to the Fennoscandian Shield. S. Suuroja et al. correlate the previous seismic reflection data (obtained before the discovery of the Neugrund structure, Gulf of Finland) with the present knowledge of this ~540 Ma old, shallow marine impact structure. A magnetic model is also presented to describe the interior structure of Neugrund. K. Suuroja outlines natural resources related to the Kärdla impact crater on Hiiumaa Island, Estonia. Except for the crystalline building stone found in the uplifted rim wall, all other resources formed after the impact but are clearly related to the morphological features caused by the impact event. Tsikalas et al. use marine seismic reflection data to correlate deposits related to the 40 km wide Mjølnir structure in the Barents Sea. On the basis of considerable thickening of the early post-impact deposits in the annular trough, a biostratigraphic age of 142 Ma is proposed to represent the minimum impact age.

Shuvalov models the formation of impact craters produced by vertical impact of asteroids and comets into the ocean. The size of a marine crater depends strongly on the water depth, and no crater forms at all if the depth of the seawater layer exceeds four times the diameter of the projectile.

Acknowledgements

The editors would like to thank the ESF IMPACT programme and the Finnish Academy for providing financial support for organizing the workshop. We thank all those who contributed to this volume by submitting their manuscripts and, in particular, those who devoted their time and effort to carry out reviews of the papers. Our colleagues from the University of Tartu, the University of Helsinki, and the Geological Survey of Finland are acknowledged for their support. We thank Christian Koeberl and Wolf-Uwe Reimold for revising the volume.

January 2002

Jüri Plado
Institute of Geology
University of Tartu
Estonia
(jplado@ut.ee)

Lauri J. Pesonen
Department of Geophysics
University of Helsinki
Finland
(lauri.pesonen@helsinki.fi)

Contents

List of Contributors

Abels, A.
Institute for Planetology, University of Münster
D-48149 Münster, Germany
abels@uni-muenster.de

All, T.
Geological Survey of Estonia
12618 Tallinn, Estonia
t.all@egk.ee

Armstrong, R.
Research School of Earth Sciences, Australian National University
Canberra 0200, A.C.T., Australia
prise.armstrong@anu.edu.au

Artemieva, N.
Institute for Dynamics of Geospheres
Leninsky Prospect 38-6, 117334 Moscow, Russia
nata_art@mtu-net.ru

Boamah, D.
Institute of Geochemistry, University of Vienna
Althanstrasse 14, A-1090 Vienna, Austria
dboamah@hotmail.com

Brandt, D.
Impact Cratering Research Group, Department of Geology,
University of the Witwatersrand
Private Bag 3, P.O. Wits 2050, Johannesburg, South Africa

Dence, M.R.
824 Nesbitt Place, Ottawa ON, K2C 0K1, Canada
mrdence@rsc.ca

Dypvik, H.
Department of Geology, University of Oslo
P.O. Box 1047 Blindern, N-0316 Oslo. Norway
henning.dypvik@geologi.uio.no

Eldholm, O.
Department of Geology, University of Oslo
P.O. Box 1047 Blindern, N-0316 Oslo, Norway

Faleide, J.I.
Department of Geology, University of Oslo
P.O. Box 1047 Blindern, N-0316 Oslo, Norway

Grieve, R.A.F.
Geological survey of Canada, Earth Sciences Sector
601 Booth Street, Ottawa, ON, K1A 0E8, Canada
rgrieve@gsc.nrcan.ca

Gurov, E.P.
Institute of Geological Sciences, National Academy of Sciences of Ukraine
55 b O. Gonchar Street, 01054 Kiev, Ukraine

Gurova, E.P.
Institute of Geological Sciences, National Academy of Sciences of Ukraine
55 b O. Gonchar Street, 01054 Kiev, Ukraine

Koeberl, C.
Institute of Geochemistry, University of Vienna
Althanstrasse 14, A-1090 Vienna, Austria
christian.koeberl@univie.ac.at

Lehtinen, M.
Finnish Museum of Natural History, University of Helsinki
FIN-00014 Helsinki, Finland
martti.lehtinen@helsinki.fi

Masaitis, V.L.
Karpinsky All-Russia Geological Research Institute (VSEGEI)
Sredny Prospect 74, 199106 St. Petersburg, Russia
vcmsts@mail.wplus.net

Naumov, M.V.
Karpinsky All-Russia Geological Research Institute (VSEGEI)
Sredny Prospect 74, 199106 St. Petersburg, Russia
mvn@mail.wplus.net

Pesonen, L.J.
Department of Geophysics, University of Helsinki
FIN-00014 Helsinki, Finland
Lauri.Pesonen@helsinki.fi

Pilkington, M.
Geological Survey of Canada
615 Booth Street, Ottawa, ON, K1A 0E9, Canada
mpilking@nrcan.gc.ca

Plado, J.
Institute of Geology, University of Tartu
Vanemuise 46, 51014 Tartu, Estonia
jplado@ut.ee

Reimold, W.-U.
Impact Cratering Research Group, Department of Geology,
University of the Witwatersrand
Private Bag 3, P.O. Wits 2050, Johannesburg, South Africa
065wur@cosmos.wits.ac.za

Shuvalov, V.V.
Institute for Dynamics of Geospheres
Leninsky Prospect 38-6, 117979 Moscow, Russia
valery@vshuvalov.mccme.ru

Sokur, T.M.
Institute of Geological Sciences, National Academy of Sciences of Ukraine
55 b O. Gonchar Street, 01054 Kiev, Ukraine

Suuroja, K.
Geological Survey of Estonia
12618 Tallinn, Estonia
k.suuroja@egk.ee

Suuroja, S.
Geological Survey of Estonia
12618 Tallinn, Estonia
s.suuroja@egk.ee

Tsikalas, F.
Department of Geology, University of Oslo
P.O. Box 1047 Blindern, N-0316 Oslo, Norway
filippos.tsikalas@geologi.uio.no

Vishnevsky, S.
Institute of Mineralogy and Petrology
3 Universitetsky Prospect, 630090 Novosibirsk-90, Russia

Wagner, R.
Anglo Technical Division
45 Main Street, Marshalltown 2107, Johannesburg, South Africa
RWagner@mhs51.tns.co.za

The Impact Cratering Record of Fennoscandia - A Close Look at the Database

Andreas Abels[1], Jüri Plado[2,3], Lauri J. Pesonen[2,4], and Martti Lehtinen[5]

[1]Institut für Planetologie, Universität Münster, D-48149 Münster, Germany.
(abels@uni-muenster.de)
[2]Geological Survey of Finland, FIN-02151 Espoo, Finland.
[3]Institute of Geology, University of Tartu, 51014 Tartu, Estonia.
[4]Department of Geophysics, University of Helsinki, FIN-00014 Helsinki, Finland.
[5]Finnish Museum of Natural History, University of Helsinki, FIN-00141 Helsinki, Finland.

Abstract. Currently thirty meteorite impact structures are identified in Fennoscandia, which corresponds to ~19 % of the known global record. Eleven of them have been discovered or confirmed during the last decade. This high number is due to intensive, determined research, including deep drilling, and high-resolution geophysical regional mapping. The ages and diameters of Fennoscandian impact structures vary considerably, but many are relatively small (<10 km) and mostly of early Paleozoic age. The latter is probably an effect of the regional geological evolution. The majority of craters were formed in complex targets composed of crystalline basement, a sedimentary cover and frequently a water pile on top. Only those that were excavated in the crystalline shield contain bodies of coherent impact melt rock or reveal indications of such. Distal impact-related deposits of some structures have locally survived, especially those that formed in platform areas with virtually uninterrupted sedimentation after impact. These deposits are useful stratigraphic markers. Apart from the confirmed structures, there are presently some sixty additional structures for which an impact origin has been suggested.

1 Introduction

Over the last decade a considerable amount of information has been accumulated on the impact cratering record of Fennoscandia. Henkel and Pesonen (1992), Wickman (1992), Puura et al. (1994) and Pesonen (1996) have presented recent accounts of this record. Wickman (1988) presented a summary of suspected and confirmed impact structures in Sweden, Pipping and Puura (1996) of confirmed sites surrounding the Gulf of Finland, and Gailius (1988) of those in Lithuania.

Masaitis (1999a) reviewed the impact structures of the Baltic states and the Fennoscandian parts of Russia and Belarus.

The present contribution is basically intended to review the most important developments in impact cratering of Fennoscandia over the last decade by presenting synoptic views of certain aspects of the impact structure record. Furthermore, we outline the historical development in this field and introduce some new ideas and suggestions for future research. Finally, we present some speculative examples of geological structures and lithologies that resemble or are at least similar to those expected for deeply eroded impact structures and associated distal ejecta deposits. The past has shown that at a first glance unrelated knowledge has been very crucial for new discoveries related to hypervelocity impact. Therefore, specialists from the classic disciplines, such as mineral exploration, sedimentology, geochemistry, geophysics or palaeontology, are a very welcome readership.

Since 1990 eleven new impact structures have been discovered or confirmed on the Fennoscandian Shield and its fringe (Table 1). For a number of craters it could be shown that they have formed in a former shallow sea (Ormö and Lindström 2000, and references therein). Dating of impact structures by means of microfossils from the post-impact crater fill has reached a precision comparable with radiometric methods (e.g., Grahn et al. 1996). The effectiveness of high-resolution geophysics has been demonstrated for the detection of strongly degraded impact structures in the crystalline shield that lack indicative surface exposures (Elo et al. 1993; Pesonen 1995; Pesonen et al. 1999a). Furthermore, the finding of impact diamonds at the Lappajärvi crater in Finland (Masaitis et al. 1998a; Langenhorst et al. 1999) and discoveries of submarine impact structures, Mjølnir in the Barents Sea (Gudlaugsson 1993; Dypvik et al. 1996) and Neugrund near the Estonian coast (Suuroja and Suuroja 2000), are outstanding new entries in the list of impact features in Fennoscandia.

In this review the concept of Fennoscandia refers to the region shown in Fig. 1, which includes the Mjølnir crater in the Barents Sea, and in Fig. 2, based on tectonic criteria. It comprises the known impact structures on the territories of Norway, Sweden, Finland, Estonia, Latvia, Lithuania, Belarus and parts of northwestern Russia (for coordinates of the structures see Table 1). This usage deviates from the physico-geographical definition of Fennoscandia introduced by W. Ramsay in 1898, which largely excluded regions with Paleozoic cover of the northwestern East European Platform and the Barents Sea (cf., Donner 1996).

2 Research History

The main crater of the Kaali strewn field on Saaremaa Island, Estonia, was the first impact structure scientifically investigated in northern Europe. Already in 1827 J.W. von Luce described uplift and fracturing of the rim and suggested that a "unique explosion of underground fire" formed the crater. In 1919 J. Kalkun (later

Kaljuvee) was the first to propose an extraterrestrial origin for Kaali (Raukas and Tiirmaa 1997). Various follow-up studies corroborated this interpretation, but it was Reinwald (1937) who succeeded in proving the impact origin through the discovery of meteorite fragments in two of the craters. Other structures that already attracted the attention of geologists in the 19th century and that were later ascribed to impacts are the Siljan ring (Wickman 1988), the Lumparn bay (De Geer 1880) and the lake Hummeln (Hisinger 1826).

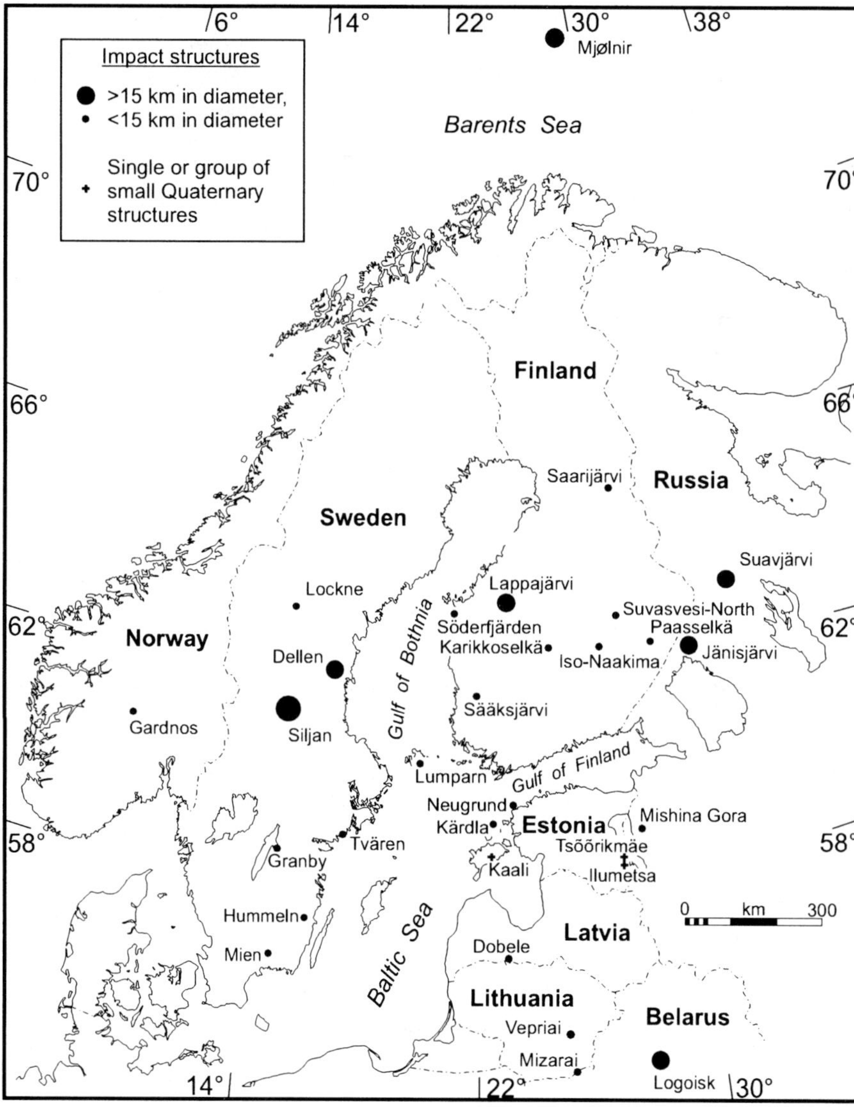

Fig. 1. Map of impact structures in Fennoscandia. See text for explanation.

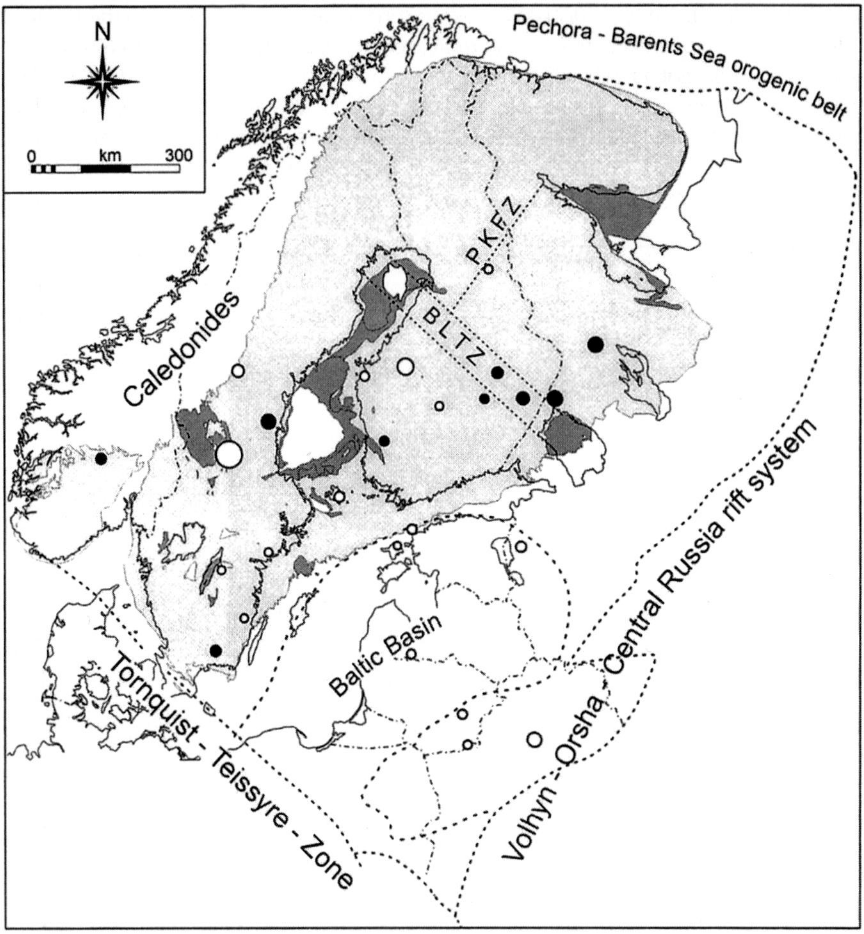

Fig. 2. Sketch map showing boundaries of Fennoscandia (after Bogdanova et al. 1996). Also shown is the outline of the crystalline shield area (light grey) with Mesoproterozoic graben structures in the exposed shield (dark grey) and Vendian-Early Paleozoic outliers (white). Not shown is crystalline rock exposed west of the present limit of Caledonian allochthon. White/black dots are known impact structures with/without Vendian-Phanerozoic sedimentary rocks in the craters (enlarged). *BLTZ*, Bothnian Bay-Ladoga Tectonic Zone; *PKFZ*, Puolanka-Kandalakša Fault Zone. Outline of Baltic Basin with present erosional limits after Poprawa et al. (1999).

Furthermore, the lakes Dellen (Svenonius 1888), Mien (Holst 1890), and Lappajärvi (Holmberg 1858) were recognized in the 19th century because of unusual "volcanic" rocks exposed on their isthmus and central islands, respectively.

At a meeting of the Geological Society of Stockholm in 1910 A.G. Högbom carefully suggested, while reporting on Meteor crater in Arizona, that the Mien and Dellen lakes could be impact structures, because of their peculiar melt rocks (Högbom 1910). Unfortunately, according to Wickman (1988), this flash of inspiration was not taken serious or went unrecognized for some 50 years. For instance, Eskola (1921) still proposed a volcanic origin for Mien, Dellen, Lappajärvi and Jänisjärvi in Russian Karelia, although he, too, noted the strange composition of the melt rocks requiring a massive contribution of country rock.

A similar fate was suffered by the Estonian astronomer E.J. Öpik, who already in 1916 noted the similarity between impact of high-velocity meteoroids and explosions and concluded that this is the reason why impact craters become circular for most angles of incidence. At least in the English-speaking community this pioneer work became recognized only much later as a significant contribution to cratering research, because it was published in Russian with a French summary in a little circulated journal (Öpik 1916).

Later, some unusual breccia occurrences (Lumparn - 1926, Neugrund - 1927, Lockne - 1940, Gardnos - 1945) and morphogeological anomalies (Tvären - 1931, Mishina Gora - ~1935, Saarijärvi - 1937) were either found or geologically investigated for the first time. However, an impact origin had not been considered or, in the case of the Norwegian Gardnos breccias, had been rejected (Broch 1945). Exceptions are the small Quaternary Ilumetsa craters in Estonia. With the experience of studies at the Kaali crater field, A.-C. Luha favoured an impact origin in 1938, the year he discovered the structures (Aaloe 1960).

With the beginning of the 1960s the paradigm in Fennoscandia shifted drastically, in particular due to the engagement of the Swedish geologists F.E. Wickman and N.-B. Svensson. In the wake of the establishment of diagnostic shock metamorphic alterations in minerals (1960-1965), Lappajärvi, Mien and Dellen were reinvestigated and their impact origin proven (Svensson and Wickman 1965; Svensson 1968a, b, 1969; von Engelhardt and Stöffler 1965; Carstens 1975; Lehtinen 1970, 1976). In addition, some new candidate structures were introduced, such as Tvären, Hummeln and Siljan (Fredriksson and Wickman 1963); the impact origin of the latter was soon confirmed on the basis of shocked quartz (Svensson 1971) and shatter cones (Svensson 1973). The impact origin of lake Sääksjärvi was recognized through the discovery of shock features in melt-bearing breccia boulders known to originate from the lake (Papunen 1969). A re-evaluation of Eskola's (1921) melt rock analyses from Jänisjärvi and map-based similarities to then confirmed craters prompted Dence (1971; written communication 2000) to propose an impact origin for this structure as well. Final proof of shock metamorphism was soon presented by Masaitis (1973).

After this pioneering work during the stimulating 'lunar era', the role of geophysics and drilling became more and more important for the discovery and confirmation of new crater structures, as most of them were found buried under sedimentary rocks. This included Kärdla, Granby, Dobele, Logoisk, Vepriai and Mizarai. Geologists became aware of the partly buried Söderfjärden crater at the Finnish west coast through early Landsat images (Talvitie et al. 1975). Lumparn owes its first impact proposal to the palaeontologist-stratigrapher Merrill (1979), who studied microfossils from a limestone deposit in the bay.

Table 1. General characteristics of impact structures in Fennoscandia

Name	(Country)	No.	Coordinates lat. N	long. E	Diameter (km)	Morph.	Imp.	Target	PL	Age (Ma, Method)	Pro.	History
Siljan	S	6	61°05'	15°00'	a: 45 b: 60 c: 52	c, e	bd, lb, sc*	sed (Or+Si+De? [~500]) cry	7	368 ±1 (Ar/Ar), ~345 (K/Ar)	-	a: ~1830 b: 1963 c: 1971
Mjølnir	N	67	73°48'	29°40'	a: 40 b: - c: (≈a)	c, m	ej	wat [200-400] sed (Pe+Tr+Ju)	1	149 - 141 (str)	ir	a: 1993 b: 1993 c: 1996
Dellen	S	5	61°50'	16°45'	a: 19 b: 22 c: 20	c, tec, l (r, u)	m, mb, lb	cry	5	89.0 ±2.7 (Rb/Sr)	-	a: 1888 b: 1910 c: 1968
Lappajärvi	F	31	63°10'	23°40'	a: 17 b: 35 c: 20	c, l (r, u), sc	m, mb, lb	sed (lMP [≥18]+Ca+Or) cry	5	77.3 ±0.4 (Ar/Ar), 73.3 ±5.3 (U/Pb)	ch	a: 1858 b: ~1955 c: 1967
Suavjärvi	R	68	63°07'	33°23'	a: 16 b: - c: -	c, e	lb, mb?	sed? cry	6	~2400 (str)	-	a: 1983 b: 1983 c: 1986
Jänisjärvi	R	36	61°58'	30°55'	a: 16 b: 18 c: 18	c, l (r, u)	m, mb, lb, sc	sed (INP?) cry	5	698 ±22 (Ar/Ar) 700 ±5 (K/Ar)	-	a: 1921 b: 1971 c: 1973
Logoisk	B	90	54°12'	27°48'	a: 8 b: 17 c: 20	c, bu	ml, mb, lb, ej	sed (lNP [220-240]+De[90] +Cr [30]) cry	2	42.3 ±1.1 (K/Ar), bstr	-	a: - b: 1979 c: 1979
Lockne	S	48	63°00'	14°49'	a: 7.5 b: 24 c: (≈a)	c, e	ej, lb, bd, mf, sc*	wat [≥200] sed (Ca+Or [~81]) cry	4	~455 (bstr)	-	a: 1940 b: 1988 c: 1990
Paasselkä	F	34	62°09'	29°24'	a: 10 × 11 b: - c: -	c?, l (r)	lb	sed cry	6	?	-	a: 1984 b: 1992 c: 1999

Table 1. (cont.)

Lumparn	F	37	60°09'	20°08'	a: 9 b: 10 c: 9	c, tec, b (r)	lb, mb, bd, sc?	(wat + sed?) cry	5	~460 - 440? (str)	-	a: 1926 b: 1979 c: 1992
Vepriai	Li	64	55°05'	24°34'	a: 7.5 b: - c: 8	c?, bu	lb, mb, sc	sed (lNP+Ca+Or+Si+De [>525])	2	155 – 165 (bstr)	-	a: - b: 1978 c: 1978
Neugrund	E	69	59°20'	23°31'	a: 7 b: 20 c: (≈a)	c, m	lb, bd	wat [50] sed (lNP+Ca[100]) cry	1	~540 (str)	-	a: 1927 b: 1996 c: 1998
Söderfjärden	F	32	62°41'	21°35'	a: 6.6 b: 11 c: (≈a)	c, bu (r)	lb, bd	cry	3	540 – 520 (bstr)	-	a: 1973 b: 1978 c: 1978
Mien	S	20	56°25'	14°52'	a: 6.5 b: 17 c: 9	c, l (r, u)	m, mb, lb	cry	5	121 ±2.3 (Ar/Ar)	-	a: 1886 b: 1910 c: 1965
Mizarai	Li	63	54°01'	23°54'	a: 5 b: - c: -	c, bu	m, lb, bd, sc*	sed (lNP+Ca) cry	5	520 – 480 (str, bstr)	-	a: 1978 b: 1978 c: 1988
Gardnos	N	42	60°40'	09°00'	a: 4.7 b: - c: 5	c, e	lb, mb	sed (lNP) cry	6	700 – 650 (str)	ch	a: 1945 b: 1990 c: 1990
Sääksjärvi	F	33	61°25'	22°23'	a: 4.5 b: 5 c: 5	s?, l (r)	ml, m?, lb, mb	cry	5	514 ±12 (Ar/Ar)	ir?	a: 1969 b: 1969 c: 1969
Mishina Gora	R	66	58°43'	28°03'	a: 2.5 × 4.0 b: 10 c: -	c?, e	lb, mb, sc	wat? sed (lNP+Ca+Or+De [460-530]) cry	4	350 – 250 (str)	-	a: ~1935 b: 1974 c: 1980
Dobele	La	65	56°35'	23°15'	a: 4.5 b: - c: -	c, bu	lb, sc	sed (Si+De+Cb)	3	240 – 170 (bstr)	-	a: - b: 1978 c: 1978

Table 1. (cont.)

Name	(Country)	No.	Coordinates lat. N	long. E	Diameter (km)	Morph.	Imp.	Target	PL	Age (Ma, Method)	Pro.	History
Suvasvesi N	F	35	62°41'	28°11'	a: 4 b: - c: -	c, tec?, l (r)	lb, mb	cry	5	560-530 *or* 260-230 (pmag)	-	a: 1986 b: 1986 c: 1993
Kärdla	E	54	58°59'	22°40'	a: 4 b: 15 c: (≈a)	c, bu	mb, lb, bd, ej	wat [<100] sed (lNP?+Ca+Or [~140])	1	~456 (bstr)	-	a: 1967 b: - c: 1980
Granby*	S	25	58°26'	14°56'	a: 2.9 × 3.5 b: - c: (≈a)	s, bu	lb	wat (25-75) sed (Ca+Or [~120]) cry	1?	~465 (bstr)	-	a: - b: 1976 c: -
Iso-Naakkima	F	61	62°11'	27°09'	a: 2 b: 3 c: < 4	s, tec?, bu	lb, bd	sed? cry	6	1200 ±100 (pmag, str, bstr)	-	a: 1989 b: - c: 1993
Tvären	S	12	58°46'	17°25'	a: 2 b: 4 c: 4	s, b (r)	lb, bd, mf	wat (75-100) sed (Ca+Or [25-80]) cry	5	~457 (bstr)	-	a: 1931 b: 1963 c: 1991
Saarijärvi	F	70	65°17'	28°23'	a: 1.5 b: 2.5 c: 3.4	s?, l (r, u?)	lb, sc*	(sed + wat)? cry	6	> 600 (bstr, str)	-	a: 1937 b: 1992 c: 1997
Karikkoselkä	F	71	62°15'	25°15'	a: 1.3 b: 2.4 c: 3	s, l (r)	lb, sc*	wat sed (Ca+Or) cry	5	450 – 440? (bstr)	-	a: 1994 b: 1994 c: 1996
Hummeln*	S	18	57°22'	16°19'	a: 1.2 b: 2 c: 1.8	s, l (r)	lb	wat (25-75) sed (Ca+Or [50-160]) cry	5	~470 - 465 (bstr)	-	a: 1937 b: 1961 c: -
Kaali	E	38	57°24'	22°40'	a: 0.11	s, l, pc, f (9)	pr, ej, lb, sc	sed (Q+Si)	1	0.00039? (C^{14}, hist?)	ir (oh)	a: 1827 b: 1919 c: 1937

Table 1. (cont.)

Ilumetsa*	E	52	57°57'	27°24'	a: 0.08	s, pc, f (2, 5?)	lb, ej?	sed (Q+De)	1?	0.0066 (C^{14}, str)	ch?	a: 1938 b: 1938 c: -
Tsõõrikmäe*	E	53	58°04'	27°27'	a: 0.04	s, pc	lb	sed (Q)	4?	0.0095 (C^{14})	-	a: 1968 b: 1984 c: -

Name: *, no shock features or projectile material documented, but in this paper accepted as impact structure
Country: *B*, Belarus; *E*, Estonia; *F*, Finland; *La*, Latvia; *Li*, Lithuania; *N*, Norway; *R*, Russia; *S*; Sweden
No.: numbering; continued from Henkel and Pesonen 1992, Pesonen 1996
Diameter: *a*, craterform (as far as preserved average rim-to-rim or edge-to-edge diameter [see text], for buried structures steepest gradient in geophysics); *b*, impact-induced structural disturbance; *c*, estimate of original rim-crest diameter
Morph.: morphology (type of crater, conditions of exposure): *c*, complex crater; *s*, simple crater; *e*, exposed; *m*, submarine; *l*, lake; *b*, bay (*r*, rim exposed; *u*, central uplift exposed); *tec*, tectonically modified; *bu*, buried under sedimentary rocks; *pc*, penetration crater; *f*, crater field (in brackets number of craters; only diameter of largest crater given)
Imp.: Impactites: observed impact lithologies (in boulder, drill core or outcrop): *ej*, ejecta outside craterform; *m*, coherent melt sheet; ml, small lenses of melt; *mf*, fragments of melt in resurge (backwash) deposit; mb, melt-bearing breccia (e.g., suevite); *lb*, lithic breccia; *bd*, breccia dyke (incl. pseudotachylites); *pr*, projectile fragments; *sc*, shatter cones (in situ *, otherwise in allochthonous breccias)
Target: *wat*, water; *sed*, sedimentary rocks; *cry*, crystalline rocks (metamorphic + magmatic); *lMP*, late Mesoproterozoic; *lNP*, late Neoproterozoic; *Ca*, Cambrian; *Or*, Ordovician; *Si*, Silurian; *De*, Devonian; *Cb*, Carboniferous; *Pe*, Permian; *Ju*, Jurassic; *Cr*, Cretaceous; *Q*, Quaternary; in [square brackets] approximate thickness and water depth, respectively, in meters
PL: preservation level: *1*, ejecta largely preserved; *2*, ejecta partly preserved; *3*, ejecta removed, rim partly preserved; *4*, rim largely eroded, crater-fill products preserved; *5*, crater-fill products partly preserved; *6*, only remnants of crater-fill preserved, crater floor exposed; *7*, crater floor removed, substructure exposed
Age: *str*, stratigraphic relationships; *bstr*, biostratigraphic dating; *pmag*, paleomagnetic dating; *hist*, historically documented event
Pro.: Projectile: *ir*, iron; *ch*, chondrite; ? indicates investigation, but interpretation uncertain or no clear signature found; *oh*, octahedrite
History: year of *(a)* discovery or first description of lithic or melt-bearing breccia, melt rock, and/or (largest) craterform; *(b)* first suggestion of impact-origin; *(c)* final confirmation of shock-features or meteoritic material

A renewed boom of impact cratering research in Fennoscandia was initiated in the mid 1980s. Major impetus for this development stems from the scientifically successful deep drilling project at Siljan (1983-1990) and, generally, from the debate on the significance of terrestrial impact cratering in the wake of the K/T boundary discoveries. International meetings held in Lappajärvi, Finland (1990) and Sigtuna, Sweden (1991) proved to be useful for exchanging experiences and initiating joint studies. Later, the European Science Foundation Scientific Network "Impact Cratering and Evolution of Planet Earth" (1993-1996) and the follow-up Scientific Program "The Response of the Earth System to Impact Processes" (since 1997) yielded a financial and coordinative basis for many efforts.

3 The Impact Structure Record of Fennoscandia

In Table 1 characteristic data of 30 sites in Fennoscandia are listed that we regard as impact structures. Note, however, that the Ilumetsa (e.g., Czegka 1997; Plado et al. 2000), Tsõõrikmäe (e.g., Pirrus and Tiirmaa 1984; Pirrus 1995), Hummeln (Svensson 1966; Lindström et al. 1999; Ormö et al. 1999) and Granby structures (Bruun and Dahlman 1980, 1982; Ormö and Lindström 2000; M. Lindström, written communication 2000) are included in the list, although no diagnostic shock features or meteoritic matter have been recognized yet. We nevertheless list them, because according to the works referred to and to our own experience no other explanation than impact is consistent with the regional geology. In our opinion, this circumstance combined with the level of investigation and documentation justifies a classification not as "confirmed", but as "accepted" impact structures.

3.1 Diameters and Morphologies

The size of an impact structure is crucially a function of the projectile's kinetic energy, target properties and erosion level (e.g., Grieve and Pesonen 1992). In Table 1 the following three diameters are defined:

(1) The present morphological diameter (D) for exposed structures. This includes rim-to-rim diameter at craters with a preserved elevated rim (e.g., Söderfjärden) and, if no rim is recognized, edge-to-edge diameter at craters which presently have a bowl-shaped (e.g., Karikkoselkä) or annular craterform (e.g., Siljan). For concentric structures, i.e., those with an inner and outer crater (Lockne, Logoisk), the diameter of the inner crater is indicated. Occasionally crater shapes deviate from circular; the eccentricity does not exceed 1.2. An exception is Mishina Gora, Russia, that reveals an oval-shaped simple crater, ~2.5 × 4 km across and filled with 796 m thick allochthonous breccia, surrounded by a ~3 km wide zone of steeply inward dipping, concentric normal faults (Shmayenok and Tikhomirov 1974; Pipping and Puura 1996). This peculiar

setting could be due to the layered target (Table 1) and an aquatic environment at impact (sections 3.2 and 3.5).

(2) The maximum lateral extent of impact-related structural disturbance, where recognized. This may include observations from geophysics, topography, remote sensing or field mapping.

(3) An estimate of the pre-erosional (original) rim-to-rim diameter, compiled from the literature or based on our own considerations (section 3.5).

The transition from simple to complex craters in Fennoscandia occurs somewhere between ~3.2 (Granby, Mishina Gora) and 4 km (Kärdla), which is more or less conform with the average value of ~4 km for impacts into crystalline targets on Earth (Grieve and Pesonen 1992). In a few cases deep erosion and/or poor exposure still prevent a definite conclusion whether a central uplift is present or not. For the Paasselkä, Vepriai and Neugrund structures a central uplift is, however, almost certain, because of their sizes. More equivocal are the deeply eroded Saarijärvi structure and Sääksjärvi, which is a poorly exposed crater.

3.2
Impact Settings and Crater-Fills

The only craters in Table 1 that were excavated in a sedimentary target, except the small Quaternary ones, are Dobele, Vepriai and Mjølnir. Most impact structures in Fennoscandia reveal evidence for a layered target at the time of impact, i.e., crystalline basement and a sedimentary cover, with or without a water column on top. This is obvious at craters with allogenic breccias containing both sedimentary and crystalline fragments (Kärdla, Lockne, Granby, Mishina Gora, Dobele, Logoisk). At the complex Siljan, Lappajärvi and, probably, Lumparn impact structures pre-impact cover rocks are preserved in ring moats, or in remnants of them that were down-faulted during the crater collapse. A 18.2 m pre-impact unit has been drilled at Lappajärvi (Pipping and Lehtinen 1992), at Siljan ~300 m are preserved, based on seismics (Vattenfall 1991), and for Lumparn at least ~72 m are known from drilling (unpubl. data). For Jänisjärvi (Masaitis 1999a) and Paasselkä (Pesonen et al. 1999a), a pre-impact sedimentary cover (siltstones) is evident from fragments within allogenic crater-fill breccias and breccia dikes, respectively. No sedimentary fragments are observed in the crater-fill breccias of Dellen (Henkel 1992), Mien (Svensson 1969), and Suvasvesi N (Pesonen 1995). This suggests a dominantly, but not necessarily purely crystalline target. This is seen at Lappajärvi, where no sedimentary fragments are observed in the breccia sequences, although Mesoproterozoic siliciclastic rocks occur in a peripheral moat (Pipping and Lehtinen 1992). The only indication of a sedimentary cover within the drilled breccias is a 2.6 m thick siltstone intercalation that contains Cambrian and Ordovician microfossils (Uutela 1998). For Sääksjärvi it may be supposed that siliciclastic rocks were part of the target in view of the late Vendian/Cambrian paleogeography and the cited radiometric impact age of ~514 Ma, but no sedimentary fragments are observed in the drilled breccias (P. Pihlaja, personal communication 2000). Furthermore, the allogenic Gardnos breccias reveal increased carbon contents, which point to the presence of carbon-bearing sediments in the target, probably represented by nearby exposed black shales, but the breccias lack such shale fragments (French et al. 1997). The amount, distribution and retention of sedimentary fragments in the crater-fill breccias may be explained by the cover rock lithology and, possibly, by the cratering process itself, although a biased recognition due to exposure and sampling density should be taken into account as well.

Evidence for a submarine setting during impact comes (1) from the character of the post-impact crater fill and (2) from the identification of target rocks formed at sea shortly before the oldest identified post-impact deposits. The post-impact fill is typically a thick, upward fining sequence, with large amounts of sedimentary breccia, deposited when the displaced water column forcefully rushed back into the newly formed depression as seen at Lockne and Tvären (Lindström et al. 1994; Sturkell 1998; Ormö and Lindström 2000). The largely buried Mishina Gora structure reveals an unusually thick (796 m) allogenic breccia body including blocks of "some tens of meters across" (Masaitis 1999a). Allogenic breccias in impact craters 3.5 - 8.0 km in diameter are usually 250 - 600 m thick (French et al. 1997). In addition, the lack of melt bodies and the vertically changing composition of the breccia resemble the deposits at Lockne and Tvären. Therefore, some kind of resurge process in an off-shore setting may also be involved at Mishina Gora. Kärdla formed in a shallow sublitoral sea, which penetrated the elevated crater through two gaps very soon after impact (Puura and Suuroja 1992). At Hummeln an inverted stratigraphic sequence of coherent crater-fill units is recognized. Lindström et al. (1999) explain this setting with an impact into a shallow sea less deep than the initial height of the elevated crater rim. Since the rim partly consisted of soft sedimentary rocks, it should have been soon overwhelmed by the exterior water pressure leading to chaotic slumping of coherent, large units into the crater. Possibly the inverted units are slumped portions of an overturned flap which was part of an initially elevated crater rim. A similar scenario may hold for Karikkoselkä, where thick coherent units (~10 - 22 m) of pre-impact silt- and claystones are situated between polymictic allochthonous breccias (Arkonsuo 2000). At Granby an overturned flap could still be preserved in the original position, but also in this crater thick sedimentary breccias (~70 m) containing large slumped blocks are observed (Bruun and Dahlman 1982; Ormö and Lindström 2000). Flow textures of clastic dikes on the shores and within Lumparn bay point to the presence of water at the time of impact and may, therefore, indicate a subaquatic impact setting as well. The dikes resemble features described from the submarine Lockne crater (Sturkell and Ormö 1997). Suuroja and Suuroja (2000) proposed a submarine setting for the Neugrund impact, based on tight paleogeographic constraints. Direct evidence is, however, not observed, as allochthonous crater-fill breccias are not accessible without drilling. Craters in Fennoscandia showing indications for impact at sea contain no coherent impact melt sheet.

In contrast, the drilled crater-fill of Söderfjärden is suggestive of a near-coastal, possibly tidally influenced on-shore setting during impact. Mechanically deformed quartz is observed up to high levels in the post-impact, marine and chronological (A. Uutela; unpublished data) crater-fill of Cambrian age. The fill contains few "sudden" coarser-grained beds as well as a thick, interbedded clay-siltstone unit at the base (Lehtovaara 1982). A possible impact melt body is seen in aeromagnetic maps. Similarly, the buried Mizarai crater reveals a rhythmically layered, sandy-clayey sequence (post-Early Cambrian) with a basal coarse-grained bed containing glassy particles. A coherent melt body has been drilled in the annular moat (Gailius 1988; Motuza 1994). As for Söderfjärden, the situation may suggest a successive back-wash of reworked ejecta into the crater, soon but not immediately after the impact (Masaitis 1999a). Drilling into the Sääksjärvi crater encountered only allogenic breccia (max. 180 m) with small pockets of melt before reaching the crater floor, but no coherent melt sheet (Pihlaja and Kujala 2000). However, melt rocks are abundantly found in glacial boulders. This may reflect the relatively increased resistivity of the massive melt rock, but could also point to a former melt body that was just eroded during the Pleistocene glaciations. At Dellen a 9 km wide sheet of impact melt (dellenite) appears mostly buried under Quaternary sediments, interpreted from aeromagnetics and known from boulders and two outcrops (Henkel 1992; Svensson 1968a). An on-shore setting for the Dellen impact is corroborated by the putative paleogeography (section 3.3). The same holds for Lappajärvi, where melt rock (kärnäite) with a drilled thickness of 145 m crops out on the central island (Pipping and Lehtinen 1992). In the center of Mien about 20 - 25 m of impact melt (mienite) has been drilled and 250 m are modelled from magnetic and gravity data (Henkel 1992). At Jänisjärvi impact melt is exposed on central islands, of some 20 m thickness, and on the southwestern shore (Feldman et al. 1981; Masaitis 1999a). Where exposed or drilled, the melt rocks overlie allogenic breccias. In summary, all these craters with preserved coherent impact melt, or with indications of such, formed on land in crystalline targets with no or a relatively thin sedimentary cover.

Impact melt is not developed at the Vepriai and Dobele craters that were excavated in sedimentary rocks only. Logoisk was excavated in a layered target, but does not contain a coherent melt sheet, but only small melt lenses and irregular bodies. The target sequence consisted of ~350 m sedimentary rocks (Table 1). At least Vepriai and Logoisk formed on-shore, as indicated by thick lacustrine crater lake sediments on top of allogenic breccias (Gailius 1988; Glazovskaya et al. 1991).

Generally the properties of the Fennoscandian craters point to a mutual relationship between the presence of sedimentary rocks in the target and the non-development of coherent impact melt bodies, which is in agreement with a study of this issue by Kieffer and Simonds (1980). With the possible exception of Mjølnir, in all Fennoscandian complex craters without evidence of a coherent impact melt body, carbonates represent a major component in the target stratigraphy. Kieffer and Simonds (1980) explain the lack of impact melt by the explosive expansion of large amounts of steamed water and of volatilised CO_2 from carbonates. This process should largely disperse initially developed impact melt. Explosive expansion of gas and steam may not only hinder the development of melt, but could also facilitate the disintegration of sedimentary rock fragments that are apparently sometimes absent in the crater-fill breccias, although a sedimentary cover was present at impact. On the other hand, Graup (1999) has recently shown that at least some of the limestone in the Ries crater target was shock melted upon impact. It occurs as a separate melt phase in the suevite, because of its immiscibility with molten silicate. Respective studies are required also for the Fennoscandian craters that reveal no coherent melt bodies but suevites, and which contain limestone sequences in the target, as assumed for Lumparn, for instance. An additional aspect that requires consideration is the penetration depth of the projectile relative to the water-sediment (Ormö and Lindström 2000) and sediment-crystalline bedrock interfaces (e.g., von Engelhardt 1997). This position has an influence on the shock metamorphism and redistribution of target rocks in specific stratigraphic levels.

3.3 Preservation of Craters

The criteria for the preservation levels (PL) listed in Table 1 are consistent with those introduced by Grieve and Robertson (1979). Note, however, that we refer PL 1 (ejecta largely preserved) and PL 2 (ejecta partly preserved) to ejecta that contribute only to the elevated rim. In this way the classification is also applicable to buried structures at which the elevated rim is eroded first but the ejecta beyond the rim remain buried and preserved. Note also that at least at the Lockne and Logoisk concentric craters, as well as at Mjølnir, elevated rims were either immediately destroyed by the seawater resurge or may never have been properly developed due to the layered targets (sections 3.2 and 3.5). The term "exposed" does not necessarily mean exposed at the surface, but refers to the relative erosion level; i.e., burial under post-impact sediments or cover by water is possible (see section 3.1).

Some of the Fennoscandian craters are exceptionally well preserved, because they were buried under sedimentary rocks soon after they came into existence. Their morphologies are virtually pristine and only the crater rim has been partly eroded. These craters are either still buried under post-impact Pre-Quaternary rocks (Vepriai, Dobele, Logoisk, Kärdla) or they have been partly exhumed, presumably not earlier than the Neogene (Neugrund, Granby, Söderfjärden). At Söderfjärden an elevated discontinuous rim is entirely exposed. Its height of ~20 - 50 m is not significantly shallower than at other well preserved craters of comparable sizes in non-glaciated areas. Other structures were likewise buried soon after formation, but their state of preservation suggests erosional action before the Neogene (Tvären, Karikkoselkä, Hummeln, probably Mishina Gora). The elevated rim has been entirely erased and also parts of the cavity are denuded. Saarijärvi (Pesonen et al. 1998) and Iso-Naakkima (Elo et al. 1993) were deeply eroded before burial, indicated by the almost total lack of allochthonous breccia bodies in drill cores. It is not entirely clear which process(es) removed the breccias, but running water or moving ice are potential agents. Removal was possibly favoured by intense chemical weathering and disintegration of the breccias indicated by abundant kaolinisation of the crater basement and fill. Partly eroded to below the true crater floor are Gardnos (French et al. 1997) and Suavjärvi (Mashchak and Naumov 1996). Siljan is the only known impact structure in Fennoscandia entirely eroded to below the true crater floor. The amount of erosion has been estimated to 1 - 1.5 km in the crystalline areas (Grieve 1988).

Whereas for most craters the post-impact erosion history and, hence, the preservation level is rather easy to understand, some structures require a more thorough discussion, because their current properties are not obvious at first glance.

The Suavjärvi impact structure (Mashchak and Orlova 1986) still contains remnants of polymict impact breccias and reveals, according to Mashchak and Naumov (1996), a ~16 km wide, circular outline in space imagery. The structure appears to have been preserved, because it was, at least, buried under a Paleoproterozoic volcanosedimentary cover (Jatuli; 2.33 - 2.1 Ga; Melezhik and Fallick 1996), which is still abundantly present in the Archean Karelian block (Sokolov and Heiskanen 1985) and on top of the impact breccias (Mashchak and Naumov 1996).

The uppermost section of the sedimentary crater-fill of the Saarijärvi crater contains ~500 - 600 Ma old microfossils (Tynni and Uutela 1985; Öhman et al. 2000). The lower portions may be older, although there appears no hiatus in a drill core sampling the whole sequence (Pesonen et al. 1998). In any case, these deposits probably once covered the entire crater and, hence, protected it. Geophysical profiling suggests a slight tilt of the whole structure (elevated in the south), and in fact a tectonic modification of the present setting is not unfounded in view of the Puolanka-Kandalakša Fault that passes Saarijärvi only 7 km to the northwest. This fault and associated branches are one of the youngest fault arrays in central Finland, probably active up to the present (Talvitie 1979; papers in Silvennoinen 1992). Post-impact tectonic activity on it (and on conjugated north and northwest-trending fractures) may have led to down-faulting of the crater-containing block and to the development of a depocenter in which Saarijärvi was buried. A prominent N-S directed fracture valley runs right into the crater.

For Iso-Naakkima, a tectonic influence on crater preservation is more evident. The deeply eroded crater is filled with Mesoproterozoic sedimentary rocks that contain abundant kaolin (Elo et al. 1993). Like for Saarijärvi, this crater-fill originally could have had a larger extension and temporarily protected the crater. In addition, Sarapää (1996) proposed protective tectonic down-faulting not only to explain the preservation of the crater and its fill, but also of authigenic kaolin several dozen metres thick within the adjacent Virtasalmi area. This author noted that all these deposits are situated southwest of the Bothnian Bay-Ladoga Tectonic Zone (Fig. 2) and that down-faulting of the respective crustal block may be due to far-reaching effects of the Sveconorwegian orogeny to the west (~1.2 - 0.9 Ga; Gaál and Gorbatchev 1987). Elo et al. (1993) attributed the bedding tilt (up to 80°) in drill cores from Iso-Naakkima to post-impact tectonic subsidence.

The most obvious influence of block tectonics on the preservation of a Fennoscandian impact crater is seen at Granby. The crater is situated within a much larger block that has been coherently down-faulted to a half-graben well within the crystalline shield area (Bruun and Dahlman 1980, 1982). Normal block tectonics favoured not only the preservation of entire craters, but also of individual parts of them. Uneven erosional levels of Dellen (Henkel 1992), Lumparn

(Winterhalter 1982) and probably Suvasvesi N (Werner et al. 2000) are effects of such displacements. Finally, protection of craters resulted also from thrusting. This holds for Lockne (Lindström et al. 1996) and Gardnos (French et al. 1997) that have been temporarily protected by Caledonian nappes, Lockne additionally by sediments prior to thrusting. An outlier of the nappe complex is still preserved within the Lockne crater.

For at least two craters, Mien and Jänisjärvi, protection by sedimentary rocks can be inferred from the paleogeographic setting. The ~121 Ma old Mien crater was likely protected from at least some Tertiary erosion because of burial under Upper Cretaceous rocks. The crater is situated right at the elevation level that marks the evident *minimum* northward extent of the Upper Cretaceous sea in southern Sweden, viz. 120 - 130 m a.s.l. (Lidmar-Bergström 1996, and references therein). Because at Mien only the central uplift has been drilled, remnants of a Cretaceous cover may still be discovered in the crater moat. The ~700-Ma-old Jänisjärvi crater also lacks direct evidence for a cover, but its location and relatively well preserved state speak in favour of it. Upper Vendian and early Paleozoic platform cover (200 - 350 m thick) crop out ~110 km south in the southern Lake Ladoga area (Fig. 2). The present northern extent of this sequence is erosional and yet there is a trend to increasing thickness towards the north (Puura et al. 1996). The beds dip to the southeast indicating post-depositional tilt and an originally flat-lying succession (Amantov et al. 1996). Therefore, these rocks extended originally much farther to the north and probably buried the crater (also Grieve and Dence 1979). The sedimentary fragments in the impact breccia (section 3.2) possibly represent a now eroded, peripheral facies of the Proterozoic fill within the Ladoga-Pasha rift basin preserved 35 km to the south (Fig. 2; Amantov et al. 1996). The Phanerozoic erosion of Jänisjärvi to below the pre-impact, crystalline-sediment interface is estimated to be some 200 m (Koistinen 1994 [measured from profile e - f]). Siljan was possibly covered by Old Red Sandstone in the Late Devonian and perhaps later, depending somewhat on the uncertain impact age (section 3.7, Table 1).

Whereas the paleogeography suggests that sediments once covered some impact craters, it likewise indicates that others have been denuded ever since their formation; viz. Lappajärvi and Dellen. These two structures suffered in particular from selective erosion by advancing Pleistocene glaciers, i.e., denudation was uneven. Weak impact breccias have been caved from the crater moats especially in the direction of the glacier movement, whereas the competent impact melt sheets were relatively less affected. Their state of preservation is similarly ascribed to PL 5 (crater-fill products partly preserved) tending to PL 4 (rim largely eroded, crater-fill products preserved).

3.4 Geophysics

Most impact structures in Fennoscandia have been investigated by geophysical methods (Table 2) and some, now confirmed, structures have even been initially discovered in this way (e.g., Kärdla, Puura and Suuroja 1992; Iso-Naakkima, Elo et al. 1993). The most commonly used methods involve gravity and magnetic potential fields, but electric, electromagnetic, seismic, and radiometric techniques have been successfully applied as well (Henkel 1992; Pesonen et al. 1999b; Tsikalas et al. 1998b, and others). Geophysics gives a fast and inexpensive method for getting an idea of the interior of an impact structure. For example, geophysics has been used to investigate the extent of authigenic brecciation affected by oxidation at Tvären (Ormö and Blomqvist 1996), to specify the shape and depth distribution of post-impact sediments and impactites at Karikkoselkä (Pesonen et al. 1999b), or to determine the depth of the fractured zone at Lockne (Sturkell and Ormö 1998; Sturkell et al. 1998).

The residual *gravity anomaly* over an impact structure depends on (1) the size, depth and morphology of the structure and (2) the density contrast between impact-related rocks and surroundings. Most of the Fennoscandian structures show negative circular gravity anomalies (Table 1) due to the high porosity and low density of impact breccias and, in some cases, crater-fill sediments. Occasionally the impact-induced anomaly is heavily distorted, however, by lateral density variations in the target rocks (e.g., Siljan, Dyrelius 1988; Lockne, Sturkell et al. 1998). In most cases, the amplitude of a gravity anomaly depends on the diameter of the impact structure (Fig. 3). The clear exceptions are Mjølnir and Lockne that show weaker anomalies than their diameters would suggest. The submarine Mjølnir crater is 40 km across, but has a negative anomaly of only -1.5 mGal (Fig. 3a). Tsikalas et al. (1998b) attributed this weak low to (1) less pervasive brecciation within the soft sediments, (2) dense centrally uplifted rocks that wipe out part of the negative anomaly surrounding it, and (3) compaction and deformation by thick post-impact sediments. At Lockne the density of impactites was elevated by hydrothermal activity that annealed fractures, mainly by calcite deposition. Moreover, the post-impact sedimentary rocks and resurge deposits have a higher density than the crystalline basement (Sturkell et al. 1998). At Siljan a ~40 km wide gravity anomaly occupies the northern periphery of the structure, but its peak of -15 mGal develops outside of it. The anomaly is partly explainable by impact-induced density differences, constrained by drill core data, but the relatively strong gravity signatures of pre-impact lithologies appear to overprint the impact signature (Dyrelius 1988, 1994). At the buried Kärdla crater the negative gravity anomaly is surrounded by a chain of positive anomalies due to the uplifted crystalline rim that is bounded externally and internally by sedimentary rocks (Plado et al. 1996).

In the crystalline shields and at deep-water targets the crater rim is usually eroded (or not formed) and not identifiable by gravity methods. In these cases an estimate of the present (and original) crater diameter by gravity alone is prone to produce flaws, since (1) the gravity diameter is extremely sensitive to the choice of the regional field, and (2) fracturing that contributes to the anomaly extends beyond the preserved cavity (Zenchenko and Tsvetkov 1999). Therefore, it is more convenient to use the half-amplitude width (HAW) for describing the range of gravity anomalies (Table 2). According to the Fennoscandian dataset (Table 1), the mean ratios of the observed HAW of the negative gravity anomaly to the present and estimated original diameters are 0.68 ± 0.15 and 0.58 ± 0.14, respectively. The level of preservation of the structure has only a minor influence

Table 2. General characteristics of impact structures in Fennoscandia

Name	(Country)	No.	Gravity				Type of the magnetic anomaly	Electric and electromagnetic surveys	Seismic surveys
			A_G (mGal)	HAW_G (km)	A_{CU} (mGal)	HAW_{CU} (km)			
Siljan	S	6	-15	(see text)	-	-	S, H (10)	MT, VLF-R	Refl
Mjølnir	N	67	-1.5	(see text)	2.5	6.0	(see text)		Refl, Refr
Dellen	S	5	-9.5	11.5	-	-	L, S, H (9)	AEM, VLF-R	
Lappajärvi	F	31	-11	13.8	-	-	L (17), D	AEM, EM	Refr
Suavjärvi	R	68	-10				L		
Jänisjärvi	R	36	-7	8.5	-	-	L (18), D, H, P		
Logoisk	B	90							
Lockne	S	48	-2.2	~8	-0.2	1.4	L (6)		
Paasselkä	F	34	-8	6.0	-	-	L (8), D		Refl
Lumparn	F	37					L (11)		Refl
Vepriai	Li	64							
Neugrund	E	69					L (5), P (7)		Refl
Söderfjärden	F	32	-6	4.2	1.5	1.3	L (5), H (1)	AEM, DC-R	Refr
Mien	S	20	-5	3.0	-	-	L (9), H (2.2)		Refr
Mizarai	Li	63	-4	2.8	-	-			
Gardnos	N	42	-	-	-	-	-		
Sääksjärvi	F	33	-6.5	3.7	-	-	L (4.5), D	AEM	
Mishina Gora	R	66	low				L		
Dobele	La	65							
Suvasvesi N	F	35	-3	2.3	-	-	S, H	AEM	

Table 2. (cont.)

Kärdla	E	6	-3; rim +2	2.3	-	-	L, P (4)		
Granby	S	67							
Iso-Naakkima	F	5	-4	2.0	-	-	L (3), S	AEM, EM	
Tvären	S	31					L (3.5), D		Refl
Saarijärvi	F	68	-1.5	1.2	0.3?	0.2	L, D	AEM, VLF-R, DC-R	
Karikkoselkä	F	36	-3.8	1.1	-	-	L (1.5)	AEM, EM	
Hummeln	S	90	-4.3	0.9	-	-	L (2)	VLF-R	
Kaali	E	48			-	-	M	GPR, DC-R	
Ilumetsa	E	34			-	-	L	GPR, DC-R	
Tsõõrikmäe	E	37							

Country: *B*, Belarus, *E*, Estonia; *F*, Finland; *La*, Latvia; *Li*, Lithuania; *N*, Norway; *R*, Russia; *S*, Sweden
No.: numbering; continued from Henkel and Pesonen 1992, Pesonen 1996
Gravity: A_G, amplitude; HAW_G, half amplitude width; A_{CU}, amplitude (central uplift); HAW_{CU}, half amplitude width (central uplift)
Type of the magnetic anomaly: *L*, magnetic low; *D*, disruption of regional magnetic trends; *S*, smoothing of regional magnetic trends; *H*, high amplitude, short-wavelength anomalies inside structure; *P*, positive anomalies due to rim wall; *M*, microanomalies due to fragments of meteorite (in brackets approximate diameter of anomaly).
Electric and electramagnetic surveys: *MT*, magnetotellurics; *VLF-R*, very low frequency resistivity; *DC-R*, direct current resistivity; *AEM*, aero-electromagnetics; *EM*, ground electromagnetics; *GPR*, Ground penetrating radar.
Seimic surveys: *Refl*, reflection seismics; *Refr*, refraction seismics.

References: (6) Dyrelius 1988; Henkel 1992; Pedersen et al. 1990; Pedersen et al. 1992, Elming and Bylund 1991; Papasikas and Juhlin 1997; Juhlin and Pedersen 1987 (67) Tsikalas et al. 1998b (5) Henkel 1992; Bylund 1974 (31) Elo et al. 1992a; Pesonen et al. 1992 (68) Mashchak and Naunov 1996 (36) Elo et al. 2000; Feldman et al. 1981 (48) Sturkell and Ormö 1998; Sturkell et al. 1998 (34) Elo et al. 1984; Pesonen et al. 1999a (37) Winterhalter 1982; (69) Suuroja and Suuroja 2000 (32) Talvitie et al. 1975; Laurén et al. 1978; (20) Henkel 1983; Henkel 1992; Åström 1998 (63) Motuza, personal communication 2000 (33) Elo et al. 1992b (66) Pipping and Puura 1996 (35) Pesonen et al. 1996 (54) Plado et al. 1996 (61) Elo et al. 1993; Järvelä et al. 1995 (12) Ormö and Blomqvist 1996; Flodén et al. 1986 (70) Elo et al. 1983; Pesonen et al. 1998 (71) Pesonen et al. 1999b (18) Ormö et al. 1999 (38) Pobul 1963 (52) Pobul 1963; Plado et al. 2000 on these ratios. The modelling shows that the major effect of erosion is expressed in a decrease of the amplitude, but minor changes are present in its diameter or HAW (Plado et al. 1999). At larger impacts denser crustal material may be centrally uplifted during collapse of the transient cavity producing a positive gravity anomaly (e.g., Pilkington and Grieve 1992). At the Fennoscandian craters only three of at least twenty complex structures reveal a gravity anomaly attributable to a central uplift (Table 2). Mjølnir (Tsikalas et al. 1998b) and Söderfjärden (Laurén et al. 1978) show weak positive features, while Lockne reveals a weak negative anomaly (Fig. 3a) due to relatively dense impact breccias and crater-fill sediments (Sturkell et al. 1998). The lack of central anomalies may be caused by (1) poor data coverage, (2) low density contrast between central uplift and surroundings, or (3) density contrasts in the target rocks that obscure an uplift-produced anomaly.

The *magnetic anomaly* of an impact structure depends on (1) the size and shape of the structure, (2) the intensity and direction of magnetisation of the impactites, with respect to the magnetisation of unshocked target rocks, (3) the local geomagnetic field, and, for aeromagnetic surveys, (4) ground clearance (distance from measurement level to source). Magnetic anomalies related to impact structures are usually more complicated than gravity anomalies, due to the large variations of magnetisations in impact-related and target rocks. A further complicating factor is that the observed magnetic signature represents a vector sum of the induced and remanent magnetisations, which depend on the direction of the present and ancient Earth's magnetic fields, respectively. Magnetic investigations have been carried out at most of the Fennoscandian structures. Many were covered by high-resolution, low-altitude (30 - 40 m) aeromagnetic (AM) mapping during routine geophysical mapping by national geological surveys. The principal magnetic characteristic observed in these AM data is a smooth magnetic low, usually distributed over all the morphological structure (e.g., Jänisjärvi, Elo et al. 2000; Neugrund, Suuroja et al., this volume; Tvären, Ormö and Blomqvist 1996). The magnetic low may either disrupt the regional anomalies (e.g., Sääksjärvi, Elo et al. 1992a; Karikkoselkä, Pesonen et al. 1999b), hide them by impact-induced magnetic features or make them smoother (e.g., Siljan, Pedersen et al. 1990; Suvasvesi N, Pesonen et al. 1996). It is frequently caused by a late- to post-impact oxidation of ferromagnetic minerals into the less magnetic forms, e.g., oxidation of ferrimagnetic magnetite to antiferromagnetic hematite (Henkel 1992, Henkel and Pesonen 1992). This phenomenon has been recognized, for instance, at Siljan, Tvären and Mien. The oxidation is most intensive in the uppermost part of a structure, where the rocks are heavily fractured, and therefore susceptible to oxygen-rich fluids. In some cases also a thick post-impact sediment infill is responsible, at least partly, for a magnetic low (e.g., Lockne, Kärdla, Söderfjärden). Besides the very low magnetization of the sediments, they hide the magnetic response from the more magnetic basement. A magnetic low may also be caused by shock-induced demagnetization (Cisowski and Fuller 1978; Pesonen et al. 1997a), or by random orientation of magnetization vectors in allochthonous crater-fill breccias. However, the latter two cases have not been thoroughly studied in the field, contributing to the still incomplete

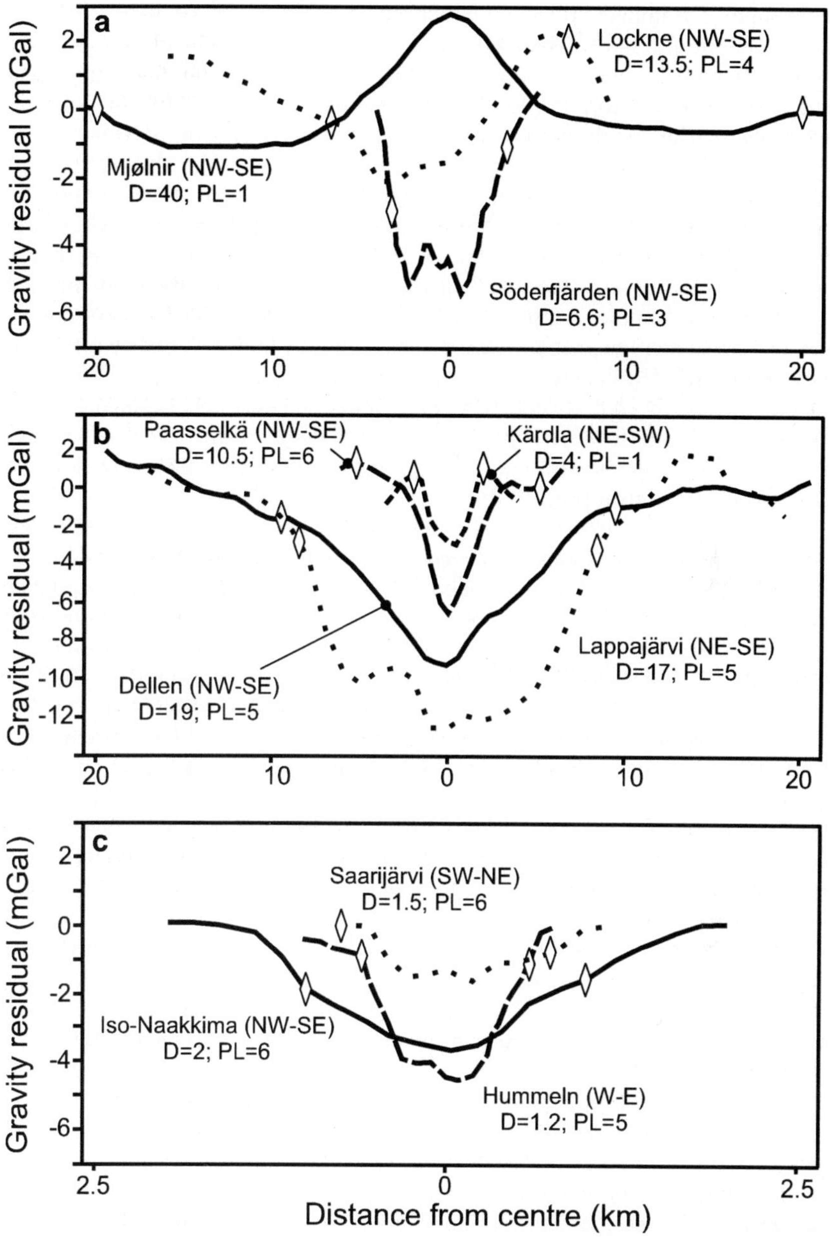

Fig. 3. Comparative representation of residual gravity profiles of some Fennoscandian impact structures. a) profiles of complex structures that show features corresponding to a central uplift, b) profiles of complex structures that do not show any gravity features related to a central uplift, c) profiles of simple structures. Note different scales. Diamonds mark a present craterform; *D*, present diameter (m); *PL*, preservation level (for definition, see text and Table 1). References to geophysical investigations of each structure can be found in the footnotes of Table 2.

understanding of impact effects on target rock magnetism (Scott et al. 1997). Five Fennoscandian structures (Mien, Dellen, Jänisjärvi, Söderfjärden, Mien, Suvasvesi N) are reported to contain central magnetic anomalies with high-amplitudes and short wavelengths compared to the surrounding magnetic lows. At Mien and Dellen near-surface melt sheets with high relative magnetizations (Henkel 1992) cause central anomalies. Suvasvesi N contains a ~650 m wide, short-wavelength, negative anomaly in its approximate center that is due to reversely magnetized impact melt within allochthonous breccias (Pesonen et al. 1996). For Söderfjärden a small buried impact melt body can be inferred (Abels 2001). The presence of impact melt, however, does not always assure the existence of magnetic anomalies. For example, Lappajärvi hosts a thick impact melt sheet, but does not show any prominent magnetic anomalies inside the overall low, due to a relatively low content of ferromagnetic minerals (Kukkonen et al. 1992). The Siljan structure reveals a central, complex, high-amplitude magnetic anomaly, but no significant magnetic low, presumably a result of deep erosion. There, the central anomaly is due to the central uplift that brought more highly magnetized granites to the surface (Pedersen et al. 1990).

Impact causes differences in the resistivities between impacted and target lithologies, which makes *electrical methods* a useful investigation technique. The net effect is an increased porosity and permeability of the target and crater-fill, and, hence, higher fluid content and electric conductivity. Measurements exploiting alternating fields, especially airborne methods, are common due to their relatively low costs and high effectiveness. Resistivity measurements utilising artificially generated direct currents have been performed at Söderfjärden (Talvitie et al. 1975), Saarijärvi (Elo et al. 1983; Turunen 1995), Ilumetsa (Plado et al. 2000) and Kaali only. In all cases a Schlumberger array was used for vertical electric sounding (VES) at several positions. Below Saarijärvi, the conductive zone, corresponding to drilled post-impact sediments and, to a lesser extent, impact breccias was clearly imaged. The conductivity at the structurally uplifted crater rim of Ilumetsa turned out to be less than that measured for sandstones in the vicinity.

Very low frequency (VLF) resistivity measurements at Siljan and Dellen show radially distributed low resistivities extending to beyond the structure's edges, an effect of fracturing (Henkel 1992). For both structures higher resistivities were measured in the central parts and attributed to uplifted target rocks. Ormö et al. (1999) performed VLF measurements in the surroundings of Lake Hummeln. They used resistivities to characterize the porosity and frequency of fissures in the bedrock by a relationship that was proposed by Henkel (1992), based on resistivity and porosity of drill-cores from Siljan. However, without the measurements on the structure, it has been impossible to identify any impact-related fracturing. At Saarijärvi, VLF measurements revealed low resistivities in particular caused by post-impact sediments.

Electromagnetic soundings or profiles of such have been performed at Siljan (Henkel 1992), Lappajärvi (Elo et al. 1992b), Karikkoselkä (Pesonen et al. 1999b) and Iso-Naakkima (Elo et al. 1993). The measurements show low resistivities in central parts of the craters, assigned to impact breccias, increased fracturing and/or post-impact sediments. Airborne low-altitude (30 - 40 m) electromagnetic (AEM) surveys were flown over a number of Fennoscandian structures, as well as aeromagnetics, in the wake of routine country-scale surveys. The acquired maps covering the impact structures show usually circular anomalies as well as high in-phase (Re) values (>500 ppm) in the absolute scale and in comparison with out-of-phase (Im) values (Re/Im > 0.5). In the case of lakes, water and bottom sediments alone cannot produce these relatively high values (in typical inland lakes Re/Im << 0.5). Therefore, a deeper lying conductive body, such as a breccia layer, must be present providing a diagnostic tool to search for impact structures. If thick post-impact sediments fill the crater, however, an impact-related signature usually cannot be inferred by AEM, as seen for Söderfjärden (Puranen et al. 1999).

A *magnetotelluric survey* has been carried out at Siljan (Zhang et al. 1988; Dyrelius 1994). No surface conductive zone was recognized, but a deeper (5 - 20 km), slightly eccentric low-resistivity zone was found inside the ring structure. It is explained by fluid migration, in particular saline water, through impact-induced fractures.

Seismic reflection studies, carried out at five Fennoscandian craters (Table 2), provide the most detailed geophysical information on the subsurface structure (e.g., thickness of breccia lenses, extent of basement fracturing, fault planes). Tsikalas et al. (1998b) demonstrated this powerful technique most recently for delineating the detailed structure of the submarine Mjølnir crater. For Siljan a present transient cavity diameter and minimum depth of 22 and 3 km, respectively, could be inferred, and it was shown that the central uplift at 9 km depth amounts still to 2 km, fading out somewhere below 15 km (Juhlin and Pedersen 1987, 1993). Refraction profiling has been used at Söderfjärden (Laurén et al. 1978), Mjølnir (Tsikalas et al. 1998b), Mien (Åström 1998) and, across the rim only, at Lappajärvi (Elo et al. 1992b).

Ground penetrating radar (GPR) measures the dielectric contrasts of the subsurface strata, which may correspond to changes in stratigraphy or lithology. In Fennoscandia it has been used for shallow subsurface investigations at the Quaternary Ilumetsa crater (Plado et al. 2000), where the radargrams reveal numerous small dislocations in the whole volume under the crater floor and elevated rim. The reflector that in the surrounding terrain corresponds to the Devonian sandstones is interrupted beneath the elevated rim.

3.5 Crater Reconstructions

Attempts to reconstruct the original morphologies of Fennoscandian impact craters have been published for Siljan (Grieve 1988) and Gardnos (French et al. 1997) on the basis of available geological constraints and previously published scaling relations. For most Fennoscandian craters, however, difficulties for the application of scaling relations arise from the layered targets that consist of a crystalline basement, differently lithified sedimentary cover rocks and sometimes a water layer on top (Table 1). Sufficiently deep water shallows the penetration depth of the projectile within the sea floor, which could result in a shallower transient crater when compared to on-shore impacts of equal total impact energy

(Ormö and Lindström 2000). Another potential effect at off-shore impacts is the immediate denudation of the elevated crater rim through a forceful resurge, especially favoured by only weakly to non-lithified rocks in the elevated crater rim, as supposed for Hummeln (Lindström et al. 1999). The vertically changing strength and possibly fabric of the target rocks are in particular responsible for the formation of a concentric crater morphology already during the crater excavation. Lockne (off-shore impact) and Logoisk (on-shore impact) are typical representatives of this crater type in Fennoscandia. If the sedimentary cover above the excavated crystalline basement is relatively thick, mega-block ledges may develop above the crystalline-sediment interface, rather than funnel-shaped outer craters. This setting could be a result of structural uplift and subsequent gravitational resetting beyond the extents of the inner crater, but without, - or at larger craters with, a subdued - wholesale collapse of the crater floor that includes the central uplift. Mishina Gora could be an example for this potential type of impact structure. Slumping can occur along stratigraphically controlled décollements which influences the final morphology of a crater as well, documented for example at Mjølnir (Tsikalas et al. 1998a). Generally, cratering in the described layered targets tends to decrease the ratio between the true depth of the final crater and its morphological diameter. Therefore the in Table 1 listed original rim diameters of the eroded craters (PL 4 - PL 7) should be regarded as rough estimates only (see also Grieve and Pesonen 1996, for critical notes on crater diameters cited in the literature).

Not only is the reconstruction of eroded crater portions a difficult task, but sometimes also the interpretation of the preserved crater portions that are not directly exposed at the surface. For instance, at the ~3.2-km-Granby crater Bruun and Dahlman (1982) interpret, on the basis of drill cores, a remarkably shallow craterform (ratio of true depth to diameter: ~0.09) when compared to the average of terrestrial simple craters (~0.28; e.g., Grieve 1998). Since the target properties, the good preservation of the crater and the probable simple craterform (Table 1; cf. Bruun and Dahlman 1982; Ormö and Lindström 2000) cannot explain this shallow depth, other factors like a shallow-oblique impact or a misinterpretation of the crater floor - perhaps the "fractured crystalline rocks" drilled in a core from the crater centre are part of an allochthonous block - should be taken into account. Similarly difficult to interpret are the sub-surface properties of the Saarijärvi crater. The partly lake-filled craterform, 1.5 km in diameter, contains a central island with exposed crystalline rocks, whereas a core drilled in only ~80 m distance reached the crystalline crater floor at ~156 m depth (Pesonen et al. 1998). For the evidence of an originally larger, deeply eroded crater (Öhman et al. 2000) and the indications for tectonic modifications (section 3.3), it is still unclear if the central elevation is due to uplift during crater collapse or post-impact displacements.

3.6 Impact Dating

Structures with preserved impact melt have been relative precisely dated by radiometric methods, although some scatter exists in view of the standard deviations and, in particular, of the results from different methods, as seen for Lappajärvi, Siljan or Dellen (review in Deutsch and Schärer 1994).

The age of an impact event may be inferred from potentially dateable remnants of a pre-impact and/or post-impact sedimentary cover in the crater or in impact lithologies. As noted before, Mashchak and Naumov (1996) reported that polymict impact breccias in the Suavjärvi impact structure are overlain by sedimentary rocks estimated at ~2.3 Ga. This would make Suavjärvi the oldest impact structure known on Earth. Biostratigraphic dating of the target and crater-fill by means of microfossils (acritarchs, chitinozoas, conodonts) has been lately shown to provide remarkably narrow time windows for the formation of those craters in Fennoscandia that were rapidly buried under sediments (e.g., Grahn et al. 1996). The Karikkoselkä crater contains a crater-fill in which thick, siliciclastic intercalations (Arkonsuo 2000) with Ordovician microfossils (A. Uutela; personal communication 2000) point to a marine setting at impact (section 3.2). As there is no evidence for a transgression in this region after the Ordovician, the age of the microfossils could represent the impact age. Circumstantial evidence at the Lumparn crater point to a pre-impact deposition of an Ordovician limestone deposit in the northern part of the structure. The deposit appears to be a mega-breccia, down-faulted during the crater collapse (unpubl. data; Merrill 1979). Indications for the presence of water at impact point also to a shelf sea setting (Caradoc to Llandovery in age), but this remains to be proven. Note that paleomagnetic dating (see below) for both Karikkoselkä (Pesonen et al. 1999) and Lumparn (L. Pesonen, unpublished data) are incompatible with the tentative estimates based on stratigraphic indications (Table 1).

Circumstantial evidence for an impact age can, in some cases, be derived from stratigraphic relationships. For example, French et al. (1997) suggested that an increased carbon-content in impactites from the Gardnos impact structure stems from a carbon-rich black shale of late Neoproterozic age (Biri Shale), which covered the target at impact. The Biri Formation may correspond to the uppermost part of the Visingsö Group in the Lake Vättern Graben dated at 707 ± 37 Ma (Kumpulainen and Vidal 1985). It evidently pre-dates the Moelv Tillite which, in turn, pre-dates the Ekre Shale (612 ± 18 Ma) in the southern Norwegian Hedmark Group. The Moelv Tillite corresponds to the Varanger glaciation dated in North-Scandinavia at 653 ± 7 Ma (Vidal and Moczydlowska 1995). Within the Gardnos structure sandy to shaly rocks with a large-fragment breccia at the base cover melt-bearing impact breccias. French et al. (1997) label the sedimentary rocks as slumped crater-fill, which implies a deposition soon, or immediately after impact. Since these rocks have no reported glacial affinity, they appear to pre-date the Moelv Tillite. Collectively this suggests an age of the Gardnos event of ~700 - 650 Ma.

Age estimates are also possible by paleomagnetics (review in Pesonen 1994). Two techniques are known and have been applied to Fennoscandian craters. These are (1) the apparent polar wander (APW) method (e.g., Siljan, Elming and Bylund 1991) and (2) the magnetic reversal method, applied when only unoriented samples are available from drill cores (e.g., Suvasvesi N, Pesonen et al. 1996). When samples are taken from outcrops and drill cores, it is essential to be able to exclude a post-impact displacement of the sampled unit, because only a slight structural tilt gives meaningless results. The measurements sometimes yield multiple remanent components, each with two possible polarities. This means that more than one "age" could be consistent with the data, even after obvious impact-unrelated components were sorted out. However, because of the usually large gap between these alternatives, a reasonable choice is sometimes possible in combination with other constraints.

3.7 Distal Impactites and Breccia Deposits

Confirmed and supposed impact-related deposits found to occur beyond the rim of some of the Fennoscandian craters comprise (1) seismically triggered breccia deposits, (2) ejected spherules and microspherules, (3) unmelted ejecta including meteoritic dust, and (4) glacially transported crater-fill impactites.

(1) The Middle Devonian impact event (~380 Ma) that formed the Kaluga crater ($D \approx 15$ km) in northwestern Russia (southeast of area shown in Fig. 2) is possibly also responsible for the Narova breccia horizon that has been encountered in coeval stratigraphic sections in the Baltic states, northwestern Russia and Belarus (Motuza 1994; Masaitis 1999b). The breccia is some tens of centimeters to several meters thick and consists of up to 10 cm large fragments of dolomites, marls and scarce sandstones embedded into dolomite clay. According to Masaitis (1999b), it is generally believed that the breccia results from sub-marine slumping through impact-triggered earthquakes.

A similar occurrence is represented by the Osmussaar breccia in northwestern Estonia (Kirsimäe et al. 2001). The up to 1.5 m thick, disturbed horizon occurs in a mixed siliciclastic-carbonate sequence (Arenig in age) and consists of fractured to brecciated sandy carbonate rocks with lithoclasts ranging in size from tens of centimetres to tens of metres. Spaces in between the fragments are filled with veins and dikes of limy mud to arenitic, siliciclastic material. In the arenites abundant grains with diagnostic shock features have been observed. Kirsimäe et al. (2001) suggested that the shocked material either originates from redeposited impactites of the nearby, Lower Cambrian Neugrund impact crater (Suuroja and Suuroja 2000), or from a yet unknown impact in the adjacent Baltic Sea that could have also triggered the generation of the carbonaceous breccia.

(2) It has been suggested (Claeys et al. 1992; Claeys and Casier 1994) that the Siljan structure is related to glassy microtektite-like spherules at the Frasnian to Famennian boundary (Late Devonian) in the Dinant Basin, Belgium. This is based on a radiometric age (Ar/Ar) of pseudotachylitic rocks from Siljan of 368 ± 1.1 Ma (Bottomley et al. 1978; recalculated in Grieve and Pesonen 1992) and the relatively close spatial association (presently ~1200 km). Unfortunately the temporal correlation is poorly constrained in view of another radiometric date (K/Ar) from Siljan of ~345 Ma (Åberg et al. 1989) and present uncertainties on the absolute age of the Frasnian-Famennian boundary (see comments in Racki 1999). However, the expectation of distal impactites in Europe is certainly reasonable in consideration of the size of Siljan ($D \approx 52$ km) and the distances related to established crater-tektite associations on Earth. According to Ūsaitytė

(2000), a largely complete sequence of the Frasnian and Famennian is known from boreholes off the coast of Lithuania (Fig. 1), approximately 650 km south-southeast of Siljan. Clearly further investigations are warranted.

Microspherules are also related to the Kaali (in addition to meteoritic matter) and probably the Ilumetsa crater fields, both of Quaternary age and located in Estonia (section 3.8).

(3) Unmelted ejecta have been preserved far outside the Kärdla (Puura and Suuroja 1992, Suuroja et al. 1999) and Lockne (Sturkell et al. 2000) impact structures, which both formed in an Ordovician shelf sea (Table 1). About Kärdla it has been encountered in various drill cores up to ~80 km from the crater center (a layer with thickness of few mm up to 2 m depending on the distance), at Lockne it is known from one locality (~30 cm thick) some ~45 km to the SSW of the crater center. Redeposition probably influenced the present extent and thickness of these layers. The impact origin of ejecta is constrained at Lockne by shocked quartz grains (Sturkell et al. 2000) and at Kärdla, in particular, by a precise stratigraphic correlation (Suuroja et al. 1999). The circumstance that textures and lithology deviate strongly from the local stratigraphy (predominately limestones) provides additional evidence. The distal deposits at Kärdla, here up to at least 30 km from the crater rim, and at Lockne have in common a lower debris layer containing allochthonous clastic fragments, cemented by a calcareous matrix that is overlain by a sandy to silty, laminated layer. At Kärdla the upper unit extends beyond the lower unit (Suuroja et al. 1999). A similar succession, i.e., silty claystone (with maximum Ir-peak) on top of mud-flake conglomerate, has been drilled 30 km northeast of the rim of the 40-km-Mjølnir impact structure in the Norwegian Barents Sea (Dypvik et al. 1996). A supposedly related Ir-peak occurs in a calcite cemented silty bed within a shale dominated succession on Svalbard, some 500 km to the northwest of the crater (Dypvik et al. 2000). The lack of a coarse-grained lower unit on Svalbard may indicate a comparable spatial situation in the environs of Mjølnir and Kärdla, i.e., the lower unit has a smaller radial extension than the upper, more fine-grained unit. Paired ejecta layers at these craters do not necessarily call for identical deposition mechanisms, but they could serve as an “at first glance”-pathfinder for moderately distal deposits related to shelf-sea impacts. In Fennoscandia the experience with previous descriptions of rocks which later turned out to be impactites indicates that a lower coarse ejecta layer may be labelled in old stratigraphic logs as “breccia”, “conglomerate” or “arkose”. Ejecta deposits are also noted by Suuroja and Suuroja (2000) to occur over 30 km from the center of the well preserved Neugrund impact structure, but these deposits still require documentation.

(4) Glacially transported impactites led to the discovery of Sääksjärvi (Papunen 1969). Kinnunen and Lindqvist (1998) described agates from these boulders and emphasized the potential of this material for the recognition of so far undiscovered impact structures. Indicative for impact are also shocked minerals and impact-derived diamonds in glaciogenic drift, like those discovered in glacial impactite boulders and loose rock from the glacial fan south of the Lappajärvi crater (Koivisto and Korhonen 1997; Masaitis et al. 1998a). Finally, boulders of sedimentary rocks that are otherwise totally eroded in the region can point to the

presence of a craterform in which they have been protected. Sites such as Tvären, Hummeln, and Lumparn have been known for this feature for more than a century. Erratic, heavily brecciated, gneissic blocks up to several meters in size have been known in the peripheral shores of the submarine Neugrund crater for more than 70 years (Suuroja and Suuroja 2000).

To our knowledge the only systematic attempt to search for distal impact signatures in Fennoscandia has been performed by Schmitz et al. (1994). A 120 m thick sequence of Silurian sediments on Gotland, Sweden, has been investigated for shocked quartz grains. The result was negative; not a single grain with planar deformation features was found. However, only bentonites and bentonite-like layers have been studied, because such units resemble very distant impact layers that can contain shocked quartz, such as the globally distributed K/T boundary clay. For deposits closer to an impact site clayey lithologies may not be the most promising, as outlined above at least for those craters formed in a shallow sea. Also for these more proximal deposits the well investigated K/T boundary sequences in the Mexican Gulf coastal plains provide qualitatively visual instruction (review in Smit 1999). Scans for shocked minerals, high-pressure polymorphs, (micro)spherules and/or iridium enrichments (among other geochemical signatures) are approaches for confirmatory evidence (review in Grieve 1997, and Montanari and Koeberl 2000). We emphazise the significance of further systematic searches in Fennoscandia. Ejecta layers could provide precise markers for local to regional stratigraphic correlation, as it has been demonstrated, for example, for Ordovician volcanic bentonites in and beyond Fennoscandia (Bergström et al. 1995).

All of the known cases of distal impactites in Fennoscandia comprise non-metamorphosed, Phanerozoic lithologies. In the cratonic basement no impact-related distal deposits have been discovered so far, although specific suspected lithologies require further investigations (section 5.3).

3.8 Small Quaternary Craterforms

At least 13 Quaternary craterforms have a probable or proven impact origin in Fennoscandia. This includes the Tor crater in Sweden ($D \approx 30$ m), the Kaali (8 craters) and Ilumetsa crater fields (at least 2 craters), as well as the single Simuna ($D \approx 8.5$ m) and Tsõõrikmäe craters ($D \approx 40$ m), all located in Estonia.

The Kaali crater field was the fourth suspected impact site on Earth, and the seventh proven at which related meteorites were found and confirmed as such (Chapter 2). The largest of the craters has a rim-crest diameter of 105 - 110 m, while the others are 13 to 39 m across (Krinov 1961; Raukas and Tiirmaa 1997; Czegka and Tiirmaa 1998; and others). The Kaali craters provide an instructive example for the approximate initial diameter at which round explosion craters form, depending on the velocity of the fragments at surface contact. The hypervelocity impact at the main crater is evident from the lack of meteorite fragments found in the smaller craters (Reinwald 1937; Czegka and Tiirmaa 1998), and the occurrence of small (two centimeter high) partial shatter cones in allochthonous dolomite breccia of the elevated rim (Krinov *in* Short and Bunch

1968). The most recent interpretations of the age of the Kaali event are 400 - 370 B.C. (Rasmussen et al. 2000) and 800 - 400 B.C. (Veski et al. 2001), both based on calibrated ^{14}C-dating of Ir-enriched peat in the Piila mire ~8 km northwest of the craters. Since charcoal is associated with the Ir-enriched layer, an earlier attempt to date the Kaali event by Aaloe et al. (1975) appears in new light. He dated charcoal found in three of the smallest craters at 540 - 930 B.C. (^{14}C). However, all these contradict an estimate of 6500 - 6720 B.C. (^{14}C), based on peat layers containing silicate spherules in various mires surrounding Kaali (Raukas 2000). This includes the Piila mire, but a horizon ~130 cm deeper than the Ir-enriched horizon. The higher age for Kaali is more compatible with investigations of drill cores from the fill of the Kaali main crater by Saarse et al. (1991); the highest radiocarbon date obtained was 1400 ± 35 B.C. Post-impact redistribution of carbon possibly make this result meaningless (Rasmussen et al. 2000), but pollen of early Subboreal age (> ~1500 B.C.) close to the crater bottom (e.g., Kessel 1981; Saarse et al. 1991) still remains difficult to reconcile with a much younger crater age. Generally one can suppose that two different events were recorded by the findings in the Piila mire, but it is still unclear which one corresponds to Kaali. Anyway, increased abundances of magnetic NiFe-spherules (>10 gm^{-3}) are observed as far as the Estonian mainland over 100 km from the Kaali crater field. The irregular distribution of these particles points to a projectile trajectory from the east-northeast (Tiirmaa 1988), whereas the arrangement of the craters with respect to their diameters points to a trajectory from the south-southeast (e.g., Krinov 1961).

The minimum age of the Ilumetsa crater field (Aaloe 1960; Czegka 1997; and others) is constrained by the ~6000 years old sediment fill of the main crater (Liiva et al. 1979) that corresponds rather well with a ~6600 years old bog layer ~6 km southwest of the craters that contains glassy spherules (Raukas et al. 2001). The main crater ($D \approx 76$ m) is excavated in Devonian sandstone, but unequivocal shock features and meteoritic fragments are unknown.

Confirming evidence for impact is also lacking at the other Quaternary craterforms Tor (Henkel et al. 1996), Simuna and Tsõõrikmäe (Pirrus and Tiirmaa 1984, 1991), all excavated in loose Quaternary rocks. The Ilumetsa and Tsõõrikmäe craters are, however, included in Table 1, because their origin is well established by analyses of the morphology, structure and crater-fill deposits (e.g., Pirrus 1995; Plado et al. 2000). The Simuna crater was possibly excavated by a fragment of the Viru-Roela bolide observed in June 1937 (Pirrus and Tiirmaa 1991). The apparent lack of meteoritic fragments in some of the small craters may be due to a stony projectile that either totally disintegrated at impact, rapidly weathered, or fragments of which are simply too difficult to recognize in the field. Although lower in strength than irons and stony-irons, we are not aware of physical evidence that excludes the formation of small craters by stony projectiles. For Ilumetsa the detection of meteoritic matter in the associated spherules may be a possibility to validate the impact hypothesis. The many efforts to identify extraterrestrial matter in peat and plants at the 1908-Tunguska explosion site in Siberia might indicate further promising approaches.

3.9 Projectiles

The projectile responsible for the Quaternary Kaali crater field in Estonia is unequivocally identified as a coarse octahedrite (IAB), as abundant fragments of it remained preserved (Reinwald 1937; Czegka and Tiirmaa 1998; and others). No fragments are preserved, as a rule, in larger impact craters. For those the identification of the projectile yields unambiguous results only in rare cases due to the inherently variable properties of the target rocks and of the projectiles themselves (review in Koeberl 1998). In Fennoscandia the earliest successful attempt was performed on melts rocks of the Lappajärvi crater, the projectile of which turned out to be chondritic (Göbel et al. 1980), probably a C-chondrite rather than a H-chondrite (Reimold 1982). Furthermore, Schmidt et al. (1997) revised an earlier work of Palme et al. (1980) and, based on additional data, proposed that the element signature in melt rocks from Sääksjärvi indicated an iron projectile. A chondritic projectile was excluded in particular because of a too high Ni/Ir ratio, and various lines of evidence indirectly suggest that the high Ni content is unrelated to the target rocks (also Pihlaja and Kujala 2000). In this respect it is, however, notable that various small ultramafic bodies occur in the vicinity of Sääksjärvi, partly less than 1 km from the inferred original crater rim (Mancini et al. 1996; Pihlaja 1994). As such bodies are frequently associated with Ni-Cu-Fe sulfide mineralization in the region, it seems to us a not too far-fetched possibility that similar bodies occurred in the central target area. Schmidt et al. (1997) also investigated impact melt from the Dellen and Mien craters in Sweden, but no clear projectile assignment could be made. Furthermore, the Mjølnir projectile is supposed to be an iron meteoroid (Dypvik and Attrep 1999) and that of Gardnos is listed by Koeberl (1998) as a chondrite.

3.10 Economic Potential

Impact events are in various cases responsible for deposits of considerable economic value including polymetallic ores, hydrocarbons and gems (review in Grieve and Masaitis 1994). In Fennoscandia minor ore occurrences in crystalline rocks, generally sulphides of Cu, Ni, Pb, Zn and/or Fe, are presumably related to the Paasselkä (Pesonen et al. 1999a), Kärdla (Sundblad 1994; Suuroja, this volume), Siljan (Johansson 1984; Wickman 1994), Lumparn (Bergman and Lindberg 1979) and Lockne (Sundblad 1994; Sturkell et al. 1997) impact structures as well as some other suspected sites. Apart from Paasselkä, in all cases at least two generations of ore deposits are known, one older than the impact event and one probably epigenitically related to it. The sulphides drilled near the center of the Paasselkä Lake could be essentially progenetic, i.e., their precipitation occurred prior to impact and they were brought to higher crustal levels by a central uplift.

Although, by present economic standards, none of the ore deposits mentioned has a sufficient significance to justify mining, their investigation is warranted, because their characterization may help to identify hydrothermal events, impact-induced and others, which post-date the cratonization of Precambrian shields. Ore deposits of considerable economic value are evidently associated with some of the large, suspected ring structures in the Fennoscandian Shield for which an impact origin has been proposed (section 5.2). Therefore, the study of these structures is also interesting from an economical point of view.

A possibly typical feature for such crater-associated hydrothermal ore deposition is to be mentioned here. In the periphery of Kärdla and Lockne compositions of lead isotopes in galena are relatively enriched in radiogenic lead (Sundblad 1994). This holds also for galena in sandstone dikes in the periphery of Lumparn (Vaasjoki 1977; Bergman and Lindberg 1979). A variable isotope composition of lead in galena is known from small-scale mines at Siljan as well (Johansson 1984; Wickman 1994). In all cases this phenomenon probably reflects remobilization of pre-impact ore and subsequent fractionation of lead isotopes via impact-induced hydrothermal activity, a process suggested by Sundblad (1994).

Worth mentioning are abundant gem-quality agates, a banded variety of chalcedonic quartz, that occur in impact breccias of the Sääksjärvi crater (Kinnunen and Lindqvist 1998). They formed by impact-related hydrothermal activity. Furthermore, Lappajärvi is of prime importance as a drinking water resource in the region (Pesonen 1996). This includes not only the open water body, but also ground water stored within crater-fill breccias and fractured basement. Lappajärvi and Söderfjärden serve as major recreation areas, which include down-hill skiing from the elevated crater rims. In Estonia the Kaali craters are under state protection, i.e., only professional studies are allowed. The craters are prominent natural monuments, annually visited by numerous tourists.

4
Preservation of Impact Structures in Fennoscandia

The discussion above has shown that every impact structure in Fennoscandia has its very own, specific properties depending on its age and size, on the target and probably the projectile, and on the post-impact geological development of the region. However, on the dynamic Earth large-scale endogenic and exogenic processes alter the crater record also as a whole (Grieve 1984; Masaitis et al. 1985). When viewing the spatial distribution of the known impact structures on the Fennoscandian Shield, it becomes obvious that, except Saarijärvi, all structures are situated in the southern part of the shield, south of 63.2°N latitude (Fig. 1). Various factors have been proposed to explain this fact (von Engelhardt 1974; Puura et al. 1994; Pesonen 1996). These include (1) a *knowledge bias* due to less detailed geological mapping of northern Fennoscandia and a lack of activity in searching for impact structures in this area, (2) a *recognition bias* due to a difference of Pleistocene glacial cover and style of glacial erosion, and (3) a *preservation bias* due to differences of the amount of erosion caused by block tectonics and/or paleogeographic extent of protecting sedimentary rocks.

(1) A knowledge bias is a possible reason, because, in general, southern Fennoscandia has a much longer tradition of geologic research than the northern part. Furthermore, a systematic search for impact structures, comparable to the campaign of the former Dominion Observatory in Canada some 30 - 50 years ago (e.g., Innes 1964), has been carried out in northern Fennoscandia at the earliest since the 1980s. A number of structures have been proposed, but detailed field studies are mostly lacking. It is believed, however, that the present state of knowledge already justifies the statement that in the north no very striking craterform structures, similar to the craters in the south containing large, roundish water bodies, are exposed. Those few circular lakes subjected to preliminary checking appear unrelated to impact, like Sompiojärvi in Finland (Henkel and Pesonen 1992; Abels 1999). Saarijärvi has been known as a geologic anomaly for more than 60 years (Pesonen et al. 1997b).

(2) A Pleistocene influence claimed on the present crater distribution was based on marked differences between glacial erosion and deposition between the northern and southern Fennoscandian Shield. However, in large areas of northern Fennoscandia glacial erosion was exceptionally weak during the Pleistocene, as indicated by the widespread preservation of pre-Quaternary weathering crusts in the so-called ice divide zone (Hirvas 1991; Olsen et al. 1996; and others). In general, glacial erosion was rather minor over the entire Fennoscandian Shield (0 - 200 m; e.g., Lidmar-Bergström 1997). Differences in glacial denudation thus cannot explain the present bias. Furthermore, the thickness of glacial deposits, possibly covering unknown impact structures, cannot be responsible, because it is, on average, much greater in the south, at least in Finland (Hirvas 1991). Only one glacial factor could contribute to the dominance of impact structures in the south. Erosion by glaciers is selective and, therefore, overdeepens already existing depressions, in particular when filled with disintegrated rocks, like breccias in impact structures. This may tend to make craters more easily perceptible as spot-like depressions in regions with stronger glacial erosion.

(3) The importance of post-impact sedimentary rocks for the preservation of impact structures has previously been recognized by various authors (e.g., Dent 1973; Grieve and Dence 1979; Masaitis et al. 1985). Most of the Fennoscandian craters were buried under sedimentary rocks and their life-time thereby increased. Furthermore, tectonic circumstances are able to influence the preservation of craters via block faulting (section 3.6). An examination of both factors with respect to possible influences on the bulk impact crater record requires a brief discussion of the post-cratonic geologic history of Fennoscandia.

Based on tectonic criteria, Fennoscandia comprises the region between the Caledonian orogen in the northwest, the Tornquist-Teissyre-Zone in the southwest, the Volhyn-Orsha-Central Russia rift system in the southeast, and the Pechora-Barents Sea orogenic belt in the northeast (Fig. 2; Bogdanova et al. 1996). In this sense Fennoscandia includes the Fennoscandian Shield with exposed Precambrian rocks and the northwestern East European Platform with largely Paleozoic cover rocks. The basement of Fennoscandia represents a patchwork of Archean and Mesoproterozoic crustal portions finally amalgamated in its westernmost part during the Sveconorwegian orogeny about ~1.2 - 0.9 Ga ago (Gaál and Gorbatchev 1987). The orogeny had far-reaching effects on the already consolidated part in the east, in particular in the form of block tectonics and diabase dike intrusions (Puura et al. 1996; Heeremans and Wijbrans 1999). Grabens developed and were filled with terrigenous sedimentary rocks, the so-called Jotnian Sandstone (Fig. 2; Kohonen et al. 1993; Amantov et al. 1996). Since these times deep erosion of the Fennoscandian basement amounted to some kilometres (Lidmar-Bergström 1995; Nironen 1997). Finally, however, denudation was replaced by a widespread transgression submerging the flat peneplain in the late Vendian (from the SE and possibly NW) and in the Lower Cambrian (from the SW and NW) continuing with variable extensions until the mid-Silurian (e.g., Flodén 1980; Uutela 1998). Major depocenters from where the adjacent regions were flooded are the Baltic Basin in the S, the Moscow Basin in the SE, and the juvenile Iapetus in the NW. In effect a sedimentary sequence of ≤500 m partly covered the shield, outliers of which are still preserved (Fig. 2). In the Upper Ordovician and Silurian the Caledonian orogeny widely affected the stable shield to the east, which reacted with block displacements. In addition, thick red beds (Old Red Sandstone) accumulated in the late Silurian-Devonian, derived from the orogeny, of which however only very few outliers east of the Caledonides are preserved. Their former widespread presence is indicated, for instance, through fission track thermochronology (e.g., Larson et al. 1999; Cederbom et al. 2000). The thickness of this detritus decreased eastwards, but still should have been 1 km or more in the Åland archipelago (Cederbom et al. 2000; at Lumparn in Fig. 1). It is unknown to what extent this molasse was lithified and, hence, how rapidly it was eroded again. However, since the Ordovician peaks the marine influence successively receded to the south of Fennoscandia, and the present shield, viz. partly the former platform, became again a terrestrial area (e.g., Puura et al. 1996; Lidmar-Bergström 1995). Minor peripheral transgressions occurred in the Late Carboniferous-Early Permian (Nikishin et al. 1996; Lidmar-Bergström 1995) and Tertiary (Tynni 1982). The big unknown is the Cretaceous, which covered southernmost Sweden, but extended possibly much farther onto the shield than evident at present (Lidmar-Bergström 1995). Normal block movements of largely unknown extent probably continued on many occasions up to the present (Puura et al. 1996; Nikishin et al. 1996; Lidmar-Bergström 1996).

The crucial question that follows from the foregoing summary is whether the post-cratonic geological history of Fennoscandia is reflected in the known impact structure record. When viewing a plot of all known crater ages, a marked concentration of eleven impacts in the Cambrian and Ordovician (of a total of 27 Pre-Quaternary craters listed in Table 1), between ~540 - 440 Ma, becomes immediately obvious (Fig. 4). Of those, eight are situated in the Fennoscandian Shield, two in the East European Platform (Kärdla, Mizarai), and Neugrund at the boundary between both (Fig. 2). Seven structures are of late Lower Ordovician to early Upper Ordovician age (note that the ages of Lumparn and Karikkoselkä are as yet based on circumstantial evidence only). This pattern does not follow the global pattern, which reveals a bias towards young (≤200 Ma) structures reflecting the cumulative trend of erosion of older structures (Grieve and Pesonen 1996). Microfossil determinations from the sediment fillings of the Lockne, Kärdla and

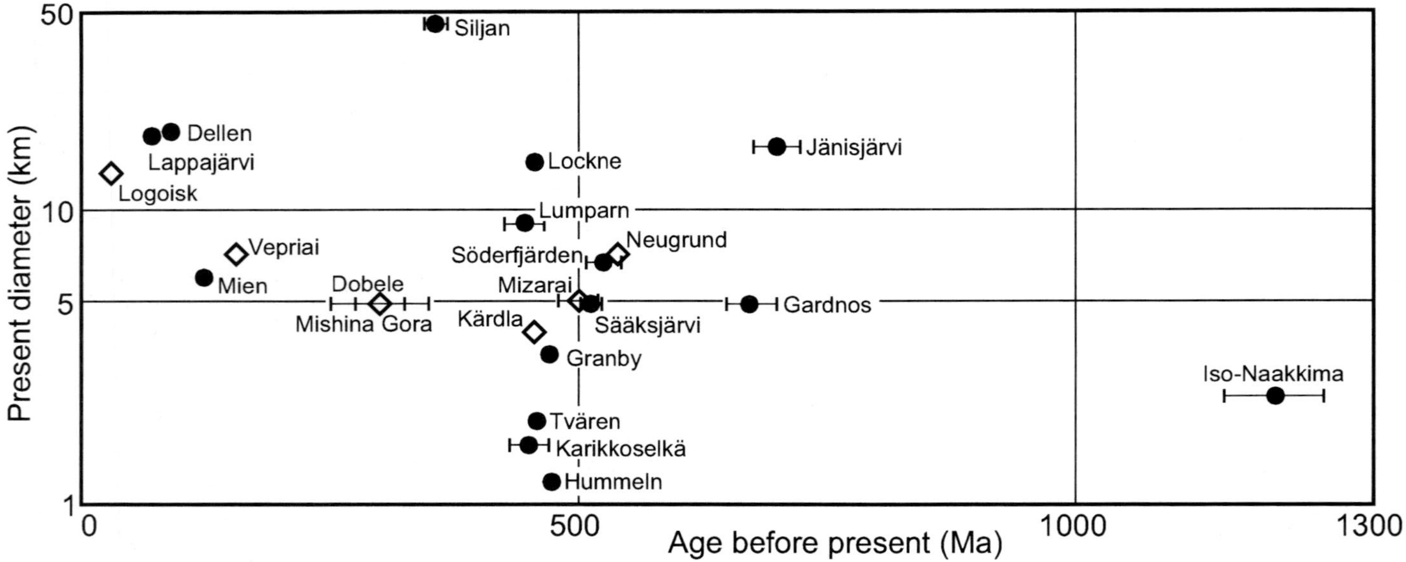

Fig. 4. Semilogarithmic plot of ages versus diameters of Fennoscandian impact structures (dots, structures in shield area; rhomboids, structures in platform area). Horizontal bars give approximate error in age estimate. Note that 'stratigraphic' ages of Lumparn and Karikkoselkä are still preliminary, and deviate from paleomagnetic results. Not shown are Suavjärvi, Saarijärvi, Paasselkä, Suvasvesi N, and Quaternary structures (see text).

Tvären craters indicate a non-contemporaneous formation (Grahn et al. 1996; Grahn 1997), as tentatively suggested by Lindström et al. (1992). Also the supposed ages of Hummeln (Lindström et al. 1999) and Granby (Grahn et al. 1996) do not match with any of these craters or, possibly, with each other. The ages of Karikkoselkä and Lumparn are presumably also in this age range, based on stratigraphic indications (section 3.2), but are still not well enough constrained to allow a reasonable statement on possible simultaneities. The same holds for the Cambrian craters Söderfjärden, Neugrund and Sääksjärvi.

In summary, the peak of Cambrian-Ordovician crater ages and the extensive transgressions of the Fennoscandian Shield, initiated in the Vendian to Cambrian, and lasting at least until the Upper Ordovician, appear to correspond. The known craters with this age formed in shelf seas or close to the advancing coast line and were rapidly buried under protecting sediments.

A similar pattern is seen for the cratering record of Laurentia, where some craters have an early Paleozoic age as well. These are Brent (~454 Ma; Grahn and Ormö 1995), Pilot (445 ± 2 Ma; Bottomley et al. 1990), Slate Islands (~436 Ma; Dressler et al. 1999), all in the shield, Calvin (~440 - 450 Ma; Milstein 1997, and references therein), Ames (~450 - 460 Ma, Repetski *in* Ormö and Lindström 2000), both in the platform, and possibly others of which the age, however, has a larger error. When comparing the crater ages of Laurentia and Fennoscandia with the remaining terrestrial record (Fig. 5), it becomes clear that, except Ilyinets in the Ukrainian Shield, the peak of early Paleozoic craters is restricted to the first two regions only. Thus, the concentration is likely an effect of the similar geological history of both. At least since the Middle Ordovician the Laurentian and East European cratons approached and, by closing the Iapetus Ocean, formed the Caledonian orogen in the Silurian. Both shields where flooded by widespread epicontinental seas (McKerrow et al. 1991; and others).

The claim that the present early Paleozoic crater record is a net effect of the burial-exhumation history of Fennoscandia demands an examination also of the differently aged impact structures in terms of their preservation relative to the geologic development. In the shield four structures have evidently a post-Ordovician age (Table 1). Three of those are also the largest craters (Siljan, Dellen, Lappajärvi) of the whole record (Fig. 4). It is clear that under equal conditions large (and hence deep) craters are degraded relatively slower than small craters. Therefore, these three large craters may be just the remaining sample of a previously larger population of which the smaller structures have been eroded. The higher age, size and deeper erosion level of Siljan relative to Lappajärvi and Dellen may reflect this selection process as well. However, the ~121 Ma old Mien crater in southernmost Sweden is one exception among the Phanerozoic craters in the shield. This crater is relatively small (Table 1), but apparently not more eroded than the 30 - 50 Ma younger and more than twice larger Dellen and Lappajärvi. One way to explain this circumstance is a temporary burial of Mien under sedimentary rocks for which indications exist (section 3.5). The assumption is in accordance with Grieve (1984), who estimated that unprotected 20 km large craters in the glaciated crystalline shields can be totally eroded in less than120 Ma, based on an integrated study of ages and preservation levels of craters

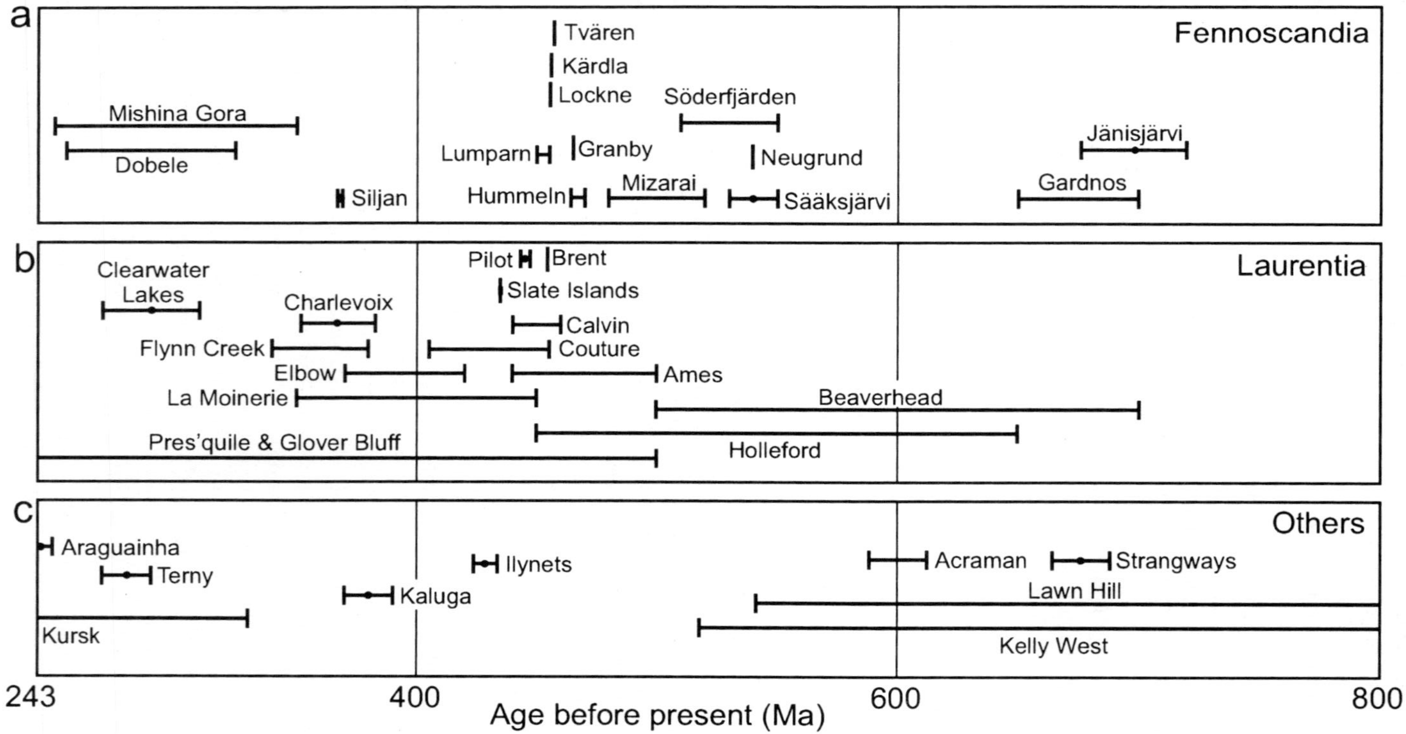

Fig. 5. Ages of impact structures in (a) Fennoscandia, (b) Laurentia and (c) the remaining global record. Note marked concentration of early Paleozoic impact ages in Fennoscandia and Laurentia. Horizontal bars give approximate estimate of error in age. Ages are compiled from various recent sources.

larger than 20 km in the Laurentian and East European Cratons. The preservation of the Precambrian structures is also explainable by burial and/or tectonic circumstances briefly summarized here (section 3.5). The Jänisjärvi crater reveals a similar preservation level and size as Lappajärvi and Dellen, but at ~700 Ma is older by almost one order. This discrepancy is likely due to burial under late Neoproterozoic and early Paleozoic platform sedimentary rocks outcropping nearby to the south (Fig. 2), presumably at least until the Cretaceous, the age of Lappajärvi and Dellen. Furthermore, Saarijärvi has been evidently covered by late Neoproterozoic to Cambrian sediments, which protected the crater from being erased, possibly in addition to protective post-impact tectonic modifications. A combined effect of burial under sediments and tectonic down-faulting also protected the ~1.2-Ga Iso-Naakkima crater for a long time. Finally, the Suavjärvi structure has survived, because it was temporarily buried, at least, under a Paleoproterozoic volcanosedimentary cover.

In summary, the character of the whole known cratering record of Fennoscandia appears reasonably explained by the geologic development of the region. Crucial factors for the preservation and destruction of pre-Quaternary impact structures in Fennoscandia are the presence or absence of a post-impact sedimentary cover and, in some cases, of protective tectonic peculiarities.

This conclusion does still not explain the observation mentioned in the introduction that the known impact structures are concentrated in southern Fennoscandia. This distribution could be indeed an indirect effect of the extent of the Vendian to early Paleozoic shelf seas and of the Caledonian foreland basin, or of a regionally different timing of Phanerozoic stripping, but a poor knowledge of the northern areas, for both impact structures and paleogeography, cannot be excluded categorically. In this respect it is an unlucky circumstance that clues on the paleogeography essentially depend on impact craters, which are often the only traps for ancient sediments within large areas. Newly discovered structures have changed paleogeographic reconstructions in Fennoscandia radically over the last decade.

The present age-size distribution of the known crater record (Fig. 4) implies the possibility to predict age limits of putative, post-cratonic impact structures (section 5.1), or of proven craters for which the age is still unknown (e.g., Paasselkä), depending on their apparent preservation levels, sizes and locations. Such predictions, however, are difficult because of (1) the aforementioned paleogeographic uncertainties, (2) a poor knowledge about the extent of tectonic faulting and tilting of crustal blocks, and of the associated variable denudation levels within the shield, (3) inaccuracies in the quantification of glacial erosion at different locations (glacier lobes *versus* interlobate zones, warm-based *versus* cold-based glacier sections, etc.), and (4) a natural variability of the phenomenology of craters even within a certain impact energy range (section 3.5). An example from the Canadian Shield elucidates the possible degree of uncertainty. The Mistastin impact crater ($D \approx 28$ km, 38 ± 4 Ma) is more deeply eroded than the similar-sized Clearwater-East impact crater ($D \approx 26$ km; 290 ± 20 Ma), despite the much higher age of the latter. This paradox may result from a difference in the amount of glacial erosion (Dent 1973), but this has not been investigated in detail so far (see also Grieve and Robertson 1979).

Finally, another note of caution is required. Our hypothesis of a geology-driven selection of the crater population cannot exclude an increased meteoroid or cometary flux between the Vendian and Ordovician, or temporal variations of the cratering rate in general. This reservation is not academic, especially in view of the work of Schmitz et al. (1996, 1997), who report from a lower-(middle) Ordovician (late Arenig-early Llanvirn) limestone sequence in southern Sweden (Kinnekulle; Fig. 6) a number of fossil meteorites and excess extraterrestrial Ir and Os. Based on these findings, these authors supposed increased accretion rates of meteorites and cosmic dust for the Lower Ordovician, possibly related to the disruption of a L chondrite parent body ~500 Ma ago. This latter date is due to recent, heavily shocked and degassed L chondrites with Ar-Ar ages of ~500 Ma suggestive of the breakup of a parent body at that time (Haack et al. 1996). Ames, Hummeln and Granby are close in age to the Kinnekulle section (Granby, with ~90 km, also close in distance) and, hence, to the postulated increased flux of cosmic matter. One can speculate that the extraterrestrial elemental signature at Kinnekulle originates from ejecta, possibly reworked, which may shed some light on the apparently not elevated cosmic ^{3}He flux in the same section noted by Patterson et al. (1998). At present, however, improved dating of various craters and search for confirmable ejecta are required before more speculations on putative relations are justified.

5 Possible Impact Features in Fennoscandia

5.1 Suspected Impact Structures

Apart from the 30 accepted and confirmed impact structures in Fennoscandia, an impact origin has been suggested at present for some 60 additional sites (Wickman 1988, 1992; Henkel and Pesonen 1992; Pesonen 1996; Abels 1999; and others). Most of these structures are poorly documented in preliminary reports or lists. We also appreciate that most sites probably turn out to be unrelated to impact or that a consensus on their formation will never be reached. However, at least the awareness of such claims is desirable for those working in the respective regions in the future. The following account is intended to show the broad variety with which impact structures could appear in the geologic record of Fennoscandia and which factors influence their appearance. The location of the examples mentioned is shown in Fig. 6.

Submarine craterform structures have been detected in the Baltic Sea by seismic surveys. The first suggested to be a possible impact structure was the Landsort Trench ~50 km north of Gotland (Wickman 1973, 1988). The feature represents a hook-shaped, 3.2 km wide morphological trench and structural graben, respectively, bounding a larger ~38 km wide, drop-shaped half-graben to

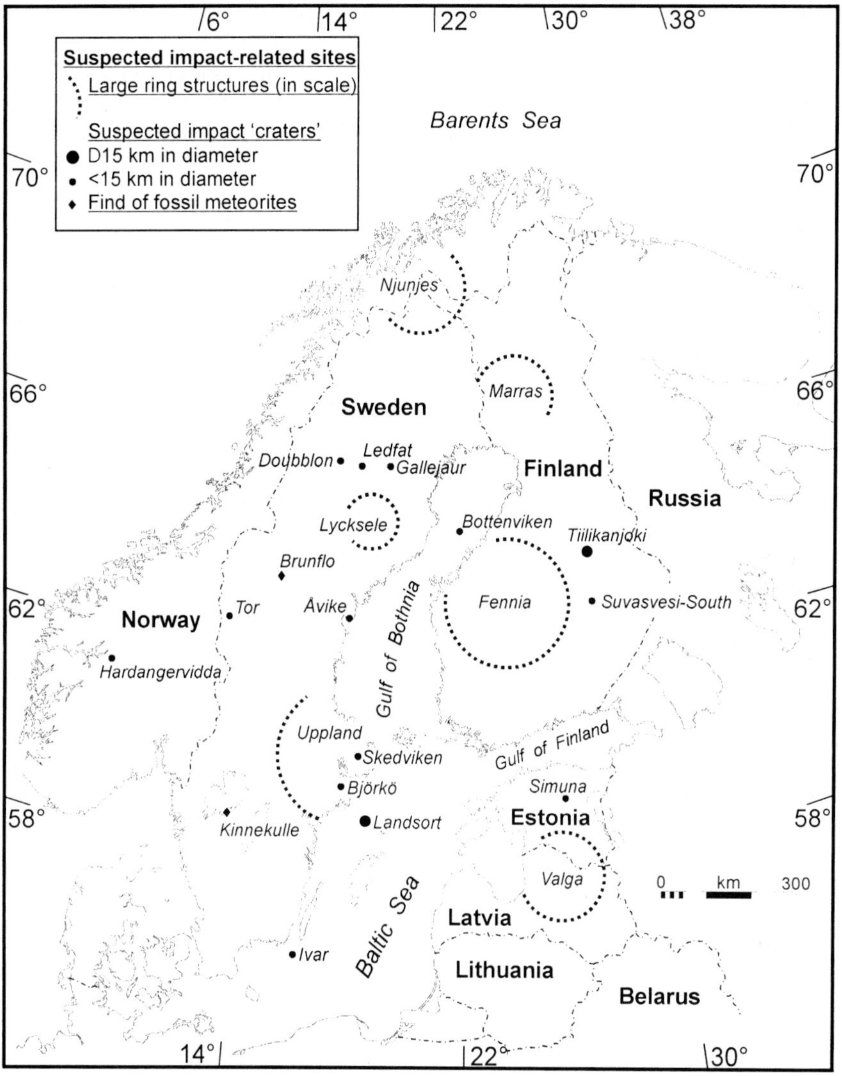

Fig. 6. Map of suspected impact-related sites and finds of fossil meteorites in Fennoscandia. See text for explanation.

the northwest. Both are situated in Precambrian crystalline basement and filled with predominantly Jotnian sandstone (~1200 Ma), buried in the SE under Cambrian and Ordovician platform cover (Flodén 1980). Wickman (1973) drew a possible analogy between the Landsort Trench and a ring graben as known from complex impact structures, e.g., Siljan. An interference pattern of tectonic faults, possibly caused by a rapakivi intrusion (Puura and Flodén 1999), is a plausible alternative, but for the isolated, roundish setting and the virtually unknown subsedimentary properties, the impact option should be kept in mind.

Further promising craterforms in the Baltic Sea are the Ivar ($D \approx 5$ km; Flodén and Bjerkeus 1994) and Bottenviken ($D \approx 5$ km; Flodén *in* Wickman 1988) anomalies. Paleogeography and seismic signatures indicate a post-Permian (Ivar) and post-Ordovician (Bottenviken) age. In consideration of a comparative study of the seismics of circular geological features by Stewart (1999) and of the regional geology, the only potential alternatives to impact structures are small-scale pull-apart basins.

On-shore suspected impact structures in Fennoscandia have been discovered in a variety of ways depending on their degree of exposure and preservation, and simply on which method was used to first survey the respective regions. This includes bedrock mapping, erratic boulder findings, airborne and ground-based geophysics, as well as topographic surveys and orbital remote sensing.

Few suspected structures have a well preserved crater-form, easily detectable in satellite imagery or on large-scale topographic maps. The 7 km wide Åvike bay at the Swedish east-coast reveals a striking circular outline and also the bathymetry is reminiscent of a central uplift surrounded by a ring moat (Söderström 1966; Henkel and Pesonen 1992). Dikes and boulders on the shores point to an eroded alkaline intrusion (the Alnö carbonatite complex is situated ~16 km to the west), but this may be coincidental, as at the Tswaing (South Africa) and Brent (Canada) impact craters. More degraded crater forms are the Suvasvesi S lake in Finland ($D \approx 4$ km), first perceived in satellite images (Lehtinen *in* Henkel and Pesonen 1992; Pesonen et al. 1996), and the shallow Skedviken depression in Sweden ($D \approx 5.5$ km), detected by an airborne electromagnetic (VLF) survey (Claesson 1985).

The Björkö probable impact structure ($D \approx 8$ km), Sweden, is an example of a long known geologic anomaly that has been only recently reinterpreted as an impact structure (Wickman 1988; Flodén et al. 1993). It comprises an irregularly shaped bay and various islands on which scarce Jotnian Sandstone crops out. These rocks define the minimum age (~1200 Ga) of the impact, as sedimentary features point to deposition in an already existing craterform now partly eroded below the floor (Flodén et al. 1993).

At structures partly or entirely eroded below allochthonous deposits, the extensive loss of a crater-like morphology makes detection difficult to impossible in commonly available remote sensing and topographic data. Furthermore, at complex impact structures the lack of horizontal layering of most crystalline targets (magmatic or metamorphic) effectively prevents the development of distinct circular signatures of the central uplift. These deeply eroded structures are likely to be detectable only through geologic mapping and drilling, new or reinterpreted, or by typical geophysical signatures. However, one reservation can be made. Irregular spot-like or ring-shaped fracture patterns, when well exposed and morphologically enhanced by differential weathering, may become perceptible at deeply eroded structures. This is because at larger impacts target rocks are fractured into blocks with sizes well above the spatial resolution of common remote sensing and topographic data, and slumping faults of the structural rim reach well below the level of the true crater floor (e.g., Ivanov et al. 1996; Juhlin and Pedersen 1987).

Finally any kind of large-scale morphologic or geologic expression of the impact gets lost and only some patches, zones or dikes with cataclastic to pseudotachylitic breccias generated in the crater basement may be preserved. Askvik (2000) described suspected breccia sites in the Hardangervidda area of southern Norway and notes similarities to exposed breccias at the Gardnos impact structure to the east (Fig. 6). Apart from exposures at the proven impact structures, drill cores of crystalline crater basements also provide illustrative material for evaluating suspected exposed features. In Fennoscandia up to meter-thick breccia dikes and fracture zones have been documented in cores from Siljan (AlDahan 1990; Metzler and Stöffler 1990), Lappajärvi (Vaarma and Pipping 1997) and Paasselkä (Pesonen et al. 1999a). Unfortunately in the absence of mineralogical or geochemical evidence, there only remains the difficult task of convincingly excluding a tectonic, sedimentary or intrusive origin of exposed breccias.

Impact craters that pre-date the major orogenies in Fennoscandia can only have survived at localities with insignificant deformation and magmatism. This is a required prerequisite for the preservation of the circular Gallejaur melt body (1.88 - 1.87 Ga) in northern Sweden, recently suggested to represent a differentiated impact melt surrounded by remnants of a breccia-filled moat (Lilljequist 2000).

After denudation of such impactites the circular lithologic arrangement of a central uplift may still remain intact, but only in presently flat-lying lithologies, either representing original stratigraphy or structural layering, for instance, tectonic nappes. An example of such a structure is Tiilikanjoki ($D \approx 15$ km) in central Finland. The feature comprises a central plug of tonalite and a collar of altered rocks, mostly amphibolites (Paavola 1997). The rocks are Archean in age and the structural elements point to a formation in the Paleoproterozoic, but the mode of emplacement is still unknown. Apart from a central uplift, it may otherwise be a tectonic dome, as well known elsewhere in eastern Finland.

If severe deformation and magmatism affects the impacted terrain, any exposed circular signature is successively lost and only portions of crater-fill lithologies and/or of the altered crater basement may still be preserved. These lithologies may even occur in their original stratigraphic sequence. The late Paleoproterozoic Letfat (Wickman 1988) and Duobblon (Wickman 1988; Lilljequist and Reimold 1999) formations in Sweden are possible examples. For the greater part of the Fennoscandian Shield, however, deformation and magmatism during the major orogenies set an effective maximum retention age for impact structures (note that Suavjärvi [Table 1] is of particular interest for analysing alternated shock metamorphic effects under geologic conditions, because according to Mashchak and Naumov [1996] it is overprinted by regional metamorphism).

A still unexploited avenue for identifying deeply eroded impact sites represents impact-induced thermal metamorphism. Various factors, like hot impact melt-bearing rocks, uplifted pre-impact geothermal gradients, release of elastic strain energy following shock and possibly shielding effects of impactites with decreased geothermal conductivity (e.g., Gibson et al. 1998; Popov et al. 1998) can induce hydrothermal systems that led to widespread metasomatic alterations in the fractured basement of an impact structure. Some of these alterations, e.g., the frequently observed exchange of sodium and potassium in impactites, are actually long known (e.g., Dence 1971). Metasomatism of a similar kind is a characteristic feature of regional metamorphic rocks in Fennoscandia, traditionally called leptites and hälleflinta. The investigation of such lithologies could be useful for a better understanding of longer-term alterations below the floors of impact craters (see AlDahan 1990, for altered fracture zones in the deep basement of Siljan). An interesting new clue is presented by Puura et al. (2000) from a study at the Kärdla crater. These authors link mineralogical changes in suevitic breccias and the sub-crater floor basement not to hydrothermal activity, but directly to the fireball-target interaction at impact. A drill core into the central uplift of the Puchezh-Katunki crater, Russia, contains highly shocked and extreme thermally metamorphosed sections that reveal partial selective melting and specific recrystallized breccias, so called coptoblastolithes (Masaitis and Mashchak 1996; Masaitis 1999a, b). Comparative and experimental investigations may show in the future if some of these alterations are only possible at the extreme p/T-conditions during impact, not observed in endogenic magmatic systems.

5.2 Large Ring Structures

The existence of huge circular structures in the Fennoscandian Shield has been claimed ever since country-scale remote sensing and geophysical data became available (Moralyov and Glukhovskií 1981; Witschard 1984; Henkel and Eriksson 1987; Korhonen 1993; Pesonen 1996). Some of these features are not generally accepted or they are simply ignored. In our opinion the structures listed in Table 3 are real, independent of their origin, because they have been inferred from various data sets (cf., references in annotation of Table 3, and text). Most authors describing these structures bring up the possibility that they represent ancient impact structures or basins, though more comprehensible impact concepts have been only presented for Lycksele, Marras, and Uppland.

Lycksele, Marras, and Uppland are characterized by central gravity highs that gradually merge outwards into arcuate lows. These are, in turn, limited by the supposed edges of the structures sometimes defined in sections by lithological contacts (Korhonen 1990; Henkel and Lilljequist 1992; Thunehed et al. 1999). In all cases either structural uplift or intrusion of relatively dense rocks could explain the central gravity high, since the structures appear to be isostatically compensated overall with respect to the regional gravity fields. At Uppland the supposedly corresponding rocks are mainly centrally exposed tonalites surrounded by lower density granites (Henkel and Lilljequist 1992), whereas at Marras and Lycksele the rocks inside the structures are also granites in the center (Korhonen 1990; Thunehed et al. 1999). At all these sites structural disturbance in the approximate centers, about 1/5 to 1/6 of the structures' overall diameters, has been interpreted from anomaly patterns in magnetic and gravity data. Njunjes is largely buried under Caledonian nappes and interpreted from regional gravity and magnetic data

Table 3. Some large ring structures in Fennoscandia for which an impact origin has been proposed

Name	Diameter (km)	Age (Ga)	Indications
Uppland	~320	1.80 - 1.50	BGLMRT
Fennia	~270	>1.89	LMPR
Njunjes	~200	Proterozoic	GM
Valga	~180	Proterozoic	GS
Marras	~160	2.22 - 1.77	GLMPRT
Lycksele	~130	1.80 - 1.26	GLMPT

Indications: *L*, distribution of lithological units; *B*, breccias; *G*, gravity; *M*, magnetics; *P*, petrophysics; *R*, orbital remote sensing; *S*, seismics; *T*, topography

References: *(Uppland)* Henkel and Lilljequist 1992, Reimold and Henkel 1997; *(Fennia)* Korhonen 1993, 1994, 1997; *(Njunjes)* Henkel and Eriksson 1987, Henkel and Pesonen 1992; *(Valga)* Pesonen 1996, unpubl. data; *(Marras)* Korhonen 1990, 1993, 1994; *(Lycksele)* Pesonen 1996; Thunehed et al. 1999

(Henkel and Eriksson 1987), and Valga has been inferred from geophysics (gravity; Pesonen 1996). Fennia comprises the Central Finland granitoid complex and is characterized by radially changing petrophysical properties (Korhonen 1997) as well as a concentric fracture system interpreted from aerial photography (Talvitie 1979). Geochemical studies suggest that the complex results from several circles of Paleoproterozoic crustal addition (Lehtinen and Huhma 1997).

The basic problem with all these features is their roughly circular shape that is, in consideration of their dimensions, rather at odds with expected plate tectonic settings. As explanations (1) meteorite impact, (2) diapiric to intrusive rise of plutons and (3) basement culminations have been advocated. Arcuate faulting and vertical displacement should result from (1) impact crater collapse and possibly post-impact isostatic adjustment, (2) updoming by forceful emplacement and possibly caldera formation and (3) updoming by lateral compressive stress or gravitational compensation of crustal loading with dense volcanic rocks, respectively. Impact-induced lower-crustal intrusion of mafic melts causing upper-crustal melting and granite generation has been suggested for the Marras and Fennia structures (Korhonen 1990, 1993). The few available age constraints suggest a formation of the Uppland and Lycksele structures after the main Svecofennian orogeny, i.e., after ~1.88 Ga.

Various authors emphasize the potential control of polymetallic ore mineralization along the edges and within some of the ring complexes (Witschard 1984; Lilljequist 1992; Korhonen 1993; Allen et al. 1996; and others). For those advocating impact, deep reaching fracturing and increased permeability, respectively, with associated thermal perturbations, are of major significance for generating the ore deposits.

In order to reach a convincing conclusion on the origin of these structures, it is essential to determine more tightly the ages of the lithologies and deformations involved. In conjunction with timing constraints set by the regional exhumation and geotectonic history, this should allow a more reliable conclusion whether the present-day geologic settings are consistent with a suddenly formed large impact structure or not. In any case, for these large, old and complex structures it is even more imperative than for other sites to find confirmatory evidence, since potential alternatives are virtually impossible to rule out with certainty.

In addition, it should be noted that the appearance of large circular features in small-scale geophysical maps and remote sensing images strongly depends on the image processing applied, the type and resolution of the data used, and even on the psychology of visual perception. Whether low-resolution data should be consulted to postulate large, subtle ring structures, is at least worth discussing, as the inherent fusion of sub-resolution and principally independent features is prone to produce perceptual artefacts. In general, such interpretations of maps or images are more convincing when the applied image processing and the sensor characteristics are also evaluated.

5.3 Potential Distal Impact Lithologies

Discoveries of spherule-bearing, putative distal impact ejecta, like those found in South Africa (3.24 - 3.46; 2.64 Ga) and Australia (2.49 - 2.63 Ga) (Simonson and Harnik 2000; Shukolyukov et al. 2000, and others), has not been announced in Fennoscandia so far. The relatively stronger metamorphism of the Fennoscandian Shield may serve as an explanation, albeit the awareness of the possibility of such impact lithologies is still the exception among geoscientists. Fennoscandian iron-formations contain abundant spherules of variable mineralogy and habitus (e.g., Paakkola and Gehör 1988) and are therefore of special interest. Vishnevsky and Raitala (2000) and Masaitis et al. (1998b) discuss the diagnostic value of impact diamonds in regionally metamorphosed Precambrian lithologies, because of their extreme resistivity against pressure and temperature.

Other lithologies of potential interest are Paleoproterozoic (Marmo and Ojakangas 1984) and Neoproterozoic diamictites (Edwards 1984) in Fennoscandia, which are generally interpreted to be glaciogenic. It has been suggested, however, that such deposits could partly represent impact-related diamictites as well, based on the apparent underrepresentation of ejecta horizons in the stratigraphic record and on the similarity to some ejecta deposits around proven impact structures (Oberbeck et al. 1993). Rampino (1994) noted a potential similarity between (1) many tillites deposited as glaciomarine debris-flows (for Fennoscandia see for example Eerola 1996) and (2) hypothetical debris-flows induced by impacts with simultaneous rainout of ejected debris. The submarine ejecta deposits of the Kärdla, Lockne and Neugrund impacts and the Narova breccia (section 3.7) represent illustrative material for a comparison with the diamictites considered diagnostic of glaciations. Since both glaciations and large-scale impacts are associated with major environmental perturbations, it is noted that Fennoscandia is a key region for the study of such perturbations associated with the formation of diamictites in the Paleoproterozoic (Melezhik and Fallick 1996).

Furthermore, interstratigraphic breccia occurrences presumably deposited in a high-relief energy, but non-glacial, environments are of interest as well. The sedimentary Rieskavaara breccia in central Finland is a striking example for such a sudden immature deposit within a Paleoproterozoic sequence (Gehör and Havola 1988). However, impact-generated and tectonic deposits are unlikely to be distinguishable, unless other evidence for impact can be discovered. This includes also tsunami layers as well as any kind of rock fall and mass flow deposit.

6 Outlook

Finally, a few probable future trends may be outlined. The investigation of the potential relationships between ore deposits in the Precambrian shield and deeply eroded impact structures is a branch of cratering research that is expected to be more significant than it was and still is. This is not only because impact researchers are increasingly forced to justify their studies on economic grounds, but also vice versa, economic geologists will hopefully become increasingly aware of the significance of impact cratering for the formation and redistribution of ores. The related study of hydrothermal systems and thermal metamorphism will possibly also provide new tools to identify deeply eroded impact structures. The study of alterations of shock features through metamorphism represents a major task for the future.

The present work has shown that the impact structure record in Fennoscandia is just a small, highly biased sample of the craters actually formed in that region. The large majority of craters has been erased from the record, although it is almost certain that some structures still await discovery, in particular those that are deeply eroded and large or buried (for a list and map of suspected and proven impacts in Fennoscandia see *http://www.gsf.fi/paleo/files/fenno.html*). Apart from the search for yet undiscovered impact structures, further attempts must be made to precisely date the known structures, not only for the benefit of improved determinations of cratering rates and of potential variations therein. For example, a detailed knowledge of the tempospatial distribution of impact structures in Fennoscandia has direct implications for the exploration on kimberlites, whose preserved facies is equally governed by the burial-exhumation history of the region.

Furthermore, the rapidly increasing knowledge of impact stratigraphy will lead to new discoveries in the stratigraphic records of Fennoscandia, in both the platform cover and the cratonic basement. Another field of study for which the Fennoscandian geology provides opportunities is the potential role of impacts for the rapid biological evolution in the Vendian to Cambrian.

Generally it is hoped that our contribution served its primary purpose of giving a comprehensible overview on some of the recent advances regarding impact phenomena in Fennoscandia, and hopefully the one or the other reader remains with an idea for future work. Studies of impacts will not only supplement our current views about the planet, but they have the potential to lift existing paradigms off their hinges.

Acknowledgements

We are grateful to M.R. Dence and M. Lindström, whose careful reviews led to considerable improvements of the present paper. L.I. Glazovskaya, R.A.F. Grieve, V.L. Masaitis, G. Motuza, M.V. Naumov, and A. Uutela provided further information on specific craters or issues. The paper contains material of a doctoral thesis (by AA) in preparation at the University in Münster, Germany.

References

Aaloe AO (1960) Ilumetsa craters in Estonian SSR (in Russian). Meteoritika 18: 26-31

Aaloe A, Eelsalu H, Liiva A, Lõugas V (1975) On the correction of the age of the Kaali meteoritic craters (in Estonian with English abstract). Eesti Loodus 18: 706-709

Abels A (1999) Possible impact structures in Finland. Interim report, GPI/IfP Münster - GSF Espoo, 46 pp

Åberg G, Collini B, Schmitz B (1989) K-Ar isotope analyses from the Siljan Ring meteorite impact structure, Sweden. Geologiska Föreningens i Stockholm Förhandlingar 111: 355-360

AlDahan AA (1990) Alteration and mass transfer in cataclastites and mylonites in 6.6 km of granitic crust at the Siljan impact structure, central Sweden. Contributions to Mineralogy and Petrology 105: 662-676

Allen RL, Lundström I, Ripa M, Simeonov A, Christofferson H (1996) Facies analysis of a 1.9 Ga, continental margin, back-arc, felsic caldera province with diverse Zn-Pb-Ag-(Cu-Au) sulfide and Fe oxide deposits, Bergslagen region, Sweden. Economic Geology 91: 979-1008

Amantov A, Laitakari I, Poroshin Y (1996) Jotnian and Postjotnian: sandstones and diabases in the Gulf of Finland. In: Koistinen T (ed) Explanation to the Map of Precambrian basement of the Gulf of Finland and surrounding area 1 : 1 mill. Geological Survey of Finland, Special Paper 21: 99-113

Arkonsuo A (2000) Impact metamorphism and the Karikkoselkä crater in Petäjävesi (in Finnish). MSc. thesis, Department of Geology and Mineralogy, University of Helsinki, 113 pp

Askvik H (2000) Breccias of assumed impact origin in the Precambrian of northern Hardangervidda, central south Norway [abs]. In: Plado J, Pesonen LJ (eds) Meteorite Impacts in Precambrian Shields. Programme and Abstracts of ESF workshop: Meteorite Impacts in Precambrian Shields. Geological Survey of Finland and University of Helsinki, Espoo: p 42

Åström K (1998) Seismic structure of the lake Mien impact structure, southern Sweden. Geophysical Journal International 135: 215-231

Bergman L, Lindberg B (1979) Phanerozoic veins of galena in the Åland rapakivi area, southwestern Finland. Bulletin of the Geological Society of Finland 51: 55-62

Bergström SM, Huff WD, Kolata DR, Bauert H (1995) Nomenclature, stratigraphy, chemical fingerprinting, and areal distribution of some Middle Ordovician K-bentonites in Baltoscandia. GFF 117: 1-13

Bogdanova SV, Pashkevich IK, Gorbatschev R, Orlyuk MI (1996) Riphean rifting and major Palaeoproterozoic crustal boundaries in the basement of the East European Craton: geology and geophysics. Tectonophysics 268: 1-21

Bottomley RJ, York D, Grieve RAF (1978) ^{40}Ar-^{39}Ar ages of Scandinavian impact structures: Mien and Siljan. Contributions to Mineralogy and Petrology 68: 79-84

Bottomley RJ, York D, Grieve RAF (1990) ^{40}Ar-^{39}Ar dating of impact craters. Proceedings Lunar and Planetary Science Conference 20: 421-431

Broch OA (1945) Gardnos breccias in Hallingdal (in Norwegian) Norsk Geologisk Tidsskrift 25: 16-25

Bruun Å, Dahlman B (1980) The Paleozoic bedrocks (in Swedish). Sveriges Geologiska Undersökning Avhandlingar 119: 59 pp
Bruun Å, Dahlman B (1982) The Paleozoic bedrocks (in Swedish). Sveriges Geologiska Undersökning Avhandlingar 120: 76 pp
Bylund G (1974) Paleomagnetism of a probable meteorite impact, the Dellen structure. Geologiska Föreningens i Stockholm Förhandlingar 96: 275-278
Carstens H (1975) Thermal history of impact melt rocks in the Fennoscandian Shield. Contributions to Mineralogy and Petrology 50: 145-155
Cederbom C, Larson SA, Tullborg E-L, Stiberg J-P (2000) Fisson track thermochronology applied to Phanerozoic thermotectonic events in central and southern Sweden. Tectonophysics 316: 153-167
Cisowski S, Fuller M (1978) The effect of shock on the magnetism of terrestrial rocks. Journal of Geophysical Research 83: 3441-3458
Claesson O (1985) The Skedviken structure. Sveriges Geologiska Undersökning, Berggrundsbyrån, Geofysikrapport no. BRAP 85 405, 24 pp
Claeys P, Casier JG (1994) Microtektite-like impact glass associated with the Frasnian-Famennian boundary mass extinction. Earth and Planetary Science Letters 122: 303-315
Claeys P, Casier J, Margolis SV (1992) Microtektites and mass extinctions: Evidence for a late Devonian asteroid impact. Science 257: 1102-1104
Czegka W (1997) The crater group of Ilumetsa (South-Estonia) (in German). Aufschluss 48: 200-210
Czegka W, Tiirmaa R (1998) The Holocene meteorite crater field of Kaali on Saaremaa (Ösel), Estonia (in German). Aufschluss 49: 233-252
De Geer G (1880) Några ord om bergarterna på Åland och flyttblocken derifrån. Geologiska Föreningens i Stockholm Förhandlingar 5: 469-484
Dence MR (1971) Impact melts. Journal of Geophysical Research 76: 5552-5565
Dent B (1973) Glacial exhumation of impact craters on the Canadian Shield. Geological Society of America Bulletin 84: 1667-1672
Deutsch A, Schärer U (1994) Dating terrestrial impact events. Meteoritics 29: 301-322
Donner J (1996) The Fennoscandian Shield within Fennoscandia. Bulletin of the Geological Society of Finland 68, Part 1: 99-103
Dressler BO, Sharpton VL, Copeland P (1999) Slate Islands, Lake Superior; a mid-size, complex impact structure. In: Dressler BO, Sharpton VL (eds) Geological Society of America Special Paper: 109-124
Dypvik H, Attrep M Jr (1999) Geochemical signals of the Late Jurassic, marine Mjølnir impact. Meteoritics & Planetary Science 34: 393-406
Dypvik H, Gudlaugsson ST, Attrep M Jr, Ferrel RE Jr, Krinsley DH, Mørk A, Faleide JI, Nagy J (1996) Mjølnir structure: An impact crater in the Barents Sea. Geology 24: 779-782
Dypvik H, Kyte FT, Smelnor M (2000) Iridium peaks and algal blooms - The Mjølnir impact [abs]. Lunar and Planetary Science 31: #1538, 2 pp (CD ROM)
Dyrelius D (1988) The gravity field of the Siljan Ring structure. In: Bodén A, Eriksson KG (eds) Deep Drilling in Crystalline Bedrock, Vol 1: The Deep Gas Drilling in the Siljan Impact Structure, Sweden and Astroblemes. Springer Verlag, Berlin, pp 85-94
Dyrelius D (1994) Geophysical studies of the Siljan impact structure [abs.]. In: Törnberg R (ed) 2nd International ESF workshop: The Identification and Characterization of Impacts, Lockne, Sweden, 1 p
Edwards MB (1984) Sedimentology of the Upper Proterozoic glacial record, Vestertana Group, Finnmark, North Norway. Norsk Geologiske Undersøkning Bulletin 394: 1-76
Eerola TT (1996) Some comments on Neoproterozoic diamictites of northern Norway [abs]. 22nd Nordic Geological Winter Meeting, Turku, Finland, abstract, 1 p
Elming SA, Bylund G (1991) Paleomagnetism and the Siljan impact structure, central Sweden. Geophysical Journal International 105: 757-770

Elo S, Jokinen T, Turunen P (1983) Geophysical investigations at the Saarijärvi clay deposit (in Finnish). Geological Survey of Finland, Report of Investigation 58: 101-116

Elo S, Laurema R, Lavikainen S, Winterhalter B (1984) Geological and geophysical observations on Lake Paasivesi and its surroundings, southeastern Finland. Geological Survey of Finland, Report of Investigation 64: 24 pp

Elo S, Kivekäs L, Kujala H, Lahti SI, Pihlaja P (1992a) Recent studies of the Lake Sääksjärvi meteorite impact crater, southwestern Finland. Tectonophysics 216: 163-167

Elo S, Jokinen T, Soininen H (1992b) Geophysical investigations of the Lake Lappajärvi impact structure, western Finland. Tectonophysics 216: 99-109

Elo S, Kuivasaari T, Lehtinen M, Sarapää O, Uutela A (1993) Iso-Naakkima, a circular structure filled with Neoproterozoic sediments, Pieksämäki, southeastern Finland. Bulletin of the Geological Society of Finland 65: 3-30

Elo S, Zhdanova L, Chepik A, Pesonen LJ, Philippov N, Shelemotov A (2000) Comparative geophysical description and modelling of Lappajärvi and Jänisjärvi impact structures, Fennoscandian shield [abs]. In: Plado J, Pesonen LJ (eds) Meteorite Impacts in Precambrian Shields. Programme and Abstracts of ESF workshop: Meteorite Impacts in Precambrian Shields. Geological Survey of Finland and University of Helsinki, Espoo: p 35

Engelhardt W v (1974) Meteorite craters (in German with English abstract). Die Naturwissenschaften 61: 413-422

Engelhardt W v (1997) Suevite breccia of the Ries impact crater, Germany: Petrography, chemistry and shock metamorphism of crystalline rock clasts. Meteoritics & Planetary Science 32: 545-554

Engelhardt W v, Stöffler D (1965) Cleavage planes in quartz as signs for impacts of large meteorites (in German). Die Naturwissenschaften 52: 489-490

Eskola PE (1921) On volcanic necks in the Lake Jänisjärvi in Eastern Finland. Bulletin de la Commission géologique de Finlande 55: 13 pp

Feldman VI, Grakovskiy LB, Karotayeva NN, Dabizha AI (1981) Jänisjärvi meteorite crater (in Russian). In: Marakushev AA (ed) Impactites. Moscow University Press, Moscow, pp 136-170

Flodén T (1980) Seismic stratigraphy and bedrock geology of the central Baltic. Stockholm Contributions in Geology 35: 240 pp

Flodén T, Bjerkeus M (1994) The proposed Ivar impact structure in the southern Baltic [abs]. In: Törnberg R (ed) 2. International ESF workshop: The Identification and Characterization of Impacts, Lockne, Sweden, 1 p

Flodén T, Tunander P, Wickman FE (1986) The Tvären Bay structure, an astrobleme in southeastern Sweden. Geologiska Föreningens i Stockholm Förhandlingar 108: 225-234

Flodén T, Söderberg P, Wickman FE (1993) Björkö, a possible Middle Proterozoic impact structure west of Stockholm, Sweden. Geologiska Föreningens i Stockholm Förhandlingar 115: 25-38

Fredriksson K, Wickman FE (1963) Meteorites (in Swedish). In: Lundholm B (ed) Svensk Naturvetenskap 16, Swedish Natural Science Research Council, Stockholm: 121-157

French BM, Koeberl C, Gilmour I, Shirey SB, Dons JA, Naterstad J (1997) The Gardnos impact structure, Norway: Petrology and geochemistry of target rocks and impactites. Geochimica et Cosmochimica Acta 61: 873-904

Gaál G, Gorbatchev R (1987) An outline of the Precambrian evolution of the Baltic Shield. Precambrian Research 35: 15-52

Gailius R (1988) Meteor Craters in Lithuania (in Lithuanian). Mokslas Press, Vilnius, Lithuania, 100 pp

Gehör S, Havola M (1988) The depositional environment of the Early Proterozoic Tuomivaara iron-formation and associated metasediments, eastern Finland. Geological Survey of Finland, Special Paper 5: 109-133

Gibson RL, Reimold WU, Stevens G (1998) Thermal metamorphic signature of an impact event in the Vredefort Dome, South Africa. Geology 26: 787-790

Glazovskaya LI, Gromov EI, Parfenova OV, Il'Kevich GI (1991) Logoisk astrobleme (in Russian). Nauka Press, Moscow, Russia, 135 pp

Göbel E, Reimold U, Baddenhausen H, Palme H (1980) The projectile of the Lappajärvi crater. Zeitschrift für Naturforschung 35a: 197-203

Grahn Y (1997) Chitinozoan biostratigraphy of the early Caradocian Lockne impact structure, Jämtland, Sweden. Meteoritics & Planetary Science 32: 745-751

Grahn Y, Ormö J (1995) Microfossil dating of the Brent meteorite crater, southeast Ontario, Canada. Revue de Micropaléontologie 38: 131-137

Grahn Y, Nõlvak J, Paris F (1996) Precise chitinozoan dating of Ordovician impact events in Baltoscandia. Journal of Micropalaeontology 15: 21-35

Graup G (1999) Carbonate-silicate liquid upon impact melting: Ries Crater, Germany. Meteoritics & Planetary Science 34: 425-438

Grieve RAF (1984) The impact cratering rate in recent time. Proceedings Lunar and Planetary Science Conference 14., Part 2, Journal of Geophysical Research 89 (Supplement): B403-B408

Grieve RAF (1988) The formation of large impact structures and constraints on the nature of Siljan. In: Bodén A, Eriksson K (eds) Deep drilling in crystalline bedrock, Vol 1: The deep gas drilling in the Siljan impact structure, Sweden and astroblemes. Springer Verlag, Berlin: 328-348

Grieve RAF (1997) Extraterrestrial impact events: the record in the rocks and the stratigraphic column. Palaeogeography, Palaeoclimatology, Palaeoecology 132: 5-23

Grieve RAF (1998) Extraterrestrial impacts on Earth: the evidence and the consequences. In: Grady MM, Hutchinson R, McCall GJH, Rothery DA (eds) Meteorites: Flux with time and Impact Effetcs. Geological Society, London, Special Publications 140: 105-131

Grieve RAF, Dence MR (1979) The terrestrial cratering record, II. The crater production rate. Icarus 38: 230-242

Grieve RAF, Masaitis V (1994) The economic potential of terrestrial impact craters. International Geology Review 36: 105-151

Grieve RAF, Pesonen LJ (1992) The terrestrial impact crater record. Tectonophysics 216: 1-30

Grieve RAF, Pesonen LJ (1996) Terrestrial impact craters: Their spatial and temporal distribution and impacting bodies. Earth, Moon, and Planets 72: 357-376

Grieve RAF, Robertson PB (1979) The terrestrial cratering record, I. Current status of observations. Icarus 38: 212-229

Gudlaugsson ST (1993) Large impact crater in the Barents Sea. Geology 21: 291-294

Haack H, Farinella P, Scott ERD, Keil K (1996) Meteoritic, asteriodal and theoretical constraints on the 500 Ma disruption of the L chondrite parent body. Icarus 119: 182-191

Heeremans M, Wijbrans J (1999) Late Proterozoic tectonic events in southern Finland, constrained by $^{40}Ar/^{39}Ar$ incremental heating and single spot fusion experiments on K-feldspars. Terra Nova 11: 216-222

Henkel H (1983) Combined gravity, magnetic and petrophysic interpretation of an impact structure [abs]. Geoexploration 21: 287

Henkel H (1992) Geophysical aspects of meteorite impact craters in eroded shield environment, with special emphasis on electric resistivity. Tectonophysics 216: 63-89

Henkel H, Eriksson L (1987) Regional aeromagnetic and gravity studies in Scandinavia. Precambrian Research 35: 169-180

Henkel H, Lilljequist R (1992) Large impacts in the Baltic Shield with special attention to the Uppland structure [abs]. Sudbury Conference on 'Large Meteorite Impacts and Planetary Evolution', Lunar & Planetary Science Institute, Contribution 790: 38-40

Henkel H, Pesonen LJ (1992) Impact craters and craterform structures in Fennoscandia. Tectonophysics 216: 31-40

Henkel H, Tiirmaa R, Fleetwood Å, Blomqvist G (1996) Tor: A postglacial impact crater in northwest Härjedalen, Sweden [abs]. Meteoritics & Planetary Science 31 (Suppl): A60

Hirvas H (1991) Pleistocene stratigraphy of Finnish Lapland. Geological Survey of Finland Bulletin 354, 123 pp, Appx
Hisinger W (1826) Notification on a deposit of fossiliferous limestone at Humlenäs in Calmar, etc. (in Swedish). Kongliga Vetenskapsacademiens Handlingar 1825: 180-187
Högbom AG (1910) Minutes of the Geological Society Stockholm 3. meeting [abs]. Geologiska Föreningens i Stockholm Förhandlingar 32: 482
Holmberg HJ (1858) Material on Finlands geognosy (in Swedish). Bidrag till Finlands naturkännedom, etnografi och statistk. Finska Vetenskapssoc., Helsinki, 4 Häftet: 118
Holst NO (1890) Rhyolithes at lake Mien (in Swedish). Sveriges Geologiska Undersökning Series C 110, 50 pp
Innes MJS (1964) Recent advances in meteorite crater research at the Dominion Observatory, Ottawa, Canada. Meteoritics 2: 219-241
Ivanov BA, Kocharyan GG, Kostuchenko VN, Kirjakov AF, Pevzner LA (1996) Puchezh-Katunki impact crater: Preliminary data on recovered core block structure [abs]. Lunar and Planetary Science 27: 589-590
Järvelä J, Pesonen LJ. Pietarinen H (1995) On palaeomagnetism and petrophysics of the Iso-Naakkima impact structure, southeastern Finland. Q19/29.1/3232/95/1. Geological Survey of Finland, Espoo, Finland, 43 pp, Appx
Johansson Å (1984) Geochemical studies on the Boda Pb-Zn deposit in the Siljan astrobleme, central Sweden. Geologiska Föreningens i Stockholm Förhandlingar 106: 15-25
Juhlin C, Pedersen LB (1987) Reflection seismic investigations of the Siljan impact structure, Sweden. Journal of Geophysical Research 92: 14113-14122
Juhlin C, Pedersen LB (1993) Further constraints on the formation of the Siljan impact crater from seismic reflection studies. Geologiska Föreningens i Stockholm Förhandlingar 115: 151-158
Kessel H (1981) How old are the bottom sediments of the Kaali lakes? (in Estonian with English abstract) Eesti Loodus 24: 231-235
Kieffer SW, Simonds CH (1980) The role of volatiles and lithology in the impact cratering process. Reviews of Geophysics and Space Physics 18: 143-181
Kinnunen KA, Lindqvist K (1998) Agate as an indicator of impact structures: An example from Sääksjärvi, Finland. Meteoritics & Planetary Science 33: 7-12
Kirsimäe K, Ainsaar L, Suuroja K, Suuroja S, Mahaney WC (2001) The Osmussaar breccia in northwestern Estonia – evidence of a ~475 Ma earthquake or an impact [abs]? In: Martínez-Ruiz F, Ortega-Huertas M, Palomo I (eds) 6th ESF - IMPACT Workshop "Impact Markers in the Stratigraphic Record", Granada, Spain: 67-68
Koeberl C (1998) Identification of meteoritic components in impactites. In: Grady MM, Hutchinson R, McCall GJH, Rothery DA (eds) Meteorites: Flux with time and Impact Effetcs. Geological Society, London, Special Publications 140: 133-153
Kohonen J, Pihlaja P, Kujala H, Marmo J (1993) Sedimentation of the Jotnian Satakunta sandstone, western Finland. Geological Survey of Finland, Bulletin 369: 34 pp
Koistinen T (ed) (1994) Precambrian basement of the Gulf of Finland and surrounding area. 1 : 1 000 000. Geological Survey of Finland, Espoo
Koivisto M, Korhonen JV (1997) Lappajärvi impact crater rocks and minerals in Quaternary sediment. In: Autio S (ed) Geological Survey of Finland, Current Research 1995-1996. Special Paper 23: 79-88
Korhonen JV (1990) The concentric structure of central Finnish Lapland, an astrobleme [abs]? In: Pesonen LJ, Niemisara H (eds) Symposium Fennoscandian Impact Structures, Programme and Abstracts, Espoo and Lappajärvi, Finland: 32
Korhonen JV (1993) Major central Fennoscandian circular structures and distribution of ore deposits [abs]. In: Stöffler D, Deutsch A (eds) 1st International ESF workshop, Nördlingen, Germany, 1 p

Korhonen JV (1994) Geophysical characteristics of two major circular structures in Finland [abs]. In: Törnberg R (ed) 2. International European Science Foundation workshop: The Identification and Characterization of Impacts, Lockne, Sweden, 1 p

Korhonen JV (1997) Petrophysics in the crustal model program of Finland. In: Autio S (ed) Geological Survey of Finland, Current Research 1995-1996. Geological Survey of Finland, Special Paper 23: 157-173

Krinov EL (1961) The Kaalijärv meteorite craters on Saarema Island. Estonian SSR. American Journal of Science 259: 430-440

Kukkonen IT, Kivekäs L, Paananen M (1992) Physical properties of kärnäite (impact melt), suevite and impact breccia in the Lappajärvi meteorite crater, Finland. Tectonophysics 216: 111-122

Kumpulainen R, Vidal JP (1985) Late Proterozoic basin evolution and sedimentation in the westernmost part of Baltoscandia. In: Gee DG, Sturt BA (eds) The Caledonian Orogen – Scandinavia and Related Areas. Part 1. J Wiley & Sons, Chichester: 213-232

Lahtinen R, Huhma H (1997) Isotopic and geochemical constraints on the evolution of the 1.93-1.79 Ga Svecofennian crust and mantle. Precambrian Research 82: 13-34

Langenhorst F, Shafranovsky GI, Masaitis VL, Koivisto M (1999) Discovery of impact diamonds in a Fennoscandian crater and evidence for their genesis by solid-state transformation. Geology 27: 747-750

Larson SA, Tullborg E-L, Cederbom C, Stiberg J-P (1999) Sveconorwegian and Caledonian foreland basins in the Baltic Shield revealed by fission-track thermochronology. Terra Nova 11: 210-215

Laurén L, Lehtovaara J, Boström R (1978) On the geology of the circular depression at Söderfjärden, western Finland. Geological Survey of Finland, Bulletin 297: 5-38

Lehtinen M (1970) New evidence for an impact origin of Lake Lappajärvi, Western Finland. Bulletin of the Geological Society of Finland 42: 89-93

Lehtinen M (1976) Lake Lappajärvi, a meteorite impact site in western Finland. Geological Survey of Finland, Bulletin 282, 92 pp

Lehtovaara JJ (1982) Stratigraphical section through Lower Cambrian at Söderfjärden, Vaasa, western Finland. Bulletin of the Geological Society of Finland 54: 35-43

Lidmar-Bergström K (1995) Relief and saprolites through time on the Baltic Shield. Geomorphology 12: 45-61

Lidmar-Bergström K (1996) Long term morphotectonic evolution in Sweden. Geomorphology 16: 33-59

Lidmar-Bergström K (1997) A long-term perspective on glacial erosion. Earth Processes and Landforms 22: 297-306

Liiva A, Kessel H, Aaloe A (1979) Age of Ilumetsa craters (in Estonian). Eesti Loodus 12: 762-764

Lilljequist R (1992) Metal ore distribution in the Uppland structure, Sweden [abs]. In: Stöffler D, Deutsch A (eds) 1st International European Science Foundation workshop, Nördlingen, Germany, 1 p

Lilljequist R, Reimold WU (1999) The Duobblon structure: a suspected impact formation [abs]. Meteoritics & Planetary Science 34 (Suppl): A76

Lilljequist R (2000) The Gallejaur structure, northern Sweden. In: Gilmour I, Koeberl C (eds) Impacts and the Early Earth. Lecture Notes in Earth Sciences, Springer Verlag, Berlin-Heidelberg 91: 363-387,

Lindström M, Flodén T, Puura V, Suuroja K (1992) The Kärdla, Tvären and Lockne craters - possible evidences of an Ordovician asteroid swarm. Proceedings of the Estonian Academy of Sciences, Geology 41: 45-53

Lindström M, Flodén T, Grahn Y, Kathol B (1994) Post-impact depositis in Tvären, a marine Middle Ordovician crater south of Stockholm, Sweden. Geological Magazine 131: 91-103

Lindström M, Sturkell EFF, Törnberg R, Ormö J (1996) The marine impact crater at Lockne, central Sweden. GFF 118: 193-206

Lindström M, Flodén T, Grahn Y, Hagenfeldt S, Ormö J, Sturkell EFF, Törnberg R (1999) The lower Paleozoic of the probable impact crater of Hummeln, Sweden. GFF 121: 243-252

Mancini F, Marshall B, Papunen H (1996) Petrography and metamorphism of the Sääksjärvi ultramafic body, Southwest Finland. Mineralogy and Petrology 56: 185-208

Marmo JS, Ojakangas RW (1984) Lower Proterozoic glaciogenic deposits, eastern Finland. Geological Society of America Bulletin 95: 1055-1062

Masaitis VL (1973) Geological Consequences of the Falls of the Crater-forming Meteorites (in Russian). Nedra Press, Leningrad, Russia, 18 pp

Masaitis VL (1999a) Impact structures of northeastern Eurasia: The territories of Russia and adjacent countries. Meteoritics & Planetary Science 34: 691-711

Masaitis VL (1999b) The Kaluga impact event and its proven and possible geological consequences. In: Gersonde R, Deutsch A (eds), Oceanic impacts and environmental perturbations. ESF-Impact workshop. Berichte zur Polarforschung 343: 60-64

Masaitis VL, Mashchak MS (1996) Recrystallisation and blastesis in shock-metamorphosed rocks of impact structures (in Russian). Proceedings of the Russian Mineralogical Society 4: 1-18

Masaitis VL, Mashchak MS, Yezerskiy VA (1985) The longevity of impact structures under different geological conditions. International Geology Review 27: 556-562

Masaitis VL, Shafranovsky GI, Fedorova IG, Koivisto M, Korhonen JV (1998a) Lappajärvi astrobleme: The first find of impact diamonds on the Fennoscandian Shield [abs]. Lunar and Planetary Science 29: #1171, 2 pp (CD-ROM).

Masaitis VL, Mashchak MS, Naumov MV, Raikhilin AI (1998b) Search for Precambrian impact structures and methods of investigation of clastic impactites (in Russian). VSEGEI Press, SPb, 34 pp

Mashchak MS, Naumov MV (1996) The Suavjärvi structure: An Early Proterozoic impact site on the Fennoscandian Shield [abs]. 27. Lunar and Planetary Science Conference, Houston: 825-826

Mashchak MS, Orlova JV (1986) The evidence of shock deformations in Early Proterozoic breccias from Lake Suavjärvi area (in Russian). Meteoritika 45: 137-141

McKerrow WS, Dewey JF, Scotese CR (1991) The Ordovician and Silurian development of the Iapetus ocean. Special Papers on Palaeontology 44: 165-178

Melezhik VA, Fallick AE (1996) A widespread positive ^{13}C anomaly at around 2.33-2.06 Ga on the Fennoscandian Shield. Terra Nova 8: 141-157

Merrill GK (1979) Ordovician conodonts from the Åland Islands, Finland. Geologiska Föreningens i Stockholm Förhandlingar 101: 329-341

Metzler A, Stöffler D (1990) Siljan impact structure, Sweden: Dyke breccias within the crater floor [abs]. In: Pesonen LJ, Niemisara H (eds) Symposium Fennoscandian Impact Structures, Programme and Abstracts, Espoo and Lappajärvi, Finland: p 48

Milstein RL (1997) The Calvin impact structure, Cass county, Michigan: Identification and analysis of a subsurface Ordovician astrobleme [abs]. In: Johnson KS, Campell JA (eds), Ames structure in north-west Oklahoma and similar features: origin and petroleum production (1995 symposium); Oklahoma Geological Survey Circular 100: 391-393

Montanari A, Koeberl C (2000) Impact stratigraphy. The Italian record. Lecture Notes in Earth Sciences 93, Springer Verlag, Berlin-Heidelberg, 364 pp

Moralyov VR, Glukhovskií SA (1981) Circular features of Precambrian shields as recognized on space photographs. Soviet Journal of Remote Sensing 3: 356-362

Motuza G (1994) Impact craters in Lithuania [abs]. In: Törnberg R (ed) 2nd International ESF workshop: The Identification and Characterization of Impacts, Lockne, Sweden, 1 p

Nikishin AM, Ziegler PA, Stehenson RA, Cloetingh SAPL, Furne AV, Fokin PA, Ershov AV, Bolotov SN, Korotaev MV, Alekseev AS, Gorbachev VI, Shipilov EV, Lankreijer A, Bembinova EY, Shalimov IV (1996) Late Precambrian to Triassic history of the East European Craton: dynamics of sedimentary basin evolution. Tectonophysics 268: 23-63

Nironen M (1997) The Svecofennian orogen: a tectonic model. Precambrian Research 86: 21-44

Oberbeck VR, Marshall JR, Aggarwal H (1993) Impacts, tillites, and the breakup of Gondwanaland. Journal of Geology 101: 1-19

Öhman T, Pesonen LJ, Raitala J, Uutela A, Tuisku P (2000) The Saarijärvi crater - older and larger than assumed [abs]? In: Plado J, Pesonen LJ (eds) Programme and Abstracts of ESF workshop: Meteorite Impacts in Precambrian Shields. Geological Survey of Finland and University of Helsinki, Espoo: p 82

Olsen L, Mejdahl V, Selvik SF (1996) Middle and Late Pleistocene stratigraphy, chronology and glacial history in Finnmark, North Norway. Norsk geologiske undersøkning Bulletin 429: 111 pp

Öpik EJ (1916) Note about the meteoritic theory of the lunar circles (in Russian with French abstract). Bulletin de la Société Russe des Amis de l'Etude de l'Univers 3 (21): 1925-1934

Ormö J, Blomqvist G (1996) Magnetic modelling as a tool in the evaluation of impact structures, with special reference to the Tvären Bay impact crater, SE Sweden. Tectonophysics 262: 291-300

Ormö J, Lindström M (2000) When a cosmic impact strikes the sea bed. Geological Magazine 137: 67-80

Ormö J, Sturkell E, Blomqvist G, Törnberg R (1999) Mutually constrained geophysical data for the evaluation of a proposed impact structure: Lake Hummeln, Sweden. Tectonophysics 311: 155-177

Paakkola J, Gehör S (1988) The lithofacies associations and sedimentary structures of the iron-formations in the Early Proterozoic Kittilä greenstone belt, northern Finland. Geological Survey of Finland, Special Paper 5: 213-238

Paavola J (1997) Geological Map of Finland, Pre-Quaternary Rocks. Sheet 3343 - Rautavaara, 1 : 100 000. Geological Survey of Finland, Espoo

Palme H, Rammensee W, Reimold WU (1980) The meteoritic component of impact melts from European impact craters [abs]. Lunar and Planetary Science 11: 848-849

Papasikas N, Juhlin C (1997) Interpretation of reflections from the central part of the Siljan Ring impact structure based on results from the Stenberg-1 borehole. Tectonophysics 269: 237-245

Papunen H (1969) Possible impact metamorphic textures in the erratics of the lake Sääksjärvi area in southwestern Finland. Bulletin of the Geological Society of Finland 41: 151-155

Patterson DB, Farley KA, Schmitz B (1998) Preservation of extraterrestrial ^{3}He in 480-Ma-old marine limestones. Earth and Planetary Science Letters 163: 315-325

Pedersen LB, Rasmussen TM, Dyrelius D (1990) Construction of component maps from aeromagnetic total field anomaly maps. Geophysical Prospecting 38: 795-804

Pedersen LB, Juhlin C, Rasmussen TM (1992) Electric resistivity in the Gravberg-1 Deep Well, Sweden. Journal of Geophysical Research 97: 9171-9182

Pesonen LJ (1994) Paleomagnetic dating of impact craters [abs]. In: Törnberg R (ed) 2nd International ESF workshop: The Identification and Characterization of Impacts, Lockne, Sweden, 1 p

Pesonen LJ (1995) The Suvasvesi N structure - A new impact crater in Central-East Finland [abs]. Annales Geophysicae, Part III, Space and Planetary Sciences 13 (Supplement III): C 741

Pesonen LJ (1996) The impact cratering record of Fennoscandia. Earth, Moon and Planets 72: 377-393

Pesonen LJ, Marcos N, Pipping F (1992) Palaeomagnetism of the Lappajärvi impact structure, western Finland. Tectonophysics 216: 123-142

Pesonen LJ, Lehtinen M, Deutsch A, Elo S, Lukkarinen H (1996) New geophysical results of the Suvasvesi N impact structure, Finland [abs.]. Lunar and Planetary Science 27: 1021-1022

Pesonen LJ, Deutsch A, Hornemann U, Langenhorst F (1997a) Magnetic properties of diabase samples shocked experimentally in the 4.5 to 35 GPa range [abs]. Lunar and Planetary Science 28: 1087-1088

Pesonen LJ, Lehtinen M, Tuukki P, Abels A (1997b) The Lake Saarijärvi structure, Taivalkoski - a new meteorite impact crater in Finland [abs]. In: Kaikkonen P, Komminaho K,

Salmirinne H (eds) Sovelletun geofysiikan XI neuvottelupäivät, Oulun Yliopisto, Oulu, pp 8-9

Pesonen LJ, Lehtinen M, Tuukki P, Abels A (1998) The lake Saarijärvi - A new meteorite impact structure in northern Finland [abs]. Lunar and Planetary Science 29: #1262, 2 pp (CD-ROM)

Pesonen LJ, Kuivasaari T, Lehtinen M, Elo S (1999a) Paasselkä: A new meteorite impact structure in eastern Finland [abs]. Meteoritics & Planetary Science 34 (Supplement): A90-A91

Pesonen LJ, Elo S, Lehtinen M, Jokinen T, Puranen R, Kivekäs L (1999b) Lake Karikkoselkä impact structure, central Finland: New geophysical and petrographic results. In: Dressler BO, Sharpton VL (eds) Large Meteorite Impacts and Planetary Evolution II: Boulder, Colorado, Geological Society of America, Special Paper 339: 131-147

Pihlaja PJ (1994) Geological map of Finland, 1:100 000. Pori. Pre-Quaternary rocks, Sheet 1143. Geological Survey of Finland

Pihlaja PJ, Kujala H (2000) Some geochemical features of impact melts, suevites and target rocks at Sääksjärvi impact structure, Finland [abs]. In: Plado J, Pesonen LJ (eds) Programme and Abstracts of ESF workshop: Meteorite Impacts in Precambrian Shields. Geological Survey of Finland and University of Helsinki, Espoo: p 50

Pilkington M, Grieve RAF (1992) The geophysical signature of terrestrial impact craters. Reviews of Geophysics 30: 161-181

Pipping F, Lehtinen M (1992) Geology, stratigraphy and structure of the Lappajärvi meteorite crater, western Finland: preliminary results of deep drilling. Tectonophysics 216: 91-97

Pipping F, Puura V (1996) Impact craters in the surroundings of the Gulf of Finland. In: Koistinen T (ed) Explanation to the Map of Precambrian basement of the Gulf of Finland and surrounding area 1 : 1 000 000, Geological Survey of Finland, Special Paper 21: 127-133

Pirrus E (1995) The ratio of depression and mound volumes: A promising criterion for the identification of small meteorite craters. Proceedings of the Estonian Academy of Science, Geology 44: 188-196

Pirrus E, Tiirmaa R (1984) Tsõõrikmägi – a meteorite crater as well? (in Estonian) Eesti Loodus 9: 566-571; 10: 638-642

Pirrus E, Tiirmaa R (1991) Did Virumaa bolide reach the Earth? (in Estonian) Eesti Loodus 4: 210-214

Plado J, Pesonen LJ, Elo S, Puura V, Suuroja K (1996) Geophysical research on the Kärdla impact structure, Hiiumaa Island, Estonia. Meteoritics & Planetary Science 31: 289-298

Plado J, Pesonen LJ, Puura V (1999) Effect of erosion on gravity and magnetic signatures of complex impact structures: Geophysical modeling and applications. In: Dressler BO, Sharpton VL (eds) Large Meteorite Impacts and Planetary Evolution II: Boulder, Colorado, Geological Society of America Special Paper 339: 229-239

Plado J, Henkel H, Olofson B (2000) Geophysical studies at the Ilumetsa impact site [abs]. In: Plado J, Pesonen LJ (eds) Programme and Abstracts of ESF workshop: Meteorite Impacts in Precambrian Shields. Geological Survey of Finland and University of Helsinki, Espoo: p 86

Pobul EA (1963) The application of geophysical methods to the study of meteorite craters in the Estonian SSR Proceedings of the Institute of Geology, Academy of Sciences of the Estonian SSR 11: 45-51

Popov YA, Pimenov VP, Pevzner LA, Romushkevich RA, Popov EY (1998) Geothermal characteristics of the Vorotilovo deep borehole drilled into the Puchezh-Katunk impact structure. Tectonophysics 291: 205-223

Poprawa P, Šliaupa S, Stephenson R, Lauzauskiene J (1999) Late Vendian - Early Palaeozoic tectonic evolution of the Baltic Basin: tectonic implications from subsidence analysis. Tectonophysics 314: 219-239

Puranen R, Säävuori H, Sahala L, Suppala I, Mäkilä M, Lerssi J (1999) Airborne electromagnetic mapping of surficial deposits in Finland. First Break 17: 145-154

Puura V, Flodén T (1999) Rapakivi-granite-anorthosite magmatism - a way of thinning and stabilisation of the Fennoscandian crust, Baltic Sea Basin. Tectonophysics 305: 75-92

Puura V, Suuroja K (1992) Ordovician impact crater at Kärdla, Hiiumaa Island, Estonia. Tectonophysics 216: 143-156

Puura V, Lindström M, Flodén T, Pipping F, Motuza G, Lehtinen M, Suuroja K, Murnieks A (1994) Structure and stratigraphy of meteorite craters in Fennoscandia and the Baltic region: A first outlook. Proceedings of the Estonian Academy of Sciences, Geology 43: 93-108

Puura V, Amantov A, Tikhomirov S, Laitakari I (1996) Latest events affecting the Precambrian basement, Gulf of Finland and surrounding areas. In: Koistinen T (ed.) Explanation to the Map of Precambrian basement of the Gulf of Finland and surrounding area 1 : 1 000 000. Geological Survey of Finland, Special Paper 21: 115-125

Puura V, Kärki A, Kirs J, Kirsimäe K, Kleesment A, Konsa M, Niin M, Plado J, Suuroja K, Suuroja S (2000) Impact-induced replacement of plagioclase by K-feldspar in granitoids and amphibolites at the Kärdla crater, Estonia. In: Gilmour I, Koeberl C (eds) Impacts and the Early Earth. Lecture Notes in Earth Sciences, Springer Verlag, Berlin-Heidelberg 91: 417-445

Racki G (1999) The Frasnian-Famennian biotic crisis: How many (if any) bolide impacts? Geologische Rundschau 87: 617-632

Rampino MR (1994) Tillites, diamictites, and ballistic ejects of large impacts. Journal of Geology 102: 439-456

Rasmussen KL, Aaby B, Gwozdz R (2000) The age of the Kaalijärv meteorite craters. Meteoritics & Planetary Science 35: 1067-1071

Raukas A (2000) Investigation of impact spherules – a new promising method for the correlation of Quaternary deposits. Quaternary International 68-71: 241-252

Raukas A, Tiirmaa R (1997) Kaali craters [abs]. In: Raukas A (ed) Impact and Extraterrestrial Spherules: New Tools for Global Correlation. IUGS-UNESCO Project 384, Excursion Guide and Abstracts: 11-15

Raukas A, Tiirmaa R, Kaup E, Kimmel K (2001) The age of the Illumetsa meteorite craters in South-East Estonia. Meteoritics & Planetary Science 36: 1507-1514

Reimold WU (1982) The Lappajärvi meteorite crater, Finland: petrography, Rb-Sr, major and trace element geochemistry of the impact melt and basement rocks. Geochimica et Cosmochimica Acta 46: 1203-1225

Reimold WU, Henkel H (1997) First results of a petrographic study of breccias from the region of the Uppland structure, Sweden [abs]. Lunar and Planetary Science 28: 1157-1158

Reinwald IA (1937) The Kaalijärv meteorite crater field (in Estonian). Loodusvaatleja 4: 97-102

Saarse L, Rajamäe R, Heisalu A, Vassilew J (1991) The biostratigraphy of sediments deposited in the Lake Kaali meteorite impact structure, Saaremaa Island, Estonia. Bulletin of the Geological Society of Finland 63: 129-139

Sarapää O (1996) Proterozoic primary kaolin deposits at Virtasalmi, southeastern Finland. Geological Survey of Finland, Synopsis, 12 pp

Schmidt G, Palme H, Kratz K-L (1997) Highly siderophile elements (Re, Os, Ir, Ru, Rh, Pd, Au) in impact melts from three European impact craters (Sääksjärvi, Mien, and Dellen): Clues to the nature of the impacting bodies. Geochimica et Cosmochimica Acta 61: 2977-2987

Schmitz B, Jeppsson L, Ekvall J (1994) A search for shocked quartz grains and impact ejecta in early Silurian sediments on Gotland, Sweden. Geological Magazine 131: 361-367

Schmitz B, Lindström M, Asaro F, Tassinari M (1996) Geochemistry of meteorite-rich limestone strata and fossil meteorites from the lower Ordovician at Kinnekulle, Sweden. Earth and Planetary Science Letters 145: 31-48

Schmitz B, Peucker-Ehrenbrink B, Lindström M, Tassinari M (1997) Accretion rates of meteorites and cosmic dust in the Early Ordovician. Science 278: 88-90

Scott RG, Pilkington M, Tanczyk EI (1997) Magnetic investigations of the West Hawk, Deep Bay and Clearwater impact structures, Canada. Meteoritics & Planetary Science 32: 293-308

Shmayenok AI, Tikhomirov SN (1974) Mishina Gora explosion structure in the Lake Chud area (in Russian). Doklady Akademii Nauk SSSR 219: 701-703

Short NM, Bunch TE (1968) A worldwide inventory of features characteristic of rocks associated with presumed meteorite impact structures. In: French BM, Short NM (eds) Shock metamorphism of natural materials. Mono Book Corporation, Baltimore: 255-283

Shukolyukov A, Kyte FT, Lugmair GW, Lowe DR, Byerly GR (2000) The oldest impact deposits on Earth - First confirmation of an extraterrestrial component. In: Gilmour I, Koeberl C (eds) Impacts and the Early Earth. Lecture Notes in Earth Sciences, springer Verlag, Berlin-Heidelberg 91: 99-115

Silvennoinen A (ed) (1992) Deep fractures in the Paanajärvi-Kuusamo-Kuolajärvi area. Geological Survey of Finland, Special Paper 13: 115 pp

Simonson BM, Harnik P (2000) Have distal impact ejecta changed through geologic time? Geology 28: 175-178

Smit J (1999) The global stratigraphy of the Cretaceous-Tertiary boundary impact ejecta. Annual Reviews of Earth and Planetary Sciences 27: 75-113

Söderström L (1966) The kimberlites of Åvike bay, on the Bothnian coast of Sweden. Geologiska Föreningens i Stockholm Förhandlingar 88: 351-360

Sokolov VA, Heiskanen KJ (1985) Evolution of Precambrian volcanogenic-sedimentary lithogenesis in the south-eastern part of the Baltic Shield. Geological Survey of Finland, Bulletin 331: 91-106

Stewart SA (1999) Seismic interpretation of circular geological structures. Petroleum Geosciences 5: 273-285

Sturkell EFF (1998) Resurge morphology of the marine Lockne impact crater, Jämtland, central Sweden. Geological Magazine 135: 121-127

Sturkell EFF, Ormö J (1997) Impact-related clastic injections in the marine Ordovician Lockne impact structure, Central Sweden. Sedimentology 44: 793-804

Sturkell EFF, Ormö J (1998) Magnetometry of the marine, Ordovician Lockne impact structure, Jämtland, Sweden. Journal of Applied Geophysics 38: 195-207

Sturkell EFF, Broman C, Forsberg P, Torssander P (1997) Impact-related hydrothermal activity in the Lockne impact structure, Jämtland, Sweden. European Journal of Mineralogy 10: 589-606

Sturkell EFF, Ekelund A, Törnberg R (1998) Gravity modelling of Lockne, a marine impact structure in Jämtland, central Sweden. Tectonophysics 296: 421-435

Sturkell EFF, Ormö J, Nõlvak J, Wallin Å (2000) Distant ejecta from the Lockne marine-target impact crater, Sweden. Meteoritics & Planetary Science 35: 929-936

Sundblad K (1994) Lead isotopic composition of galena associated with the Early Paleozoic meteorite impact structures at Lockne (Sweden) and Kärdla (Estonia) [abs]. In: Törnberg R (ed) 2nd International ESF workshop: The Identification and Characterization of Impacts, Lockne, Sweden, 1 p

Suuroja K, Suuroja S (2000) Neugrund structure - the newly discovered submarine Early Cambrian impact crater. In: Gilmour I, Koeberl C (eds) Impacts and the Early Earth. Lecture Notes in Earth Sciences, Springer Verlag, Berlin-Heidelberg 91: 389-416

Suuroja K, Puura V, Suuroja S (1999) Kärdla crater (Hiiumaa Island, Estonia) – The result of an impact in a shallow epicontinental sea [abs]. In: Gersonde R, Deutsch A (eds) Oceanic impacts and environmental perturbations. ESF-Impact workshop. Berichte zur Polarforschung 343: 89-91

Svenonius F (1888) Andesite from Norra Dellen in Helsingland (in Swedish). Geologiska Föreningens i Stockholm Förhandlingar 10: 262-285

Svensson N-B (1966) Lake Hummeln. A possible astrobleme in southern Sweden. I. The bottom topography. Sveriges Geologiska Undersökning Ser. C 608, Årsbok 60/3: 1-18

Svensson N-B (1968a) The Dellen Lakes, a probable meteorite impact in Central Sweden. Geologiska Förenigens i Stockholm Förhandlingar 90: 314-315

Svensson N-B (1968b) Lake Lappajärvi, Central Finland: a possible meteorite impact structure. Nature 217: 438
Svensson N-B (1969) Lake Mien, southern Sweden - A possible astrobleme. Geologiska Förenigens i Stockholm Förhandlingar 91: 101-110
Svensson N-B (1971) Probable meteorite impact crater in central Sweden. Nature Physical Science 229: 90-92
Svensson N-B (1973) Shatter cones from the Siljan structure, Central Sweden. Geologiska Föreningens i Stockholm Förhandlingar 95: 139-143
Svensson N-B, Wickman FE (1965) Coesite from Lake Mien, southern Sweden. Nature 205: 1202-1203
Talvitie J (1979) Seismo-tectonics in Finland. Geologiska Föreningens i Stockholm Förhandlingar 100: 247-253
Talvitie J, Pernu T, Raitala J (1975) The circular Vaasa structure in the Baltic Shield, western Finland. Department of Geophysics, University Oulu, Contribution 59, 15 pp
Thunehed H, Elming S-Å, Pesonen LJ (1999) Lycksele structure in northern Sweden: Result of an impact? In: Dressler BO, Sharpton VL (eds) Geological Society of America Special Paper 339: 109-124
Tiirmaa R (1988) Distribution of pulverized mateoritic matter in the Kaali crater field. Proceedings of the Estonian Academy of Science, Geology 37: 43-46
Tsikalas F, Gudlaugsson ST, Faleide JI (1998a) Collapse, infilling and postimpact deformation at the Mjølnir impact structure. Geological Society of America Bulletin 110: 537-552
Tsikalas F, Gudlaugsson ST, Eldholm O, Faleide JI (1998b) Integrated geophysical analysis supporting the impact origin of the Mjølnir structure, Barents Sea. Tectonophysics 289: 257-280
Turunen P (1995) VLF-R and DC measurements at Saarijärvi in Taivalkoski region (in Finnish). Geological Survey of Finland Report of Investigation Q 19/3553/1995/1, 14 pp, Appx
Tynni R (1982) The reflection of geological evolution in Tertiary and interglacial diatoms and silicoflagellates in Finnish Lapland. Geological Survey of Finland, Bulletin 320: 40 pp
Tynni R, Uutela A (1985) Late Precambrian shale formation of Taivalkoski in northern Finland (in Finnish). Geologi 4-5: 62-65
Ūsaitytė D (2000) The geology of the southeastern Baltic Sea: a review. Earth Science Reviews 50: 137-225
Uutela A (1998) Extent of the northern Baltic Sea during the Early Paleozoic era - New evidence from Ostrobothnia, western Finland. Bulletin of the Geological Society of Finland 70: 51-68
Vaarma M, Pipping F (1997) Pre-Quaternary rocks of the Alajärvi and Evijärvi map-sheet areas (in Finnish). Explanation to the maps of Pre-Quaternary rocks 1 : 100 000, Sheets 2313 and 2314, Geological Survey of Finland, Espoo, 83 pp
Vaasjoki M (1977) Rapakivi granites and other postorogenic rocks in Finland: their age and the lead isotopic composition of certain associated galena mineralizations. Geological Survey of Finland, Bulletin 294: 64 pp
Vattenfall (1991) Scientific summary report of the deep gas drilling project in the Siljan ring impact structure. Vattenfall - Swedish State Power Board, 238 pp
Veski S, Heinsalu A, Kirsimäe K, Poska A, Saarse L (2001) Ecological catastrophe in connection with the impact of the Kaali meteorite about 800 - 400 BC on the island of Saaremaa, Estonia. Meteoritics & Planetary Science 36: 1367-1376.
Vidal G, Moczydlowska M (1995) The Neoproterozoic of Baltica - stratigraphy, palaeobiology and general geological evolution. Precambrian Research 73: 197-216
Vishnevsky S, Raitala J (2000) Impact diamonds as indicators of shock metamorphism in strongly-reworked Precambrian impactites. In: Gilmour I, Koeberl C (eds) Impacts and the Early Earth. Lecture Notes in Earth Sciences, Springer Verlag, Berlin-Heidelberg 91: 229-247
Werner SC, Janle P, Pesonen LJ (2000) Suvasvesi N – modeling gravimetric and aeromagnetic data of a Finnish impact structure [abs]. In: Plado J, Pesonen LJ (eds) Programme and

Abstracts of ESF workshop: Meteorite Impacts in Precambrian Shields. Geological Survey of Finland and University of Helsinki, Espoo: p 38

Wickman FE (1973) The origin of the Landsort Trench in the Baltic. Geologiska Föreningens i Stockholm Förhandlingar 95: 339-341

Wickman FE (1988) Possible impact structures in Sweden. In: Bodén A, Erikkson GK (eds) Deep Drilling in Crystalline Bedrock, Vol 1: The Deep Gas Drilling in the Siljan Impact Structure Sweden and Astroblemes, Springer Verlag, Berlin: 298-327

Wickman FE (1992) Förelagna impaktstrukturer i Norden och närliggande områden. Astronomisk Tidskrift 25: 49-62

Wickman FE (1994) The Siljan ring impact structure: possible connections with minor ores in its neighborhood. GFF 116: 145-146

Winterhalter B (1982) The bedrock geology of Lumparn Bay, Åland. Geological Survey of Finland, Bulletin 317: 115-130

Witschard F (1984) Large-magnitude ring structures on the Baltic Shield - Metallogenic significance. Economic Geology 79: 1400-1405

Zenchenko Y, Tsvetkov V (1999) Formation of a flattened subsurface zone around meteorite craters. In: Dressler BO, Sharpton VL (eds) Large Meteorite Impacts and Planetary Evolution II. Geological Society of America Special Paper 339: 223-227

Zhang P, Rasmussen TM, Pedersen LB (1988) Electric resistivity structure of the Siljan impact region. Journal of Geophysical Research 93: 6485-6501

Re-examining Structural Data from Impact Craters on the Canadian Shield in the Light of Theoretical Models

Michael R. Dence

824 Nesbitt Place, Ottawa, ON, K2C 0K1, Canada. (mrdence@rsc.ca)

Abstract. A broad consensus now exists on hypervelocity impact crater mechanics, based on field observations, theory and computer modeling. Refinement requires closer attention to key parameters observable in better preserved terrestrial craters. Among them, eleven craters on the Canadian Shield, ranging in size over 1.5 orders of magnitude, have the advantage of having formed in crystalline rock targets with relatively uniform physical properties. This compensates in part for the general lack of structures for direct determination of displacements.

By re-examining previously published data, earlier models for simple craters are modified by comparing the apparent shock attenuation for rocks underlying the lowermost melt rocks at the Brent crater with computations of shock decay. The full complement of craters is then used to summarize features that change with crater size, including topography and structure, gravity and other geophysical parameters, the development and emplacement of pseudotachylites, as well as the distribution of melts rocks and associated allochthonous breccias. This leads to the following:

- a satisfactory hypervelocity impact model for crystalline rock targets can be developed by taking into account calculated changes in the rate of shock wave attenuation between the near field and far field; observation harmonizes with experiment and calculation if the far field decay exponent is -2.
- the model provides calculations of apparent displacements, nominal projectile size and impact energy.
- crater size is controlled by the extent of Grady-Kipp fracturing, which, with increasing impact energy, is progressively less extensive down axis than the imprint of shock metamorphism.
- by influencing the efficiency of Grady-Kipp fracturing, gravity is important in the early stages of crater formation in addition to governing excavation and late stage structural modifications.

1 Introduction

It is over fifty years since the New Quebec crater was introduced to science as a probable meteorite impact crater (Meen 1950), leading to the systematic exploration of the Canadian Shield for additional ancient craters (Beals et al. 1956; Dence 1965). The Canadian discoveries, in turn, helped stimulate the worldwide search for impact sites, resulting in a steadily increasing number of confirmed locations (Dence 1972; Grieve 1991; Grieve and Shoemaker 1994). It is thus appropriate at this time to re-examine the craters on the Canadian Shield and compare the diverse data set that has been amassed over the subsequent half century with current understanding of hypervelocity impact crater mechanics, as summarized, for example, by Melosh (1989).

For this review, data are drawn primarily from representative well-preserved craters on the Canadian Shield. These are listed in Table 1, with references to data sources used here. In each case the crater diameter given is an estimate of the final diameter prior to erosion rather than that of the transient cavity. In general, the diameters assigned are based on residual evidence for a structural rim and for the gravity anomaly associated with each crater as well as collaborative evidence from pre-existing or post-impact sedimentary rocks and similar data where they exist. Such data are also used to estimate the depth of erosion of each crater, with results that vary from negligible to several hundred metres. Detailed arguments for the estimates preferred here require a case-by-case analysis and will be presented elsewhere. In most instances the estimates used lie within 10 % of generally accepted values in the literature, but for a few, such as Manicouagan, the difference is greater (see, for example, Grieve 1991; Grieve and Cintala, 1992 and references therein). Craters not discussed are either less completely studied or are formed in layered targets. Similarly, the largest and oldest Canadian crater, Sudbury, is not included in this review as modification by subsequent tectonic events and other factors introduce complexities that require special consideration. Its distinctive features will be discussed separately.

The better-preserved Canadian craters range in size over about 1.5 orders of magnitude. Most are reasonably well exposed at the surface and many have been investigated in the subsurface by geophysical methods and by diamond drilling, most notably through the program carried out by the former Canadian Dominion Observatory between 1955 and 1967. They furnish terrestrial examples of the progression observed on other planetary bodies and may be compared with recent computer simulations (O'Keefe and Ahrens 1993, 1999; Melosh and Ivanov 1999).

By concentrating on craters on the Shield, advantage is taken of the broadly uniform properties of the crystalline rocks, which form the target materials. This minimizes structural complications due to variations in composition and physical properties, in contrast to the situation with craters formed in layered target materials. On the other hand, crystalline rocks have the distinct disadvantage of generally lacking pre-existing structures by which displacements can be

Table 1. Representative Canadian craters in Precambrian igneous and metamorphic rocks

Crater name	Diameter (km) [b]	Structure	Age (Ma) [c]	Form	References [d]
West Hawk Lake [a]	2.75	Simple	100 ± 50	Bowl-shaped	Halliday and Griffin 1967; Short 1970
New Quebec	3.0	Simple	1.4 ± 0.1	Bowl-shaped	Meen 1950; Dence 1965; Currie 1965
Brent	3.8	Simple	450 ± 3	Bowl-shaped	Dence 1965, 1968; Dence and Guy-Bray 1972; Grieve 1978
Deep Bay	9.5	Complex	100 ± 50	Flat-floored	Innes et al. 1964; Dence et al. 1968
Nicholson Lake	12.5	Complex	<400	Central peak	Dence et al. 1968
Clearwater East	20	Complex	290 ± 20	Central peak	Dence 1964, 1965; Dence et al. 1968
Lake Mistastin	28	Complex	38 ± 4	Central peak	Taylor and Dence 1969; Grieve 1975
Slate Islands [a]	30	Complex	<350	Central peak	Robertson and Grieve 1977
Clearwater West	32	Complex	290 ± 20	Peak ring	Dence 1964, 1965; Dence et al. 1977
Charlevoix	52	Complex	357 ± 15	Peak ring	Rondot, 1968, 1989; Robertson 1975; Robertson and Grieve 1977
Manicouagan	80	Complex	212 ± 2	Multi-ring	Willmore 1963; Dence 1964, 1965; Dence et al. 1977

[a] Target rocks mainly greenstone (West Hawk) or metasedimentary and volcanic (Slate Islands)

[b] Conservative estimate of original diameter; other, generally larger, estimates are given in the literature (cf. Grieve 1991; Grieve and Cintala, 1992)

[c] After Grieve 1991

[d] Main sources for data used in this review

determined directly. However, the imprint of shock metamorphism can be used as a proxy for such structural markers, thereby indirectly providing information on net displacements resulting from the impact.

In the first part of this paper the comprehensive drilling results at the Brent crater are used as the principal example in reviewing the main structural elements of simple craters. Particular attention is given to the down axis limit of allochthonous brecciation, which is interpreted, following the terminology of Melosh (1989), as the Grady-Kipp fracturing limit (GKL), and its relationship to the extent of shock metamorphism. Then follows a review of the use of shock metamorphism as a method of estimating displacements in the breccias and underlying parautochthous rocks, based on apparent rates of shock pressure attenuation. Previously, Dence et al. (1977) employed experimental data from the Piledriver buried nuclear explosion in granodiorite at the Nevada test site (Borg

1972) as a standard. This is here complemented by the calculations of Ahrens and O'Keefe (1977) to derive a more complete picture of shock decay and movements during crater formation.

In the second part there is a brief discussion of some of the factors which complicate the application of our understanding of simple craters to craters of larger size and more complex form. Among the more interesting parameters that change with crater diameter are crater morphometry, structure, geophysical signature, distribution of pseudotachylite veins and of allochthonous melt rocks and breccias. More detailed attention is given to relationships between crater size and the level of shock metamorphism in the parautochthonous rocks underlying the limit of Grady-Kipp fracturing.

2 Main Structural Elements of Simple Craters

The principal structural elements of simple craters in Canada, as revealed by geophysics, surface mapping and drilling, have been described a number of times (e.g., Beals et al. 1963; Dence 1965, 1968; Dence et al. 1977; Grieve 1978; Halliday and Griffin 1967; Short 1970). The gross structure of the craters conforms reasonably well to that anticipated from theoretical arguments and earlier work at Barringer Meteor Crater (Shoemaker 1960): a simple, bowl-shaped depression with a lens of brecciated rock below it and, where preserved, a raised rim of uplifted country rock. In every case general erosion and Pleistocene glaciation have effectively removed any *in situ* ejecta, though very small quantities have been found as glacial float in the vicinity of the youngest, the New

Fig. 1. (opposite) Simplified radial sections of Brent crater as a representative simple hypervelocity impact crater. (a) Transient cavity at maximum depth. The bolide is shown in black as a sphere impacting vertically. Also shown down axis is the limit of the near field where shock attenuation is low (cf. Fig. 2) changing to the far field region of higher attenuation. The volume directly melted by shock is outlined. Two representative curved flow lines show typical directions of movement for shocked material during excavation. The transient cavity is lined with a thin layer of impact melt overlying an allochthonous breccia consisting of clasts disrupted by Grady-Kipp fracturing. The breccia is underlain by a parautochthon of disturbed rocks that, near the surface, are uplifted to form the transient rim. Dimensions accord with calculations given in the text. (b) Present configuration with outline of the final rim before erosion and sedimentation. Four drill holes that completely intersect the breccia lens are shown. The star at the collar of central hole B1-59 is 220 m below the estimated position of the surface at the time of impact. Sedimentary fill consists of Ordovician grits, evaporites, shales and limestones that filled the crater soon after formation. The lower boundary of the breccia is the primary shear plane along which the allochthon and adjoining parautochthon forming the transient cavity walls slumped towards the centre (arrow indicates direction). Some higher shears within the breccia are also shown, as are layers of impact melt near the top and the residual pool of melt at the base (black). The primary shear conforms closely to a spherical surface with radius approximately twice the depth of the breccia in the centre of the crater. Within the breccias, gray horizons are layers with mainly fragmented impact melt, and jagged lines indicate shear zones.

BRENT

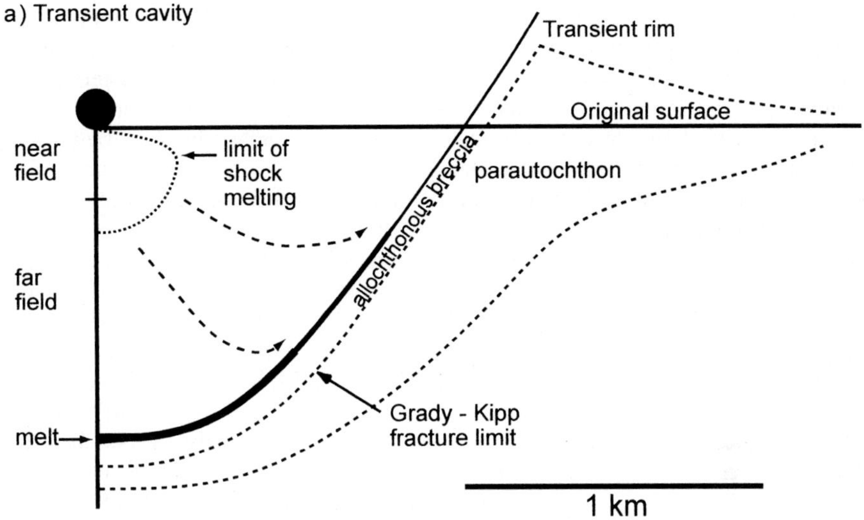
a) Transient cavity
Transient rim
Original surface
near field
far field
limit of shock melting
parautochthon
allochthonous breccia
Grady - Kipp fracture limit
melt
1 km

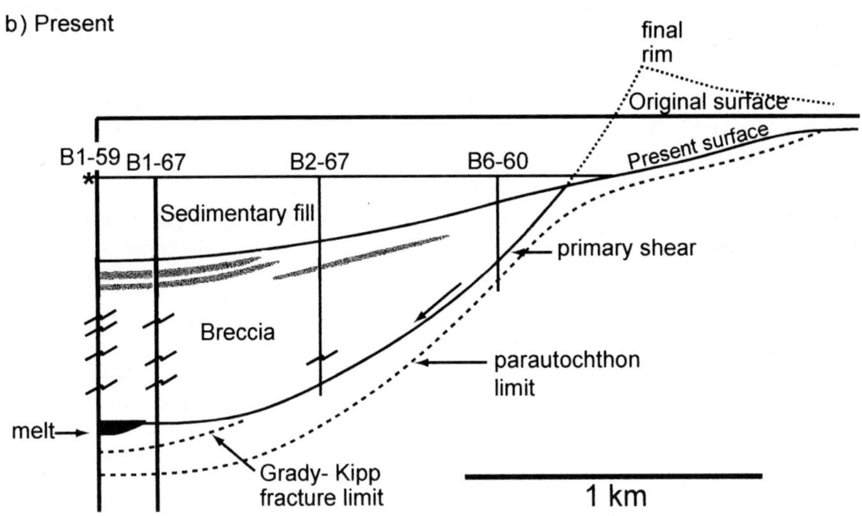
b) Present
final rim
Original surface
Present surface
B1-59
B1-67
B2-67
B6-60
Sedimentary fill
primary shear
Breccia
parautochthon limit
melt
Grady- Kipp fracture limit
1 km

Quebec Crater (Currie 1965). Fragments in breccias are derived from target rocks and are associated with melted rocks that are also derived from the country rock and contain, in some cases, minor additions of meteoritic material. This geometry has generally been understood in terms of an impact model in which target material near the point of impact is melted or vapourized, and at greater distance is stressed above the Hugoniot elastic limit and broken into fragments. A proportion of this material is ejected and the remainder displaced to form a lining to the growing cavity. Some low velocity ejecta mantles the crater rim, but the bulk of the rim consists of uplifted country rocks.

Once the breccias were analyzed in greater detail, with emphasis on the distribution of melted rocks and grades of shock metamorphism, it was realized that the final form of the breccia lens was predominantly the result of slumping (Dence 1968). The transient cavity formed as the first stage in the growth of the crater closely resembles computer calculations, such as those of O'Keefe and Ahrens (1993). Its configuration (Fig. 1a) can be reconstructed from the structure of the deepest part of the breccias observed in drill holes B1-59 and B1-67 (Fig. 1b). By extrapolation from the sections underlying the zone of melt rock, it can be deduced that the transient cavity was lined with a relatively thin layer of mobilized melt underlain by a thicker sheet of allochthonous breccia. Where preserved in the centre, this lining of melt plus breccia is almost 100 m thick. The transition down the drill core from allochthonous breccia to underlying basement rocks is relatively abrupt as, in contrast to the breccias, the latter retain overall continuity, though they are locally strongly fractured with pockets of monomict breccia. This is also seen in surface exposures of the eroded crater margin where the disruption is related to the displacements that occurred as the rocks marginal to the cavity were raised to form an uplifted rim. It is assumed that, before deep erosion, the rim at Brent was similar to that preserved at the New Quebec crater (Dence 1965, 1968). The disturbed zone around the crater is a parautochthon of moved, but only locally fractured or fragmented, country rocks.

It is proposed that the clasts in the melt and underlying breccia result from Grady-Kipp fragmentation, due to interactions of direct and refracted shock waves, with further disintegration resulting from turbulent motion along the expanding crater floor. The transition from breccia to basement may, thus, be said to mark the limit of Grady-Kipp fracturing and is so designated in Fig. 1a. The parautochthon lies beyond the Grady-Kipp limit and its outer boundary with undisturbed country rock is generally gradational, though at some craters it may be marked by a pattern of ring fractures (Dence et al. 1977).

The final form of the crater as shown in Fig. 1b results from slumping and subsequent erosion and sedimentation. The walls of the transient cavity are deduced to have been unstable and to have failed rapidly along a series of shear surfaces. The lowermost shear is here labeled the primary shear and can be said to be an example of a superfault in the sense of Spray (1997). The resulting mass of mainly brecciated rock has the form of a sphere segment. It comprises disrupted sheets of the allochthonous breccia and melt, lenses of suevitic breccia which may in part be fallout, and slabs of weakly shocked country rock which slid off the adjacent parautochthon. The character of the primary shear changes along its trace. As seen in hole B6-60 at its upper part near the rim, it is a well defined boundary between allochthonous breccia and the residual parautochthonous wall rocks. Where intersected near the bottom of drill hole B2-67, it is a thin zone of crushed and partly melted rock between an overlying sheet of largely intact but displaced parautochthon and weakly shocked basement. It ends in the middle of the residual melt rock outlined by holes B1-67 and B1-59. Within the breccia mass there are additional shear zones indicating that the mass slumped as a series of large slabs of the order of 100 - 150 m thick (Dence and Guy-Bray 1972).

To reconstruct the motions during crater modification in greater detail, an estimate of displacements can be made by comparing the final distribution of shocked materials with those predicted from theoretical and experimental models of shock wave attenuation. The section below the melt zone in drill hole B1-59 was subdivided by Dence (1968) into six zones of progressively decreasing shock metamorphism (Table 2 and Fig. 2). Beneath the zone of complete recrystallization, where heat from the overlying melt combined with the residual heat of shock metamorphism has obliterated most indicators of shock, lies a zone where some shock indicators are preserved, followed, in turn, by rocks which retain a full record of their shock history. The main shock metamorphism indicators at this level include feldspars with partial development of maskelynite and quartz with lower than normal birefringence and refractive indices (Dence 1968). There are, thus, traces of a complete range of shock metamorphism in the

Table 2. Brent crater drill hole 1-59: zones of shock metamorphism from melt zone to bottom of hole

Lithology	Depth below est. point of impact (m)	Shock zone	Estimated shock pressure (GPa)
Upper melt rocks, v. fine grained, many clasts	1043-1048	vii	>60
Central melt rocks, medium grained, few clasts	1048-1077	vii	60-150
Lower melt rocks, fine grained, many clasts	1077-1086	vii	60+
Breccia, qtz, matrix re-crystallized, f'spar clasts deformed	1086-1095	vi	40-60
Breccia, qtz recrystallized, f'spar clasts with many pfs, partly isotropic	1095-1102	v	25-40
Breccia, qtz with mainly π pdfs, low RIs and BR, f'spar with pfs, low BR	1102-1109	v	20-25
Breccia, qtz with ω + other pdfs, RIs slightly low, f'spars with few pfs	1109-1113	iv	15-20
Breccia, qtz with mainly ω pdfs, normal RIs, f'spars with rare pfs	1113-1130	iii	10-15
Breccia, qtz with basal + ω pdfs, f'spars normal	1130-1150	ii	7-10
Fractured, locally brecciated gneiss, qtz with rare pdfs, f'spars normal	1150-1190	i	5-7
Fractured gneiss, qtz and f'spars normal	1190-1286	o	<5
End of hole	1286		

Qtz, quartz; *f'spar*, feldspar (commonly andesine + mesoperthite); *pdfs*, planar decorated features (in quartz); *pfs*, planar features (in feldspar); *RIs*, refractive indices; *BR*, birefringence.

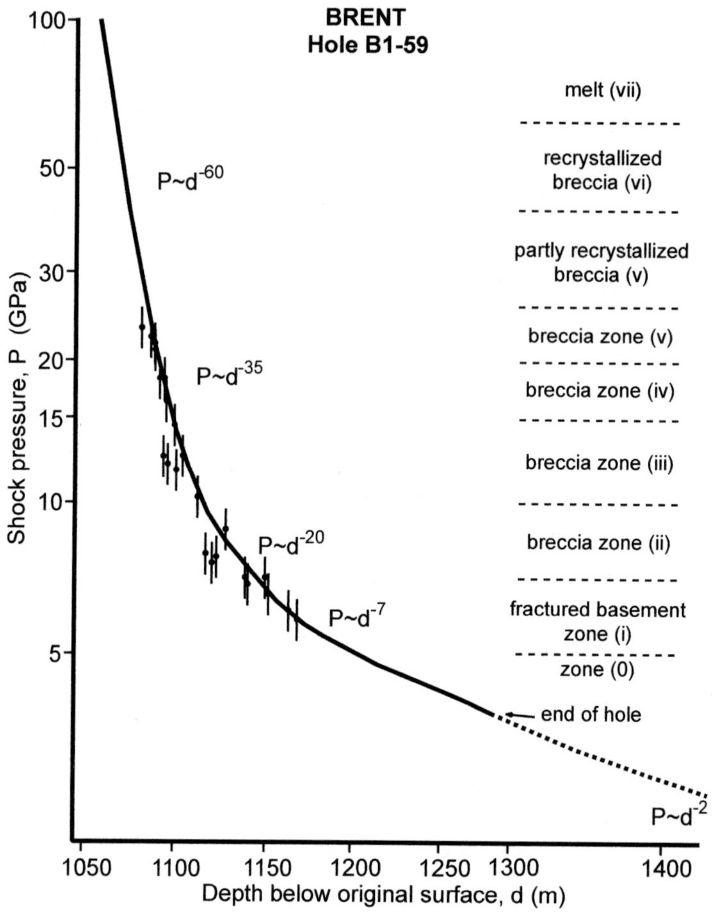

Fig. 2. Detail of shock metamorphism in the lowest approximately 300 m of Brent drill hole B1-59. Depth *d* below the original surface (taken as 220 m above the collar of the hole at the present surface, as in Fig. 1) is plotted against estimated shock pressure (GPa). Shock zones (ranging from o to vii) are taken from Dence (1965, 1968). Estimates of shock pressures with error bars are derived from measurements of shock deformation in quartz by Robertson and Grieve (1977). The apparent shock pressure attenuation is shown as a smooth solid curve with shock pressure, P, decreasing approximately as depth, d^{-60} near the melt zone, changing to d^{-7} at the lowest measurements of shocked quartz with a further change to d^{-2} inferred within zone 0.

section below the melt, with the lower, well preserved portion corresponding to stages I and II (shock pressures up to ~30 GPa) in the classification of Stöffler (1966), Stöffler and Langenhorst (1994) and similar schemes (French 1998).

Subsequently, more detailed observations were made to calibrate the unrecrystallized part of this section by averaging measurements of shock-induced planar deformation features in quartz (Robertson 1975; Robertson and Grieve 1977). Their results give ranges of shock pressures to each shock zone, as summarized in Table 2, and indicate that zones (i) to (v) defined by Dence (1968) at Brent range from <5 to >23 GPa. While the absolute precision of the calibration is probably of the order of ± 3 - 5 GPa, the method gives a faithful depiction of the relative intensity of shock metamorphism in rocks of similar lithology. Hence, it can be used to assess the rate of change of shock metamorphism with depth or distance from the centre.

Robertson and Grieve (1977) recognized that, over the interval they measured, the average apparent rate of shock pressure attenuation is high (exponent ~20). The exponent of shock pressure decay is even higher within the breccias (~35) but decreases abruptly to ~7 in the lower part of the breccias near the transition to the underlying parautochthon. They place the change in rate at 925 m below the collar of the drill hole, where the shock pressure had dropped to ~7.5 GPa, an estimated 1145 m below the original surface. That depth is 59 m below the base of the melt zone and about 5 m above the level originally identified by Dence (1965) as the bottom of the breccia within the crater and now equated with the approximate limit of GKL. Thus defined, the GKL at Brent occurs at an estimated shock level of 7 GPa.

Alternatively, as shown in Fig. 2, the apparent rate of shock pressure attenuation can be viewed as changing smoothly and extending beyond the measured section upwards into the melt zone and downwards into zone 0. The exponent of the apparent rate of decay ranges from ~ -60 near the melt to ~ -2 over the last 94 m of the drill core where shock effects are absent.

3 Shock Pressure Attenuation and Displacements in Breccias and Parautochthon

The data given in Table 2 can be used to estimate the net displacement that has occurred in rocks below the melt zone at Brent by comparing the apparent shock pressure attenuation with a reliable standard. In the original analysis along these lines (Dence et al. 1977), the standard for comparison was experimental data from the 64 kt TNT equivalent Piledriver underground nuclear explosion in granodiorite at the Nevada Test site (Borg 1972). There, when instrumented stations were restored to their original position, the indicated exponent of shock wave attenuation was -2. However, in applying this result to natural impact craters no allowance was made for effects close to the surface and a single attenuation rate was assumed from the point of impact. Indeed, the data seemed to give a better fit when higher rates of attenuation were modeled, so it was argued that shock pressure attenuation may change with increasing crater size. In the case of Brent a rate with exponent -2.5 was considered to give the best fit to the data and even higher rates were judged to be necessary to fit similar data from the complex Charlevoix crater to an impact model (Grieve and Robertson 1977).

In returning to this problem using the calculations of Ahrens and O'Keefe (1977), previously neglected near-surface effects are taken into account (see also Melosh 1989). They show that, in the near field close to the point of impact, the on-axis peak pressure attenuation rate has an exponent between -0.15 and -0.3 for a wide range of impact conditions. In the far field at a depth along axis of typically one to two projectile diameters, the exponent of shock pressure decay changes to between -1.5 and -3. Ahrens and O'Keefe (1977) described the centre line pressure in terms of:

$$\log_{10}(P - Mbar) = a\log_{10}(R/R_o) + b \tag{1}$$

where P is the centre line pressure in bars, R is the distance down the centre line from the point of impact, and R_o is the initial meteoroid radius in metres. In the Ahrens and O'Keefe (1977) calculations for a velocity of 15 kms^{-1}, a is -0.222 in the near field and -1.97 in the far field. For b the corresponding values are 0.285 and 1.399. The change from near to far field occurs at $P \sim 140$ GPa.

The Ahrens and O'Keefe formula for the far field can be rounded and transcribed in the form:

$$P = 2512(R/R_o)^{-2} \tag{2}$$

where P is now in Gigapascals (1 GPa = 10 kbar). To apply this formula to Brent crater, their specific instance of a gabbroic anorthosite projectile impacting a similar target material is accepted as a reasonable approximation to the physical properties of the country rocks and probable bolide. The predominant country rocks at Brent are high grade quartz-feldspar gneisses, and geochemical analysis of the melt rocks has indicated that the impacting meteorite was probably chondritic (Grieve 1978). In addition, a shock pressure must be chosen at which the apparent attenuation matches the theoretical exponent of -2 in formula (2). The depth at which this occurs is the limit of net post-impact movement due to the shock wave. In Dence et al. (1977) this pressure was taken to be the same at Brent as at the Piledriver experiment, namely 2 GPa, corresponding to a depth of 1,600 m below the point of impact at which no net displacement was inferred to have occurred at Brent. However, an impact crater formed by energy equivalent to the 64 kt Piledriver nuclear explosion would have a diameter of about 480 m (using formula (3) given below). Then, by formula (1), the GKL for such a crater would be about 2.4 GPa. If the gradient from the GKL to the point of no displacement is similar at Brent, where the GKL (given above) is 7 GPa, the shock pressure at which the apparent and actual attenuation rates are similar is ~5 GPa. The corresponding depth, R, is 1,220 m below the point of impact. This is shown in Fig. 3 as the conjunction of the measured apparent shock attenuation curve redrawn from Fig. 2 and the actual shock attenuation curve derived from Ahrens and O'Keefe (1977).

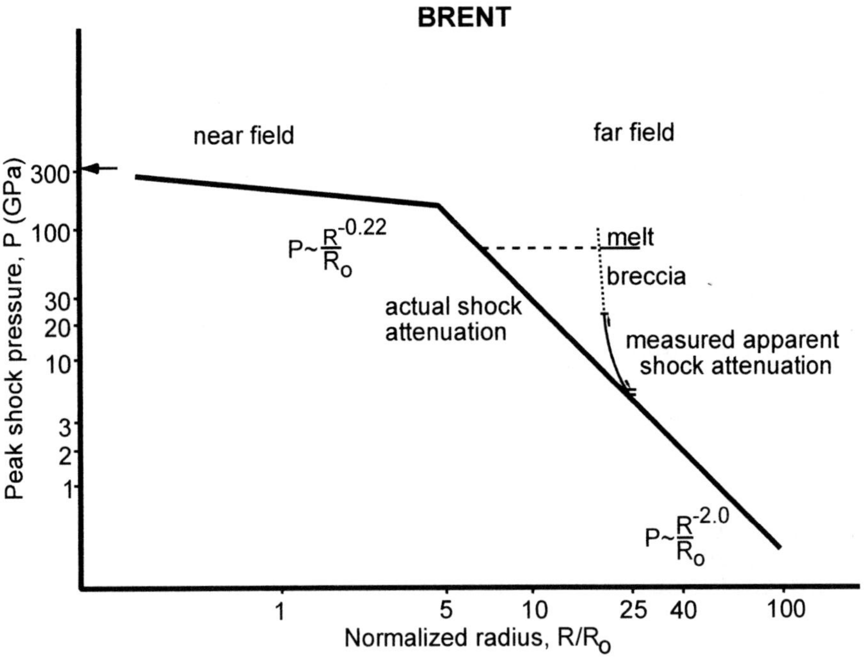

Fig. 3. Comparison of observed apparent shock pressure attenuation in the Brent central drill hole B1-59 (from Fig. 2) with calculated actual attenuation in the near and far fields generalized after Ahrens and O'Keefe (1977). Peak shock pressure along the centre line (GPa) is plotted against R/R_o where R is the distance (depth) from the impact point and R_o is the radius of the bolide, a nominal stony meteorite impacting vertically at 15 kms^{-1}. The arrow at the vertical axis indicates the maximum pressure at the point of impact. The measured part of the Brent data is shown as a solid curve, with extrapolation to the melt zone dotted. The intersection of the observed and theoretical curves occurs at a shock pressure of 5 GPa where $R/R_o = 22.4$. As R is 1220 m, R_o is calculated as 53.4 m (see text). A horizontal dashed line at $P = 60$ GPa joins the actual ($R/R_o = 6.5$) and observed ($R/R_o = 20$) attenuation curves at the base of the impact melt, providing an estimate of its downward displacement.

From this result a nominal size for the stony projectile that formed the Brent crater, given a vertical impact at 15 kms^{-1}, can be derived. By formula (2), R/R_o at $P = 5$ GPa is 22.4, so, for an observed $R = 1{,}220$ m, the radius of the projectile is $R_o = 54.3$ m. Further, the downwards displacement of the impact melt and underlying breccia can be calculated by comparing the present position of any level with the original depth calculated using formula (2). For example, the base of the melt, taken as corresponding to a pressure of 60 GPa, is now observed at a depth of 1,086 m below the point of impact ($R/R_o = 20.0$). Extrapolating along the horizontal dashed line in Fig. 3 to the equivalent pressure on the actual shock attenuation curve gives $R/R_o = 6.5$. It follows that the original depth down axis of the rocks that became the base of the impact melt was 352 m prior to compression and displacement by the shock wave. The exponent of apparent decay of -60

(Fig. 2) indicates that this level of the target was compressed vertically to 1/30 of its original thickness and spread into a disk more than 5 times its original lateral extent.

The energy of the impact can also be derived as 3×10^{17} J, assuming the impacting body has a similar density (3,900 kgm^{-3}) to the high pressure phase of the gabbroic anorthosite used in the Ahrens and O'Keefe calculations. Further, if the relationship between kinetic energy and crater diameter is of the form

$$D = xE^{1/3.4} \tag{3}$$

then, with $D = 3.8$ km and $E = 3 \times 10^{17}$ J, the value of x is 2.75×10^{-5}. This may be compared with $x = 1.96 \times 10^{-5}$ derived by Dence et al. (1977) by extrapolation from nuclear explosion data. The larger value in the present calculation may be viewed as the limiting case in which there is greater efficiency of energy transfer than is likely to occur in nature or experiment.

4
Some Size-dependant Parameters of Canadian Craters

4.1
Topographic Form and Structure

The Canadian craters in crystalline rocks (Table 1) with original diameters <4 km across (West Hawk Lake, New Quebec, Brent) are bowl-shaped and representative of craters with simple structure. At larger diameters, the form of the craters is successively flat-floored, with a topographically subdued central uplift (Deep Bay), bowl-shaped with a pronounced central uplift (Nicholson Lake, Clearwater East, Lake Mistastin, Slate Islands), combined central peak and ring (Clearwater West, Charlevoix) and transitional to multi-ring structures (Manicouagan).

Although erosion introduces considerable uncertainty in some cases, the deepest crater is probably Deep Bay where the relatively flat centre is an estimated 570 m below the original surface. Larger craters are shallower, the most conspicuous change being at the two Clearwater Lake craters where the larger is at least 120 m less deep than the adjacent smaller crater. There is a corresponding decrease in the ratio of depth to diameter in accord with the view that, with increasing size, gravity is increasingly important in determining final crater form.

4.2
Gravity and Other Geophysical Parameters

Early in the investigation of circular topographic features in Canada great emphasis was placed on the presence of a corresponding circular negative Bouguer gravity anomaly as diagnostic of impact (Innes 1961; Beals et al. 1963). For some, the case was strengthened by an associated low seismic velocity and subdued magnetic anomalies. The prevailing view at the time was that the negative anomalies were due to fractured, low density material and would be characteristic of all impact sites. This proved to be a reliable criterion in the case of craters up to about 30 km across. However, larger structures, such as Manicouagan, which was the subject of geophysical investigations as early as 1953 (Willmore 1963), had ambiguous geophysical signatures, difficult to reconcile with the general model. In these craters the negative anomalies are concentrated near the margins and the gravity and seismic signatures at the centre are similar to those of the surrounding regional rocks. This has been attributed to higher density rocks being raised in the central uplift (Sweeney 1978), but the generality of the feature in large craters suggests that the main mechanism is a relative absence of fracture porosity in the central uplift. Whereas the central regions of craters with $D < \sim 30$ km have either a concentration of low density allochthonous breccias or fractured and locally crushed parautochthonous rocks of the central uplift, those of larger craters are shocked but lack porosity, as discussed below.

4.3 Pseudotachylites

Pseudotachylites are generally included as one of the distinguishing features of impact structures (French 1998), though the literature is dominated by two occurrences, the type locality at Vredefort (Reimold and Colliston 1994) and the analogous Sudbury breccias (Thompson and Spray 1994). In the smaller Canadian craters veins of finely crushed rock, found in parautochthonous rocks and occasionally with allochthonous breccias, are generally narrow (<1 cm thick) and occur predominantly in the shear zones in which they were generated. An example is the primary shear zone intersected by drill hole B2-67 at Brent (Fig. 1b) and higher shears within the breccia lens (Dence and Guy-Bray 1972). Whether complete frictional melting has taken place in these and similar instances remains an open question and needs further investigation. Drill core from the Deep Bay (Dence et al. 1968) and East Clearwater (Dence 1964, 1965) craters provide similar examples of relatively inconspicuous veins within the crushed and fractured rocks of their central uplifts. This seems to be the general case in craters less than 30 km across.

However, at the 32 km West Clearwater crater and still larger Charlevoix and Manicouagan craters undoubted pseudotachylite veins are observed in the parautochthonous central uplifts. Those at West Clearwater and Charlevoix are typically 1 - 2 cm thick. At the former they are found in core from two holes within 2 - 3 km of the centre (Dence 1964) and at Charlevoix they have been recorded predominantly within 4 km of the central peak of Mont des Éboulements (Rondot 1989). Rondot describes them as clast-bearing veins with groundmass of fluidal texture that have invaded large blocks of shocked but otherwise mildly fractured rocks of the parautochthon. He distinguishes them from veins and zones of strongly crushed rock, which he calls mylolisthénites, that occur predominantly at the margins of the central uplift and in the annular graben of down-faulted Paleozoic rocks. Similarly, at Manicouagan, the writer observed excellent exposures of cross-cutting pseudotachylites up to 10 cm thick within the mildly fractured anorthosite massif of Mont de Babel at distances up to 10 km from the centre (Dence 1964, 1965). Here some veins retain glass within the matrix and have locally melted adjoining shocked feldspar (in part maskelynite) to form a tight bond with the host. The result, most notably at the two largest craters, is a tough rock that, despite shock metamorphism at an estimated 20 - 30 GPa and widespread alteration of constituent feldspars to zeolites (prehnite at Charlevoix, thomsonite at Manicouagan), has proved as resistant to glacial erosion as the strongest regional rocks. The interpretation offered is that many cracks have been sealed by the migration and injection of hot frictional melt from the shear zones where it was generated during a stage when the host rock was dilated. This would be the case if, during formation, the central uplift rose above its present position, and possibly well above the original surface, before subsiding and consolidating in its final configuration.

4.4
Distribution of Melts Rocks and Associated Allochthonous Breccias

Canadian crater studies have contributed empirically in several ways to the present understanding of melt rock formation and distribution (Dence 1971; Grieve 1978; French 1998). It is now well known that hypervelocity impact into crystalline rocks of near average crustal composition produces substantial volumes of melt rocks as well as associated breccias of variably shocked country rocks. In simple craters the portion not ejected is concentrated in the centre. As discussed above, at the 3.8 km Brent crater the melt plus underlying allochthonous breccia had a thickness of about 100 m at the transient crater stage (Fig. 1a). In complex craters the allochthonous melt rocks and breccias occur as annular sheets around the central uplift (Dence 1971; Grieve et al. 1977). In some instances (Deep Bay, Clearwater East), later sedimentary cover has preserved a reasonably complete assemblage. At others, such as West Clearwater, Charlevoix and Manicouagan, erosion has removed considerable information about the initial proportions of melt and breccia at different radial distances from the centre. Even so, it is clear that, as the central uplift is formed, there is an accompanying radial movement of allochthonous melt and breccia away from the centre.

We can go further by noting that at West Clearwater the thickest parts of the preserved melt sheet occur on the outer margins of the islands, which form the ring. Here, the melt rocks overlie and have thermally metamorphosed small patches of pre-impact Palaeozoic limestone which have been brought down by the formation of the outer peripheral trough (Dence et al. 1977). Thus, in this case, the motion of the melt sheet and underlying breccia carried a large proportion of their combined volume beyond the rim of the transient cavity. Much the same picture emerges at Manicouagan where only a few exposures of melt rock are found within 8 km of the centre and a thick section of melt rock overlaps residual patches of Paleozoic limestone in the peripheral trough. There are other, less readily apparent, changes which are nonetheless important in understanding how crater mechanics change as size increases. One is considered in the next section.

5 Shock Metamorphism in Parautochthonous Basement Rocks

In complex craters the parautochthon forms the central uplift and includes rocks underlying the surrounding layers of allochthonous breccias and shock melted rocks. Unlike the chaotic juxtaposition of all levels of shock metamorphism in clasts in these layers, parautochthonous rocks show a broadly regular distribution of shock metamorphism. Where outcrops permit, zones of shock metamorphism can be mapped that typically decrease radially from moderately strong in the centre to weak towards the margin (Dence 1968; Robertson 1975; Robertson and Grieve 1977). Grieve and Cintala (1992) noted further, from a survey of data in the literature that, with increasing crater size, rocks from the centre show a relatively weak trend from about 25 GPa in 2 km craters to more than 40 GPa in those ~100 km across.

This intriguing result is extended here by including data from additional Canadian craters (Table 3), which fill some gaps, notably for craters with diameters between 5 and 20 km.

The data recorded in Table 3 and plotted in Fig. 4 are based on the method of averaging 25 or more measurements of shock grade for each sample as used at Brent, discussed above, and at the Slate Islands and Charlevoix craters (Robertson and Grieve 1977). The calibration of shock zones at Brent (Fig. 2) is used here to estimate shock grades from less complete data from other craters. The results indicate lower levels of shock in the centre of the parautochthon at small craters and a pronounced gradient to higher levels in larger craters.

Table 3. Estimated central shock pressure at top of parautochthon (Grady-Kipp limit; GKL), and present diameter for selected Canadian craters; sample type and location in structure also given

Crater name	Diameter (km) [a]	Central shock pressure at GKL (GPa) [b]	Sample type	Position
Brent [c]	3.8	7	Drill core	Base of breccia at centre
Deep Bay	9.5	12.5	Drill core	Central peak ~1 km from centre
Nicholson L	12.5	12.5	Surface	Central peak ~1 km from centre
Clearwater E	20	15	Drill core	Central peak ~1.5 km from centre
Mistastin [c]	28	18	Surface	Central peak ~2 km from centre
Slate Islands [c]	30	20	Surface	Central peak at centre
Clearwater W [c]	32	20-25	Surface Drill core	Central peak at centre ~1 km from centre
Charlevoix [c]	52	23	Surface	Central peak at centre
Manicouagan [c]	80	30-35	Surface	Central peak ~2 km from centre

[a] As in Table 1, estimated final diameter of some craters is lower than is given elsewhere (Grieve 1991; Grieve and Shoemaker 1994)

[b] Shock pressures are based on zones of shock metamorphism calibrated as in Table 2

[c] Structure included in Table 5 of Grieve and Cintala (1992)

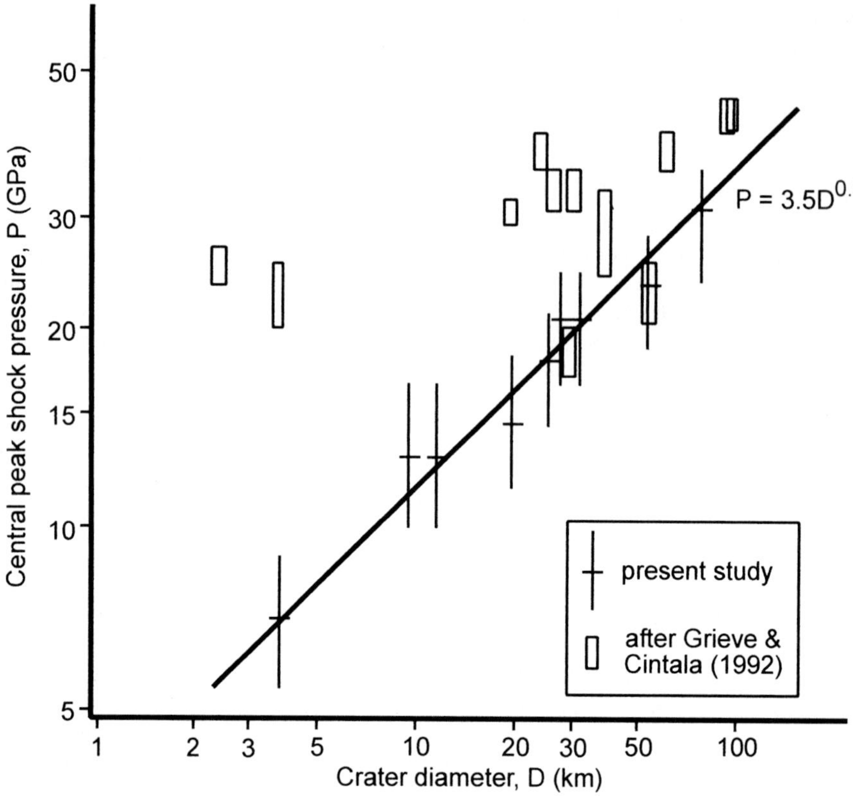

Fig. 4. Variation of estimated central peak shock pressure (GPa) at top of the parautochthon with diameter of present crater (Table 3). Data are compared with an earlier data set compiled by Grieve and Cintala (1992). The relationship between shock pressure at the centre of the parautochthon P and crater diameter D is calculated as $P = 3.5\ D^{0.5}$ (solid line). Pressures are taken at the interface between allochthonous breccia and underlying basement at the only simple crater (Brent) or at the top of the central uplift in complex craters. Size is the estimated final crater diameter (km). In most cases the horizontal bar approximates the range for these diameters given in the literature; similarly the vertical bar is an indication of estimated error in assigned shock pressure, as well as the effects of erosion and variation in sample distance from the geometric centre of each structure.

The present result may be expressed in the form:

$$P = aD^n \tag{4}$$

where P is the central shock pressure (GPa) at the GKL, D is the final diameter of the crater (km), $a = 3.5 \pm 0.3$ and $n = 0.5 \pm 0.2$.

Although the deviation in the exponent n reflects considerable scatter in the data, the trend is clear. However, there are substantial differences between the present result and that of Grieve and Cintala (1992). They have listed the maximum recorded shock stress levels at the centre of each crater in their Table 5, and in some cases appear to have included measurements on clasts within allochthonous breccias rather than restrict them to material from the highest available levels of the parautochthon. This is certainly the case for Brent, where their list uses the highest values measured for shocked but not recrystallized quartz in the allochthonous breccias underlying the impact melt in drill hole B1-59. The data in Table 2 of this study demonstrate that there is a continuum of decreasing shock levels from the melt, partly obscured by the overprint of thermal metamorphism. For the present analysis, the appropriate horizon for comparison with central peak data from complex craters is the boundary between allochthonous breccias and parautochthon at Brent. This, as defined earlier, is the GKL at 7 GPa, and it is this latter value that is adopted in Table 3 and Fig. 4.

In most diagrams of the initial stages of impact, fracturing and brecciation are usually depicted as extending well beyond the zone of recognizable shock metamorphism (Melosh 1989; French 1998). The present results demonstrate that this is generally not so in the case of terrestrial craters formed in crystalline rocks. It is evident that Grady-Kipp fragmentation controls the size of the transient cavity and is progressively less extensive relative to the extent of shock metamorphism with increasing crater size, i.e., impact energy. Larger craters will have smaller transient cavities relative to energy deposited, as recent computer models of complex crater formation suggest (O'Keefe and Ahrens 1999). It seems that, in more energetic events, the energy of refracted waves travelling down axis is dissipated or absorbed at approximately twice the rate of the decay of the direct shock wave. Why this is the case needs further consideration, but it is probable that an important factor is the work the refracted wave expends against gravity as it fractures the target rocks. Other implications for the nature of the transient crater in large impacts need to be assessed along with considerations concerning melt volumes discussed by Grieve and Cintala (1992).

6 Discussion

The general progression, which emerges from these and other observations, is of two principal changes in morphology and structure in impact craters formed in igneous and metamorphic rocks. The first is the well-known change from simple to complex form at a diameter of ~4 km. The second is the change in complex crater form at about 30 km where central peak is replaced by peak plus ring morphology. Those in the range >4 and <30 km have central uplifts which become increasingly prominent with increase in crater size and, except for some high resolution methods, resemble simple craters in their geophysical signatures. They probably include the deepest terrestrial craters. In this size range it is inferred that the peak does not rise substantially beyond its present position, is not strongly compressed and so retains significant fracture porosity. The second transition at about 30 km is marked by the development of peak ring structure. Fracture porosity is lower in the centre than towards the margin, which is characterized by the presence of a peripheral trough. Craters with these characteristics are inferred to result from the central peak initially rising above the original level and then subsiding to form the ring and residual central peak. Pseudotachylites are injected into dilated fractures to seal much fracture porosity. Also, as the peak rises, breccia and melt rocks lining the transient cavity and already sliding towards the periphery, gain additional momentum. The result is a concentration of melt rocks and associated breccia towards the periphery of the crater. In the largest craters, the centre may rise and fall more than once leading to multiple ring structure.

A general objective of this review is to contribute to the question of how available data constrain impact crater relationships as a function of energy of formation. The direct analysis of down axis displacements possible in the case of Brent and other simple craters, where drill core data are available, cannot be applied in similar fashion to complex craters. Where drill core is available, a general drop in shock level with depth is observed (Dence et al. 1977), but disruption and tilting by block faulting and rotation during central uplift degrades the usefulness of data on shock metamorphism relationships. In the few cases where shock zones have been mapped in terms of radial rather than vertical distance from the centre (Robertson and Grieve 1977), the results cannot be expressed in terms of R/R_o without making additional assumptions.

If, for example, it is accepted that formula (3), or a similar relationship between energy and diameter, applies to larger craters, R_o can be calculated for a nominal meteorite composition and velocity. The depth to GKL at the transient cavity stage, with a maximum shock level calculated from formula (1) can then be derived to give a measure of the "stratigraphic" uplift of the central peak. In the example of the 52 km Charlevoix crater such calculations give $R_o \approx 1$ km for a stony meteorite impacting at 15 kms^{-1} and a net vertical "stratigraphic" uplift in the centre of approximately 10 km. As a result, the ratio of depth to diameter is little more than 1 : 5, compared with the 1 : 8 to 1 : 10 ratio typically recorded from craters formed in stratified rocks. However, the greater strength of crystalline rocks and lack of bedding planes that facilitate slip make direct comparisons difficult with structures formed, for example, in shale and limestone targets. An intermediate case is that of targets of predominantly massive orthoquartzites and sandstones, as at Gosses Bluff (Milton et al. 1996). There, the depth to diameter ratio is about 1 : 6, supporting a low ratio for craters formed in crystalline targets.

Further exploration of models for large complex craters will be considered elsewhere. One aspect is to consider the total vertical movement of the central uplift. Such an estimate can be made from the assumption that the central peak rises only to its present level in small central peak craters, but progressively higher than the point of impact in large ($D > 30$ km) craters. Preliminary calculations suggest that in craters of the order of 100 km across the total vertical distance traversed during central peak formation may be comparable to the final crater diameter.

7 Conclusions

A number of general conclusions can be drawn from this compilation and re-examination.

- A satisfactory hypervelocity impact model for crystalline rock targets can be developed by combining observations from deep drilling at Brent and theoretical calculations of shock wave attenuation.
- By taking into account calculated changes in rate of attenuation between the near field and far field, an exponent of far field shock wave attenuation of -2 harmonizes observation with experiment and calculation.
- The model allows calculations of apparent displacements, projectile size and impact energy, recognizing that the results are for an ideal, highly efficient case unlikely to be matched in nature or experiment.
- Crater size is controlled by the extent of Grady-Kipp fracturing which, down axis, is progressively less extensive with increasing crater size than the imprint of shock metamorphism.
- Gravity, by influencing the efficiency of Grady-Kipp fracturing, is important in the early stages of crater formation as well as governing excavation and late stage structural modifications. This should be considered in extending these results to craters on other planets.

Acknowledgements

I thank Jüri Plado, Lauri Pesonen and their colleagues for the warm hospitality I enjoyed at the Lappajärvi meeting and assistance with this contribution to the proceedings. The paper has benefited considerably from perceptive reviews by John Spray and Lauri Pesonen, and comments from the editor. I am grateful to Andrew Dalby and Patrick Lyons of Carleton University, Ottawa, for their essential contributions to the production and electronic transmission of the figures.

References

Ahrens T, O'Keefe J (1977) Equations of state and impact-induced shock-wave attenuation on the moon. In: Roddy DJ, Pepin RO, Merrill RB (eds) Impact and Explosion Cratering, Pergamon Press, New York, 639-656

Beals CS, Ferguson GM, Landau A (1956) A search for analogies between lunar and terrestrial topography on photographs of the Canadian Shield, Part I, II. Journal Royal Astronomical Society of Canada 50: 203-211, 250-261

Beals CS, Innes MJS, Rottenberg JA (1963) Fossil meteorite craters. In: Kuiper GP, Middlehurst BM (eds) The Solar System, University of Chicago Press, Chicago 4: 235-285

Borg IY (1972) Some shock effects in granodiorite to 270 kilobars at the Piledriver site. American Geophysical Union Geophysical Monograph 16, 293-311

Currie KL (1965) The geology of the New Quebec Crater. Canadian Journal of Earth Science 2: 141-160
Dence MR (1964) A comparative structural and petrographic study of probable Canadian meteorite craters. Meteoritics 2: 249-270
Dence MR (1965) The extraterrestrial origin of Canadian craters. Annals New York Academy of Science 123: 941-969
Dence MR (1968) Shock zoning at Canadian craters: Petrography and structural implications. In: French BM, Short NM (eds) Shock Metamorphism of Natural Materials, Mono Book Corp, Baltimore, pp 169-184
Dence MR (1971) Impact melts. Journal of Geophysical Research 76: 5552-5565
Dence MR (1972) The nature and significance of terrestrial impact structures. Proceedings of the 24th International Geological Congress, Section 15: 77-89
Dence MR, Guy-Bray JV (1972) Some astroblemes, craters and cryptovolcanic structures in Ontario and Quebec. International Geological Congress 24th Montreal, Guidebook A65: 61 pp
Dence MR, Innes MJS, Robertson PB (1968) Recent geological and geophysical studies at Canadian craters. In: French BM, Short NM (eds) Shock Metamorphism of Natural Materials, Mono Book Corp, Baltimore, pp 339-362
Dence MR, Grieve RAF, Robertson PB (1977) Terrestrial impact structures: Principal characteristics and energy considerations. In: Roddy DJ, Pepin RO, Merrill RB (eds) Impact and Explosion Cratering, Pergamon Press, New York, pp 247-275
French BM (1998) Traces of Catastrophe: A Handbook of Shock-Metamorphic Effects in Terrestrial Meteorite Impact Structures. LPI Contribution No. 954, Lunar and Planetary Institute, Houston, 120 pp
Grieve RAF (1975) Petrology and chemistry of the impact melt at Mistastin Lake Crater, Labrador. Bulletin Geological Society of America 86: 1617-1629
Grieve RAF (1978) The melt rocks at Brent crater, Ontario, Canada. Proceedings of the 9th Lunar and Planetary Science Conference, pp 2579-2608
Grieve RAF (1991) Terrestrial impact: The record in the rocks. Meteoritics 26: 175-194
Grieve RAF, Cintala MJ (1992) An analysis of differential impact melt-crater scaling and implications for the terrestrial impact record. Meteoritics 27: 526-538
Grieve RAF, Shoemaker EM (1994) The record of past impacts on Earth. In: Hazards Due to Comets and Asteroids, University of Arizona Press, Tucson, pp 417-462
Grieve RAF, Dence MR, Robertson PB (1977) Cratering processes: As interpreted from the occurrence of impact melts. In: Roddy DJ, Pepin RO, Merrill RB (eds) Impact and Explosion Cratering, Pergamon Press, New York, pp 791-814
Halliday I, Griffin AA (1967) Summary of drilling at the West Hawk Lake crater. Journal Royal Astronomical Society of Canada 61: 1-8
Innes MJS (1961) The use of gravity methods to study the underground structure and impact energy of meteorite craters. Journal of Geophysical Research 66: 2225-2239
Innes MJS, Pearson WJ, Geuer JW (1964) The Deep Bay Crater. Publications Dominion Observatory, Ottawa 31: 21-51
Meen VB (1950) Chubb crater, Ungava, Quebec. Journal Royal Astronomical Society Canada 44: 169-180
Melosh HJ (1989) Impact Cratering. A Geologic Process. Oxford University Press, Oxford, 245 pp
Melosh HJ, Ivanov BA (1999) Impact crater collapse. Annual Reviews Earth and Planetary Science 27: 385-415
Milton DJ, Glikson AY, Brett R (1996) Gosses Bluff - a latest Jurassic impact structure, central Australia. Part 1: geological structure, stratigraphy, and origin. AGSO Journal of Australian Geology and Geophysics 16: 453-486
O'Keefe JD, Ahrens TJ (1993) Planetary cratering mechanics. Journal of Geophysical Research 98: 17,011-17,028

O'Keefe JD, Ahrens TJ (1999) Complex craters: Relationship of stratigraphy and rings to impact conditions. Journal of Geophysical Research 104: 27,091-27,104

Reimold WU, Colliston WP (1994) Pseudotachylites of the Vredefort Dome and the surrounding Witwatersrand Basin, South Africa. In: Dressler BO, Grieve RAF and Sharpton VL (eds) Large Meteorite Impacts and Planetary Evolution. Boulder, Colorado, Geological Society of America Special Paper 293: 177-196

Robertson PB (1975) Zones of shock metamorphism at the Charlevoix impact structure, Quebec. Bulletin Geological Society of America 86: 1630-1638

Robertson PB, Grieve RAF (1977) Shock attenuation at terrestrial impact structures. In: Roddy DJ, Pepin RO, Merrill RB (eds) Impact and Explosion Cratering, Pergamon Press, New York, pp 687-702

Rondot J (1968) Nouvel impact météoritique fossile? La structure semi-circulaire de Charlevoix. Canadian Journal of Earth Science 5: 1305-1317

Rondot J (1989) Géologie de Charlevoix. Ministère de l'Énergie et des Ressources Québec, MB 89-21, 606 pp + 19 cartes

Shoemaker EM (1960) Penetration mechanics of high velocity meteorites, illustrated by Meteor Crater, Arizona, International Geological Congress XXI, Norden, Pt 18: 418-434

Short NM (1970) The anatomy of a meteorite impact crater: The West Hawk Lake structure, Manitoba, Canada. Bulletin Geological Society of America, 81: 609-648

Spray J (1997) Superfaults. Geology 25: 627-630

Stöffler D (1966) Zones of impact metamorphism in the crystalline rocks of the Nördlinger Ries crater. Contributions to Mineralogy and Petrology, 12: 15-24

Stöffler D, Langenhorst F (1994) Shock metamorphism of quartz in nature and experiment: I. Basic observation and theory. Meteoritics, 29: 155-181

Sweeney JF (1978) Gravity study of great impact. Journal of Geophysical Research 83: 2809-2815

Taylor FC, Dence MR (1969) A probable meteoritic origin for Mistastin Lake, Labrador. Canadian Journal of Earth Science 6: 39-45

Thompson LM, Spray JG (1994) Pseudotachylytic rock distribution and genesis within the Sudbury impact structure. In: Dressler BO, Grieve RAF and Sharpton VL (eds) Large Meteorite Impacts and Planetary Evolution: Boulder, Colorado, Geological Society of America Special Paper 293: 275-287

Willmore PL (1963) The seismic investigation of the Manicouagan-Mushalagan Lake area in the province of Quebec. Publications Dominion Observatory, Ottawa 27: 325-336

Popigai Crater: General Geology

Victor L. Masaitis

Karpinsky Geological Institute, Sredny Prospekt 74, 199106 St. Petersburg, Russia.
(vcmsts@duxnet.spb.ru)

Abstract. The 35.7-Ma-old and ~100-km-wide multi-ring Popigai is a well-preserved impact structure at the Anabar shield, central Arctic Siberia. Because the structure is exposed, but only slightly eroded, a variety of impactites are preserved and accessible for petrological studies. The present contribution outlines the geological and petrological features of Popigai.

The Popigai structure formed 35.7 Ma ago (Bottomley et al. 1997) as a result of the collision of an asteroid-sized body with a bulk composition similar to that of an ordinary chondrite. The collision generated an impact crater about 100 km in diameter, which is filled with strongly shocked, brecciated, and melted rocks - lithic impact breccias, suevites and tagamites (see sketch map and cross-section in Pilkington et al., this volume). Today the impact site has only been slightly modified by erosion, and is imprinted on the flat or hilly landscape of the northern part of Siberian platform as a round-shaped depression that extends to a maximum depth of about 150 - 200 m below the surrounding topography, despite the presence of crater-hosted Pliocene and Quaternary deposits. Several semi-circular chains of table-shaped uplands and a semi-circular drainage system emphasize the concentric structural features within the impact crater. Topographic features that are concentric with respect to the crater are distinguished as far as 50 km from the rim of the crater depression (Masaitis et al. 1975, 1980).

The impact structure (71°38'N, 111°11'E) is located at the northeastern margin of the Anabar shield, where Archaean gneisses, schists and minor Lower Proterozoic igneous rocks are partly covered by gently north-east dipping sedimentary sequences of Upper Proterozoic, Cambrian, Permian, and Cretaceous age. Most of these rock sequences are siliciclastic and carbonate sediments with a total thickness of approximately 1 - 1.5 km. Dolerite sills and dykes of Triassic age that form part of the Siberian traps are also present, especially to the north of the impact structure.

Geological mapping, accompanied by geophysical investigations and drilling (up to 1.5 km deep), demonstrates that the structure is complex. The relief of the true bottom of the crater shows a central depression, surrounded by a peak ring (diameter about 45 km), which gradually passes outwards into an annular trough (outer diameter about 60 km). The trough is surrounded by a flat annular terrace (width about 20 km).

Several principal types of impact lithologies are recognized in the Popigai crater and its vicinity. The brecciated and cataclastic crystalline rocks that comprise the annular uplift and the bottom of the annular trough are regarded as authigenic (autochthonous or parautochthonous) monomict breccias. These rocks, and their rock-forming minerals, display evidence of shock transformations under pressures of up to 8 - 10 GPa (Masaitis et al. 1998).

The polymict allogenic breccia is subdivided into two varieties: blocky breccia and fine-grained psammitic breccia with rare large clasts (so-called coptoclastites). The second variety originated mainly as a result of the fragmentation of friable coal-bearing Cretaceous sandstones. Both breccia types may contain up to 10 vol% impact glass particles or isolated bombs of chilled impact melt. The blocky allogenic polymict breccia contains all types of sedimentary and crystalline target rocks fragments in a matrix of the same comminuted material, which is mineralized in places. Large blocks, as well as small fragments of gneisses and other rocks enclosed in polymict allogenic breccia, can be unshocked or moderately shocked, but locally exhibit features of shock metamorphism formed as a result of compression that may have been as high as 40 - 45 GPa. The blocky polymict breccias occur predominantly at the base of the crater fill sequence, and its thickness may reach several hundred meters. The fine-grained polymict breccia (coptoclastites) dominates in the upper part of the sequence. Both breccia types can grade into one another.

Suevites and tagamites predominantly comprise chilled or partly crystallized impact melt. Numerous clasts of shocked gneisses, sedimentary rocks and their minerals, make up 10 - 20 vol% of tagamites and up to 90 vol% of suevites. Larger inclusions of shocked crystalline rocks within tagamites may be about 50 - 100 m across. Suevites contain chilled impact melt as bombs and fragments; while in tagamites this partly crystallized material constitutes the groundmass. Suevites and tagamites occur in the form of lenses, and sheet-like bodies that alternate with one another locally. The thickness of individual tagamite bodies was established by drilling and may reach 600 - 1000 m (Masaitis et al. 1999). The impact melt was predominantly a product of the melting of crystalline target rocks; consequently, it has inherited chemical compositions similar to those of the target garnet-biotite gneisses. When compared with the composition of the initial gneisses, the tagamites display enrichment of siderophile elements that result from their contamination with dispersed projectile matter.

Both the tagamites and suevites are diamond-bearing. The impact diamonds formed by the solid transformation of graphite in the crystalline target rocks due to shock compression above 35 GPa. The diamonds are polycrystalline aggregates of microcrystals. These aggregates inherit the morphology of precursor graphite and have an average diameter of about 0.5 mm, but the largest aggregates are up to 10 mm in size (Masaitis 1998; Masaitis et al. 1972, 1998).

The interior of the Popigai crater shows complicated structural patterns. Flat relief characterizes the central part of the crater, which coincides with its ring central structural depression. Numerous lakes are located in this swampy area.

Drilling shows that alternating fine-grained breccia (coptoclastites) and suevites are distributed in the upper part of the crater fill. Brecciated basement gneisses that comprise the surrounding peak ring (or annular uplift) are exposed on the top of a semicircular chain of hills. The southeastern part of this uplift is buried under a cover of impactites (Mashchak and Selivanovskaya 1988). The uplift shows a gentler inner slope and a steeper outer slope. Alternating lens-like bodies of breccias, tagamites and suevites occur on the outer slope (Masaitis et al. 1999). On the inner slope suevites contain fewer impact glass bombs and particles, and alternate with coptoclastites (Raikhlin 1996). These contrasts were caused by distinctly different mechanisms of deposition of ejected material on either slope of the uplift. On the outer slope it was deposited from the ejecta moving mainly along ballistic trajectory; the inner slope was covered by material settling from collapsed column of gas, dust and fragments (Masaitis et al. 1998).

Suevites and tagamites are mainly exposed within the annular trough. A drilling profile across the trough demonstrates that it is filled by a composite sequence of allogenic polymict blocky breccia, tagamite and suevite, with the latter two dominating in the upper part of this sequence. The relief of the true bottom of the trough is irregular, with numerous blocky uplifts (Masaitis et al. 1999). Several varieties of polymict breccia can be distinguished in the lower part of the sequence. These varieties are based on their dominant clast type, including crystalline and sedimentary rocks, their type of matrix, and content of impact glass.

The tagamite sheet, which caps this sequence and overlaps allogenic breccia along the western edge of crater, has an undulating base. The maximum thickness of the overlapping tagamite sheet at this location is about 600 m. The tagamite sheet consists of two varieties of consolidated melt: high-temperature and low-temperature fractions, which are distinct in their petrography (Masaitis 1994; Masaitis et al. 1998). The tagamite sheet thins and splits near the rim towards southwest of the structure. The rim (the outer edge of the annular trough) grades outwards into a disturbed annular zone (annular terrace), surrounding this trough. Numerous arcuate thrusts faults, faults, small irregular folds, gradually decrease in offset and amplitude, respectively, with increasing distance outwards from the rim.

Shallow radial troughs are present on the inner edge of the annular terrace. These radial troughs were excavated in the target rocks, particularly in the sedimentary rock units, and are filled with allogenic polymict breccia, suevite and tagamite.

Remnants of ejected breccias and melted rocks are typically located on the outer terrace, and are found as far as 20 - 25 km from the crater rim. The ejected blocks of gneiss and quartzite, and small spots of suevite are scattered on the underlying undisturbed sedimentary and crystalline. These blocks can be up to 20 - 30 meters in diameter. Thin tagamite sheets can also be found in this zone.

Many aspects of the Popigai impact structure are of great value for comparative planetology, geology and the study of economic minerals (Melosh 1989; Grieve and Masaitis 1994; Deutsch et al. 2000a, b). The large diameter and complex inner structure, similar to peak-ring complex craters on the surface of other planets are among its distinguishing features. The distal ejecta layer found in Italy in Late Eocene sequences, which contain clasts of shocked quartz and has elevated iridium contents, is probably linked to the Popigai impact event (Montanari and Koeberl 2000). Geochemical data show that Late Eocene clinopyroxene-bearing spherules from deep-sea sediments are linked with Popigai ejecta as well (Whitehead et al. 2000).

The Popigai impact structure has escaped significant erosion, and large extended outcrops are available for study. A wide spectrum of impact lithologies, among them a large amount of impact melt rocks occur within the crater. The total volume of preserved impact melt is estimated to be equal to about 1750 km^3 (Masaitis 1994).

Various forms of disturbed target rocks, impactites, and lithic breccias are well exposed both inside and outside the crater depression. Data from numerous drillcores allow for 3-dimensional modeling of the morphology of these rock units.

Finally, most of the tagamites and suevites are diamond-bearing (Grieve and Masaitis 1994; Masaitis et al. 1998), which has initiated great economic interest in Popigai. Impact diamonds also occur in the placer deposits. The total amount of diamonds in the Popigai crater exceeds the resources of all other diamond-bearing provinces of the world combined.

A significant amount of data regarding the Popigai impact crater, its structural patterns, and mineralogical, petrological, and geological features of its rocks has been acquired during the last thirty years of research. No doubt, the scientific potential of research has not yet been exhausted, and new data will be obtained during future investigations. However, the development of the mineral resources of this remote area requires finding a solution to the numerous important barriers that exist, including those related to economics, technology, ecology, and sociology.

Acknowledgements

The bulk data regarding the Popigai crater geology were obtained in the course of field works and sample and drillcore studies carried out for many years together with my colleagues at VSEGEI, St.-Petersburg, and geologists of the Polar Exploration Expedition, Khatanga. I would like to thank them all, in particular, M.V. Mikhailov, T.V. Selivanovskaya, M.S. Mashchak, A.I. Raikhlin, G.I. Shafranovsky, A.N. Danilin, V.A. Maslov, M.V. Naumov, L.M. Zaretsky, N.A. Donov, V.T. Kirichenko. J. Whitehead and D.D. Badyukov provided thorough and constructive reviews of the manuscript, which led to considerable improvements; their help is also acknowledged.

References

Bottomley R, Grieve R, York D, Masaitis V (1997) The age of the Popigai impact event and its relation to events at the Eocene/Oligocene boundary. Nature 388: 365-368

Deutsch A, Langenhorst F, Masaitis VL (2000a) Der Popigai-Krater - Eine Schatzkammer in Sibirien. Forschung 3/4: 36-41

Deutsch A, Masaitis VL, Langenhorst F, Grieve RAF (2000b) Popigai, Siberia - well preserved giant impact structure, national treasury, and world's geological heritage. Episodes 23: 3-11

Grieve RAF, Masaitis VL (1994) The economic potential of terrestrial impact craters. International Geology Review 36: 105-151

Masaitis VL (1994) Impactites from Popigai crater. In: Dressler BO, Grieve RAF, Sharpton VL (eds) Large Meteorite Impacts and Planetary Evolution. Geological Society of America Special Paper 293: 153-162

Masaitis VL (1998) Popigai crater: origin and distribution of diamond-bearing impactites. Meteoritics & Planetary Science 33: 349-359

Masaitis VL, Futergendler SI, Gnevushev MA (1972) Diamonds in the impactites of the Popigai meteorite crater (in Russian). Proceedings of the All-Union Mineralogical Society 101: 108-113

Masaitis VL, Mikhailov MV, Selivanovskaya TV (1975) The Popigai meteorite crater (in Russian). Nauka Press, Moscow, 124 pp (English translation: NASA TT-F-16900 1976)

Masaitis VL, Danilin AN, Mashchak MS, Raikhlin AI, Selivanovskaya TV, Shadenkov EM (1980) The geology of astroblemes (in Russian). Nedra Press, Leningrad, 231 pp (English translation: NASA TM-77743 1984)

Masaitis VL, Mashchak MS, Raikhlin AI, Selivanovskaya TV, Shafranovsky GI (1998). Diamond-bearing impactites of the Popigai astrobleme (in Russian). VSEGEI Press, St. Petersburg, 180 pp

Masaitis VL, Naumov MV, Mashchak MS (1999) Anatomy of the Popigai impact crater, Russia. In: Dressler BO, Sharpton VL (eds) Large Meteorite Impacts and Planetary Evolution II. Geological Society of America Special Paper 339: 1-17

Mashchak MS, Selivanovskaya TV (1988) Breccias and impactites of the southeastern rim of the Popigai astrobleme (in Russian). Meteoritika 47: 178-188

Melosh HJ (1989) Impact cratering: A geologic process. Oxford University Press, Oxford, 245 pp

Montanari A, Koeberl C (2000) Impact stratigraphy: The Italian Record. Springer Verlag, Lecture Notes in Earth Sciences 93, Berlin-Heidelberg, 364 pp

Raikhlin AI (1996) Suevites from the Popigai impact crater: internal structure and condition on formation. Solar System Research 30: 14-18

Whitehead J, Papanastassiou DA, Spray JG, Grieve RAF, Wasserburg GJ (2000) Late Eocene impact ejecta: geochemical and isotopic connections with the Popigai impact structure. Earth and Planetary Science Letters 181: 473-487

Geophysics and Petrophysics of the Popigai Impact Structure, Siberia

Mark Pilkington[1], Lauri J. Pesonen[2], Richard A.F. Grieve[3] and Victor L. Masaitis[4]

[1]Geological Survey of Canada, 615 Booth Street, Ottawa, ON, K1A 0E9 Canada. (MPilking@NRCan.gc.ca)
[2]Department of Geophysics, University of Helsinki, P.O. Box 4, FIN-00014 Helsinki, Finland
[3]Earth Sciences Sector, 601 Booth Street, Ottawa, ON, K1A 0E8 Canada
[4]Karpinsky Geological Institute, Sredny Prospekt 74, 199106 St. Petersburg, Russia

Abstract. The 100-km-diameter Popigai structure was formed 35.7 Ma ago in a target consisting of ~1 - 1.5 km of Proterozoic, Cambrian and Permian sedimentary rocks overlying Archean crystalline basement. The structure is characterized by a gravity anomaly low of -35 mGal amplitude, one of the largest magnitude gravity anomalies associated with a terrestrial impact structure. Superimposed on the gravity low is a concentric ring-shaped high, ~45 km in diameter, that coincides with uplifted Archean basement. Magnetic data indicate a -300 nT amplitude simple anomaly low over the structure. Both the gravity and magnetic signatures of Popigai differ from those of most other impact structures of comparable size, being distinguished by the lack of a central circular gravity high related to a central structural uplift or high-amplitude magnetic anomalies caused by thick melt/suevite deposits.

Two-dimensional forward modelling of an E-W profile through the crater is initialized using existing and recently-acquired petrophysical data and a structural cross-section based on geologic mapping and drill hole information. Petrophysical measurements indicate that lithologies making up the crater fill have susceptibilities nearly two orders of magnitude less than those determined for the basement. Crater fill densities are also considerably lower, with values up to 0.4 gcm^{-3} less than the surrounding target rocks. Gravity data modelling indicates the presence of low-density crater fill plus an extensive region of reduced-density fractured basement. The magnetic data also suggest a significant volume (down to ~5 km depth) of fractured basement below the true crater floor.

1 Introduction

The 100-km-diameter Popigai structure was excavated within a target consisting of a 1.0 - 1.5 km thick layer of Proterozoic, Cambrian and Permian sedimentary rocks overlying Archean basement rocks of the Anabar Shield (Masaitis et al. 1975, 1980). Basement lithologies (Fig. 1) consist of mainly biotite-garnet gneisses and plagiogneisses of the Khapchansk and Upper Anabarsk groups. Both groups are migmatised and intruded by charnockitic granites, and are overlain by flat-lying Proterozoic and Cambrian quartzites, dolomites and limestones. Permian sandstones and argillites occur to the north of the crater. Impact lithologies crop out mostly within the structure itself. Coarse, polymict, allochthonous breccia is exposed at the outer edge of the crater, mainly in the north, east and southeast. This unit ranges from megabreccia, comprising fragments up to several hundred metres in size, to breccia containing centimetre- to metre-sized inclusions of both crystalline and sedimentary target rocks. Shock features are particularly common in gneissic fragments. The exposed central part of the structure consists mainly of suevitic deposits and fine-grained polymict breccia (coptoclastite). The suevites contain rock fragments, impact glass, and glass bombs. Several types are recognised based on the dominant clast lithology (Masaitis et al. 1980). Within the suevite are exposures of tagamite (impact melt rock) that commonly form topographic highs. Tagamites are extensively developed in the western, south-western, and southern parts of the structure. They contain up to 25% clasts and have similar chemical compositions as the suevites. A mixture of Khapchansk lithologies matches the average tagamite composition well, however, suevites generally require a greater component of sedimentary lithologies (Masaitis et al. 1980).

Figure 2 shows a cross-section through the structure based on geological and geophysical mapping and numerous drill holes (up to 1.5 km in depth). A prominent structural feature is the annular uplift at ~45 km diameter. In the west and north-west (Fig. 1), the uplift is expressed in the form of an arcuate outcrop and consists of shocked, fractured and brecciated Archean basement rocks. In the south-east, this feature is buried beneath ~1 km of tagamites and suevites (Masaitis et al. 1980). Here, the crater floor is at its deepest, reaching depths of >2 km within the annular trough radially exterior to the uplift. A poorly developed 10 - 15 km diameter central uplift peak is inferred from a weak magnetic high at the centre of the structure, noted by Masaitis et al. (1975). Thick lenses of allochthonous breccia that overlie Archean basement constitute part of the crater fill. At the crater edges, breccia has been deposited on top of the Proterozoic and Cambrian sedimentary formations. The bulk of the crater fill, however, comprises near-horizontal alternating suevite and tagamite sheets and lenses with thicknesses from 10 - 30 to 400 - 600 m (Masaitis 1998). This sequence is overlain, in places, by thin (<200 m) layers of coptoclastites containing lenses of suevites. Popigai is a well-preserved impact structure with most of the crater-fill lithologies present. The crater rim and associated uplifted rocks are partly preserved (Masaitis et al. 1980).

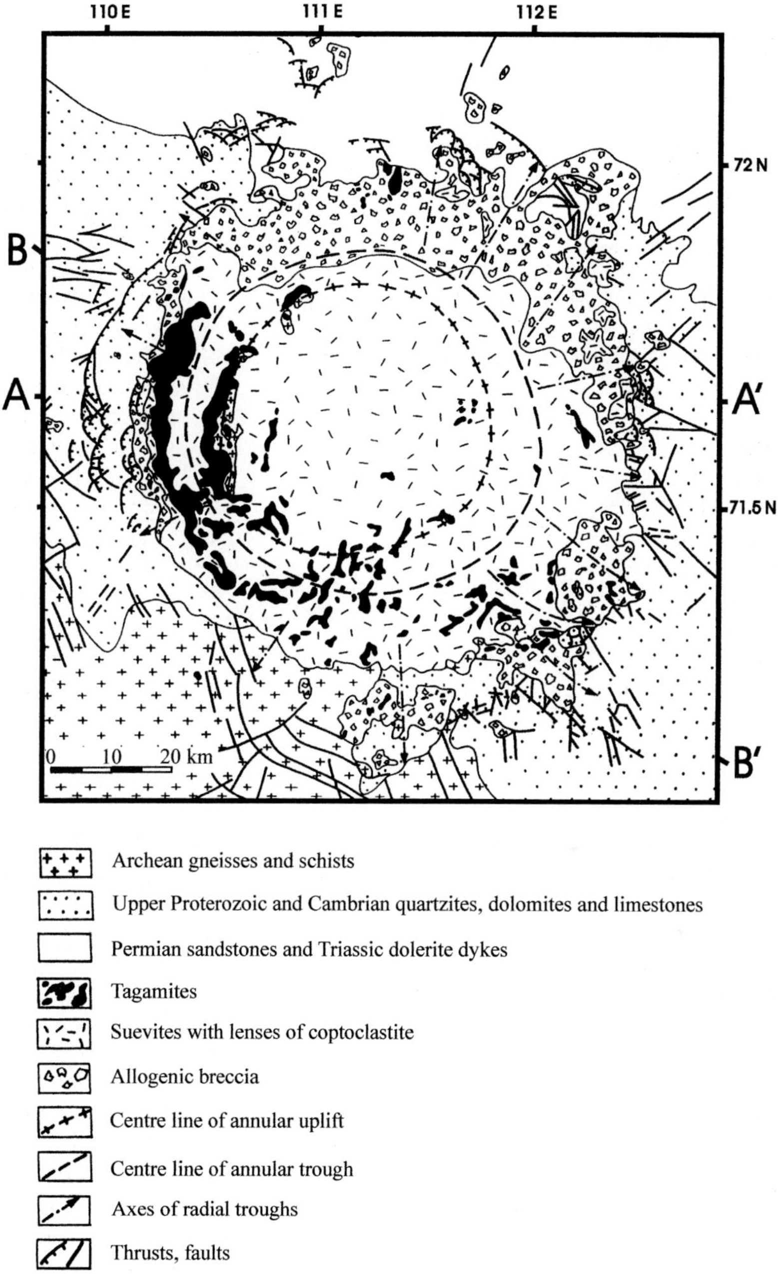

Fig. 1. Geology of the Popigai impact structure, after Masaitis (1994). A-A' denotes the position of E-W profile used for modelling. B-B' is the cross-section shown in Fig. 2.

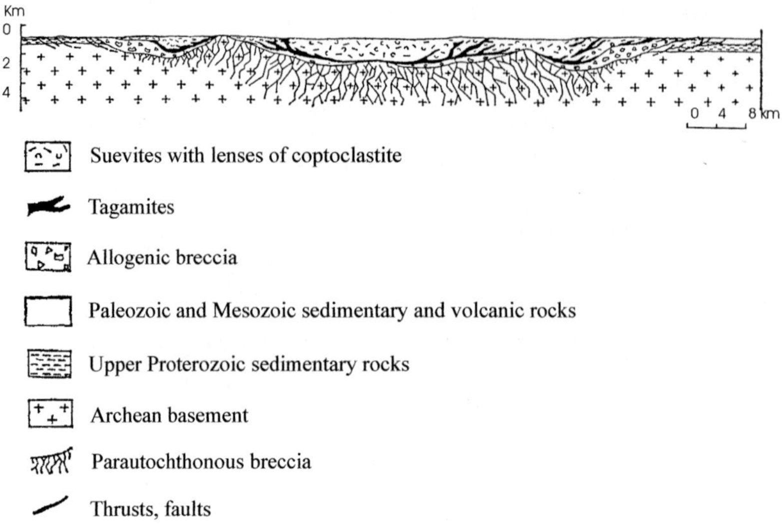

Fig. 2. Geologic cross-section through Popigai impact structure, after Masaitis et al. (1980). Location of cross-section given in Fig. 1. Vertical exaggeration 11x.

2 Gravity

The observed gravity field over Popigai reaches a minimum of -52 mGal with respect to the general level of the regional field of -17 mGal in the vicinity of the crater. This results in a gravity anomaly of -35 mGal associated with the impact structure itself. This is one of the largest magnitude gravity anomalies resulting from a terrestrial impact structure (cf. Pilkington and Grieve 1992). The anomaly over Popigai is essentially a gravity low, with a lateral extent of 80 - 110 km and with minimum values occurring in the southern portion of the structure. Strong NNW-SSE anomalies (~20 mGal) related to Archean crystalline basement structures/lithologies influence the periphery of the crater anomaly. The gravity map shown in Fig. 3a is based on an 8-km spaced data set of the Former Soviet Union (made available by GETECH Ltd.), so it has limited resolution compared to the published map of Masaitis (1998), based on data collected in 1977 by Shulgin. However, the latter, higher resolution data are displayed in Masaitis (1998) after an undocumented regional component has been removed. These data are also restricted in areal coverage to mainly the structure itself and were not available for use in this study. The 8-km data were interpolated to a 2-km interval for plotting as grids and profiles.

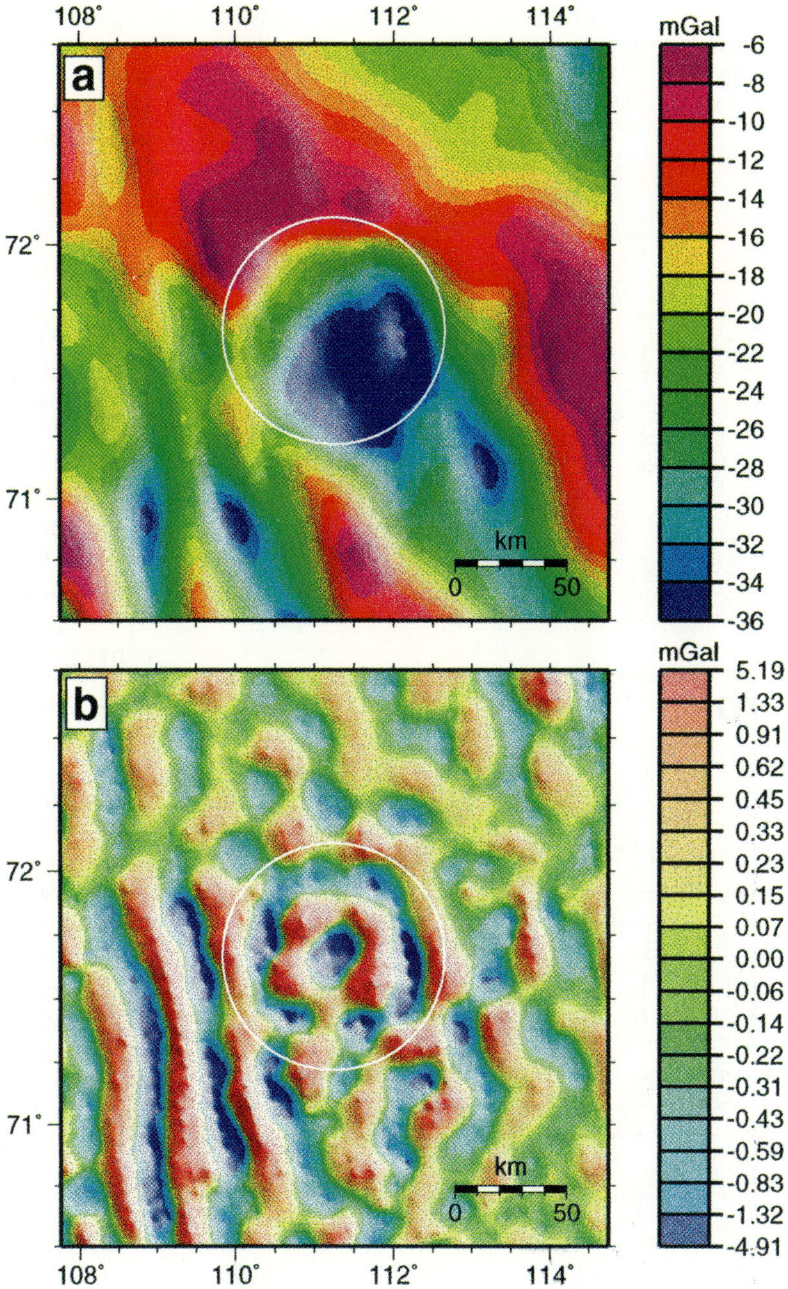

Fig. 3. (a) Gravity field over the Popigai region (data courtesy of GETECH Ltd.). White circle represents estimated crater rim at 100 km diameter. (b) Gravity field data, high-pass filtered to retain wavelength components <50 km (data courtesy of GETECH Ltd.).

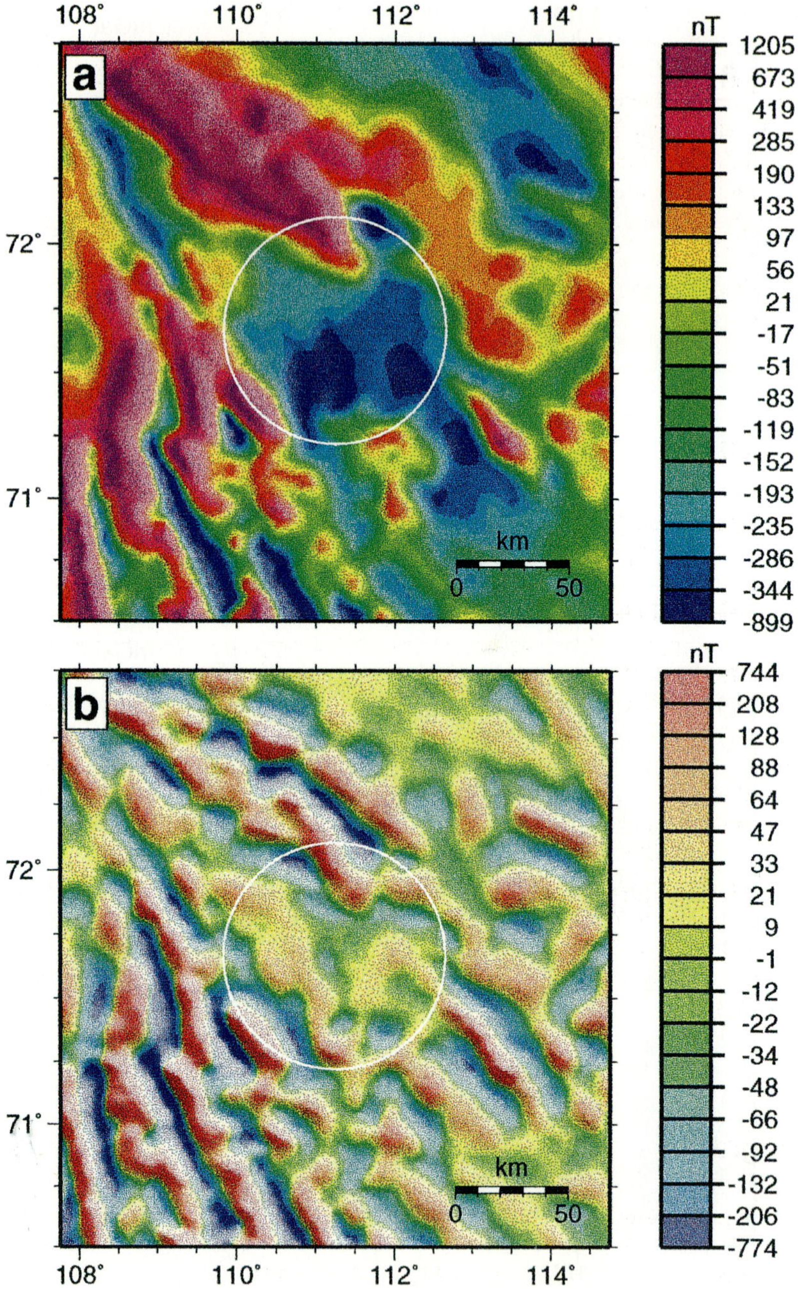

Fig. 4. (a) Magnetic field over the Popigai region. White circle represents estimated crater rim at 100 km diameter. (b) Magnetic field data, high-pass filtered to retain wavelength components <50 km.

Enhancements to the data in Fig. 3a reveal that the simple gravity low is more complex than it initially appears. Figure 3b shows a high-pass filtered (wavelengths <50 km) version of the gravity field over Popigai, which indicates that the crater-related anomaly consists of a central low flanked by a gravity high (best developed in the east and west) surrounded by lower values. A cut-off wavelength of 50 km was chosen to most effectively remove the first-order effect of the crater (the simple gravity low) and emphasize gravity variations caused by density changes within the structure. Figure 3b is similar to the published map of Masaitis (1998). The (discontinuous) gravity high at ~45 km diameter is coincident with the annular uplift of shocked and brecciated basement rocks (Fig. 1).

3 Magnetics

The magnetic field over the Popigai region (Fig. 4a) is based on data from a 2.5 km grid compilation of the former Soviet Union (Hittelman et al. 1991). These data were digitized from U.S.S.R. Ministry of Geology maps and are available from the National Geophysical Data Center (NGDC), Boulder, Colorado. The 2.5-km data was interpolated to a 1-km grid for further processing. The Popigai structure is marked by a magnetic low of up to 300 nT in amplitude (with respect to the regional background) compared to NNW-SSE-trending high amplitude (>1000 nT) anomalies in the west and more ovoid, less intense, NW-SE-trending anomalies to the east. These trends, most likely related to basement structure/lithology, are similar to observed gravity trends (Fig. 3a), however, a one-to-one correspondence is lacking. The magnetic low over Popigai shows some internal texture which is highlighted on high-pass filtered maps (Fig. 4b) of the area. The low gradients and amplitudes of these internal features, plus their similarity to the regional magnetic/gravity trends, suggest that deep (below the crater floor) magnetization variations related to basement are their cause. In contrast to the gravity field, no anomalies related to crater lithologies are apparent within the structure itself, apart from the extensive magnetic low.

4 Topography

The Popigai structure is characterized by a circular topographic low of >150 m depth with an average diameter of 90-100 km (Fig. 5). Data shown here are from the GLOBE database available from NGDC that corresponds to an approximately 1-km grid spacing (Hastings and Dunbar 1998). The Popigai depression is relatively flat in the north and east but is marked by ridges parallel to the crater circumference in the west. The outer of the two ridges corresponds to exposures of

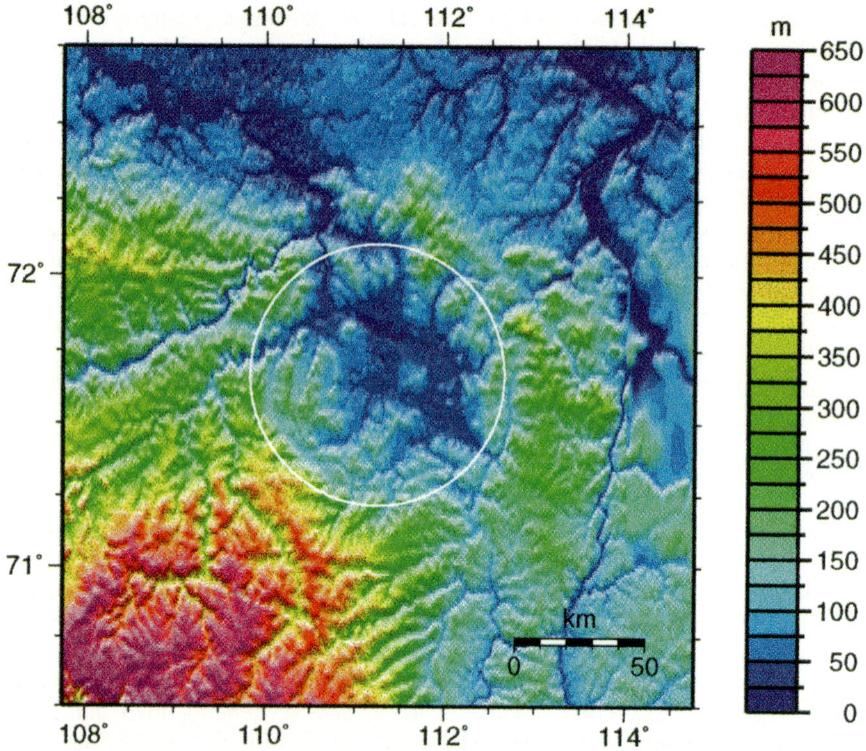

Fig. 5. Topography of the Popigai region. Data from GLOBE (Hastings and Dunbar 1998). White circle represents estimated crater rim at 100 km diameter.

tagamites whereas the inner one is composed of tagamite and exposed basement. The ridges reach heights of >100 m above the floor of the depression. In the south, a more complex topographic variation reflects the more random distribution of tagamite exposures plus the effects of local drainage.

5 Comparative Geophysical Signature of Popigai

Terrestrial impact structures of similar diameter (D) to Popigai are characterized by certain types of geophysical signatures (Pilkington and Grieve 1992; Henkel 1992; Grieve and Pilkington 1996). The expected gravity signature is a circular negative anomaly, with an area coincident with the diameter of the structure and a relative central gravity high having a diameter that extends to <0.5 D. The negative anomaly is mainly caused by fracturing and brecciation of target rocks beneath the parautochthonous crater floor and fragmentation and redistribution of target lithologies within the allochthonous breccia deposits. Low-density impact melt sheets, where present, may also contribute to the gravity low. The central gravity high is caused by the uplift of deeper, denser crustal material during crater formation. Additionally, the compressive regime within the central uplifted volume may also reduce the initial increased porosity associated with impact and lead to higher densities compared to the outer portions of the crater. Popigai has the expected magnitude of gravity anomaly (-35 mGal) for a structure of its size (Fig. 3a). For structures with $D > 30$ km, a maximum negative anomaly of ~30 mGal is expected (Pilkington and Grieve 1992). Popigai, however, is distinct in not having a central gravity high, in contrast to similar sized structures, e.g. Manicouagan, Canada ($D = 100$ km). The presence of a subtle annular gravity high that coincides with the mapped annular uplift is also a unique feature of the Popigai gravity anomaly, although enhanced versions of other crater gravity anomalies are not currently available for comparison.

Impact structures are generally distinguished by a circular magnetic low or subdued zone (Pilkington and Grieve 1992; Henkel 1992). High-amplitude central magnetic anomalies tend to be also present at larger structures ($D > 40$ km). The processes of brecciation, alteration, shock, and possible infill by non-magnetic sediments contribute to the development of the magnetic lows. However, a significant contribution to these anomalies is caused by basement rocks beneath the impact structures, where reduced magnetizations are likely caused by a combination of shock demagnetization (Hargraves and Perkins 1969; Pohl et al 1975; Cisowski and Fuller 1978) and the partial demagnetization of magnetic carrier minerals by the impact-induced transient stress wave as it travels away from the point of impact (Scott et al. 1997). The central high-amplitude magnetic anomalies are due either to more magnetic material being brought up in the central uplift or strongly magnetized zones within impact melt rocks or suevite deposits. The magnetizations causing these anomalies are usually remanent, being formed at the time of impact. Thermal, chemical and to a lesser degree, shock, remanent magnetizations have been documented at several impact sites (Grieve and Pilkington 1996). The final character of the magnetic signature will also be influenced by the degree of erosion and the target lithologies, especially when layered (Plado et al. 1999). Popigai is significant in the larger than expected magnitude of its magnetic low (~300 nT) and the lack of any central magnetic anomalies (Fig. 4a).

Both the gravity and magnetic fields of Popigai are unlike those of most other impact structures of comparable size, e.g., Morokweng, South Africa ($D = 70$ km); Manicouagan, Canada, ($D = 100$ km); Chicxulub, Mexico ($D = 180$ km). Popigai is distinguished by the lack of a central circular gravity high related to a central structural uplift or high-amplitude magnetic anomalies caused by thick melt/suevite deposits.

6
Petrophysics

Petrophysical measurements of impact lithologies and target rocks in the vicinity of Popigai are given in Tables 1-4. Positional information is only available for the recent International Popigai Expedition (IPEX) data (Table 1) that were collected in the west of the structure along the Rassokha river. Other rock properties for target lithologies in the Popigai region are only available from Savrasov (1967) and Masaitis et al. (1975), given in Tables 2 and 3, respectively. The mainly gneissic Archean basement has a consistent density of 2.76 - 2.83 gcm^{-3}. Overlying sedimentary formations show a much larger density range, e.g., 0.82 gcm^{-3} for Cambrian carbonates. Permian sandstones show consistently lower densities (2.2 - 2.4 gcm^{-3}) compared to Proterozoic and Paleozoic sediments. Without sample locations, it is not possible to determine whether density variations are spatially related within a given lithology. Susceptibilities within each of the units show much larger variations but Archean gneisses are distinguished by values more than one order of magnitude higher than those of the sedimentary units. No magnetic remanence measurements are available for the target rocks.

Table 1. Petrophysical data for samples collected during the International Popigai Expedition

	Lithology			NRM	Q	D_{NRM}	I_{NRM}
		(gcm^{-3})	($\times 10^{-3}$ SI)	(mAm^{-1})		(°)	(°)
PM1	suevite	1.855	0.230	3.3	0.5	137	53
PM2	gneiss	2.620	0.270	13	0.9	220	30
PM3	carbonate	2.728	-0.005	0.1	-	variable	variable
PM4	suevite	1.925	0.340	30	2.0	85	86
PM5	gneiss	2.886	0.310	2.1	0.15	249	-20
PM6	gneiss	2.570	0.405	120	8.2	200	-15
PM7	carbonate	2.638	-0.012	0.1	-	95	38
PM8	tagamite	2.578	0.348	18	1.3	30	72
PM9	tagamite	2.560	0.790	95	3.1	46	66
PM10	tagamite	2.555	0.510	60	2.8	60	74
PM11	tagamite	2.538	0.448	48	2.7	179	42
PM12	tagamite	2.510	0.480	64	3.1	168	85
PM13	tagamite	2.471	0.365	21	1.6	160	64
PM14	tagamite	2.510	0.550	55	2.4	170	86
PM15	tagamite	2.528	0.350	13	0.9	50	77
PM16	tagamite	2.550	0.340	12	1.1	20	74
PM17	tagamite	2.508	0.405	23	1.5	260	62
PM18	tagamite	2.540	0.370	19	1.5	68	45
PM19	tagamite	2.420	0.290	16	1.4	185	81
PM21	gneiss	2.920	0.330	1.2	0.08	42	50
PM22	gneiss	2.880	0.380	15	0.5	variable	variable

, density; , magnetic susceptibility; *NRM*, intensity of the natural remanent magnetization; *Q*, Koenigsberger ratio; D_{NRM}, declination of NRM; I_{NRM}, inclination of NRM

Table 2. Density and susceptibility data for target rocks of the Popigai region (Savrasov 1967)

Lithology and age	(gcm^{-3})	(× 10^{-3} SI)
Upper Anabarsk (Archean crystalline basement)	2.76	22.6
Khapchansk (Archean crystalline basement)	2.80	13.8
Proterozoic quartzites, dolomites, limestones	2.70 - 2.75	-
Cambrian dolomites, limestones, marls	2.55 - 2.65	-
Ordovician limestones, marls, argillites	2.40 - 2.50	-
Silurian limestones and dolomites	2.50 - 2.55	-
Permian sandstones and aleurolites	2.20 - 2.40	-
Triassic dolerites	2.90 - 2.92	-

, density; , magnetic susceptibility

Table 3. Density and susceptibility data for target rocks of the Popigai region (Masaitis et al. 1975)

		ρ (gcm^{-3})		χ (× 10^{-3} SI)	
Lithology	N	Range	Mean	Range	Mean
Archean crystalline rocks	32	2.26-3.66	2.83	0-34.0	0.880
Proterozoic quartzites	22	2.34-2.62	2.51	0-1.13	0.125
Cambrian limestones and dolomites	26	2.00-2.82	2.67	0	0
Permian sandstones and aleurolites	7	2.01-2.52	2.23	0-0.125	0.063
Triassic dolerites	7	2.86-3.47	2.91	3.39-12.0	8.29

, density; , magnetic susceptibility

Table 4. Density and susceptibility data for impact lithologies at Popigai (Masaitis et al. 1975)

		ρ (gcm^{-3})		χ (× 10^{-3} SI)	
Lithology	N	Range	Mean	Range	Mean
Impact-metamorphized crystalline rocks from allochthonous and autochthonous breccias	70	2.14-2.88	2.57	0.00-6.53	0.375
Pseudotachylites	16	1.67-2.58	2.39	0.125-3.64	0.880
Inclusions of impact-metamorphosed and partly melted crystalline rocks in tagamites	45	2.02-2.46	2.43	0.00-2.07	0.440
Massive tagamites	82	2.10-2.92	2.48	0.063-2.01	0.500
Porous tagamites	13	1.62-2.76	2.11	0.125-0.500	0.315
Vitroclastic suevites (sintered)	15	2.08-2.83	2.41	0.00-1.38	0.440
Vitroclastic ("tuff-like") and lithovitro-clastic suevites	14	1.87-2.43	2.13	0.125-0.628	0.315
Granovitroclastic (small fragment) suevites	17	1.87-2.97	2.12	0.375-6.53	1.13
Granovitroclastic (large fragment) suevites	15	2.06-2.66	2.31	0.250-5.27	1.57
Coptoclastites (fine grained cement of the allochthonous breccia)	8	1.40-2.26	2.07	0.250-0.690	0.565

, density; , magnetic susceptibility

Impact-related lithologies were sampled during IPEX (Table 1) and their properties conform in general to earlier measurements (Table 4) of Masaitis et al. (1975). In particular, susceptibilities from both studies show a considerable decrease compared to the crystalline target rocks. All impact lithologies, breccias, tagamites and suevites, have susceptibilities $<1.0 \times 10^{-3}$ SI compared to Archean basement values of $>20.0 \times 10^{-3}$ SI. The former susceptibilities are in the range of paramagnetic rocks in which the (weak) magnetization is carried by mainly silicate minerals and not by the ferrimagnetic iron oxides (cf. Henkel 1994; Korhonen et al. 1997). These weakly magnetic lithologies only produce small anomalies (<<100 nT), if any, and are usually distinguished as lows or subdued zones when juxtaposed with ferrimagnetic rocks. Carbonates within the allogenic breccia were found to be diamagnetic with small negative susceptibilities (Table 1).

IPEX samples were oriented, allowing remanent magnetization directions and magnitudes to be determined (Table 1). Tagamites and suevites have remanence directions showing some scatter, although inclinations are generally steep and tend to reinforce induced magnetizations in the direction of the Earth's field (inclination 85°N; declination 5°W) at Popigai (Fig. 6). Koenigsberger ratios (Q) show little variation for the tagamites, with a mean value of 1.95. Q values larger than unity are a characteristic of impact melt rocks from several impact structures, as is a small directional scatter of remanence directions (Pilkington and Grieve 1992).

Densities of the few basement blocks sampled within the allogenic breccia (mean 2.77 gcm^{-3}) fall in the upper range of samples from Masaitis et al. (1975). Tagamite measurements from IPEX are distinguished by a small variation in density, with a mean value of 2.52 gcm^{-3}, similar to the massive tagamites (mean 2.48) sampled by Masaitis et al. (1975). The various types of suevites show large

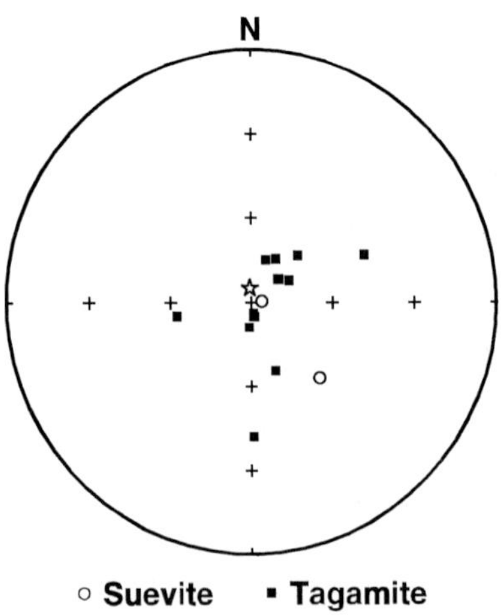

Fig. 6. Remanent magnetization directions for Popigai tagamites and suevite samples. Star denotes Earth's magnetic field direction.

density variations, with a range of 1.87 to 2.97 gcm^{-3} for all samples. The coptoclastites are characterized by the lowest densities (mean 2.07 gcm^{-3}; Table 4).

7 Modelling

Modelling of the geophysical data over Popigai provides constraints on the structure of the crater, particularly in the vertical dimension, and can corroborate existing geologic models based on surface geologic mapping and isolated drillhole data. In the absence of deep borehole information, geophysics is the only source of information on the deep structure of the crater. Here, we use two-dimensional models, based on forward modelling of profile data extracted from the available gridded gravity and magnetic data (Figs. 3a, 4a). The geologic cross-section of Masaitis et al. (1980) is used as the starting model. Initially, individual model units must be distinguished on the basis of physical properties. These units need not, however, correspond to lithologic divisions.

We look at several initial density and susceptibility distributions, distinguished mainly by the different physical property data sets available for the impact related lithologies (Tables 1 and 4). For all gravity models, Archean basement is taken to be 2.8 gcm^{-3}, the average of the Savrasov (1967; Table 2) and Masaitis et al. (1975; Table 3) densities. Proterozoic and Paleozoic sedimentary formations show a significant difference between Tables 2 and 3. We set initial values of the Proterozoic carbonates to 2.65 gcm^{-3}, and Paleozoic sediments to 2.60 gcm^{-3}, both rough averages of values in Tables 2 and 3. Allogenic breccia is treated as one unit since its stratigraphic location above the crater floor is reasonably well-known. The remainder of the crater fill material (tagamites, suevites, coptoclastites) are grouped together under one density/susceptibility unit since these lithologies are interleaved and their distributions only poorly constrained by drill hole intersections.

The profile chosen for modelling is oriented E-W, passes through the crater centre and crosses the annular uplift. The cross-section of Masaitis et al. (1980) is not coincident with this profile but oriented NW-SE and passes to the south of the centre of the structure (Fig. 1). For the input model geometry, the geologic section (Fig. 2), which does not pass through the centre of the structure, was scaled up to the diameter of the crater (100 km) along the E-W modelling profile. Observed gravity values were extracted from the gridded data (Fig. 3) at 2 km intervals. Several radial profiles through the structure were used to estimate a regional background field of -20 mGal which was subtracted from the profile data. Fig. 7 shows the calculated gravity effects of three initial models with the same model geometry but different density distributions. Model 1 is based on densities from Masaitis et al. (1975) and reproduces the overall amplitude level of the anomaly. The relative gravity highs over the annular uplift, however, are much larger than observed. We note that using a two-dimensional approximation to the true circular

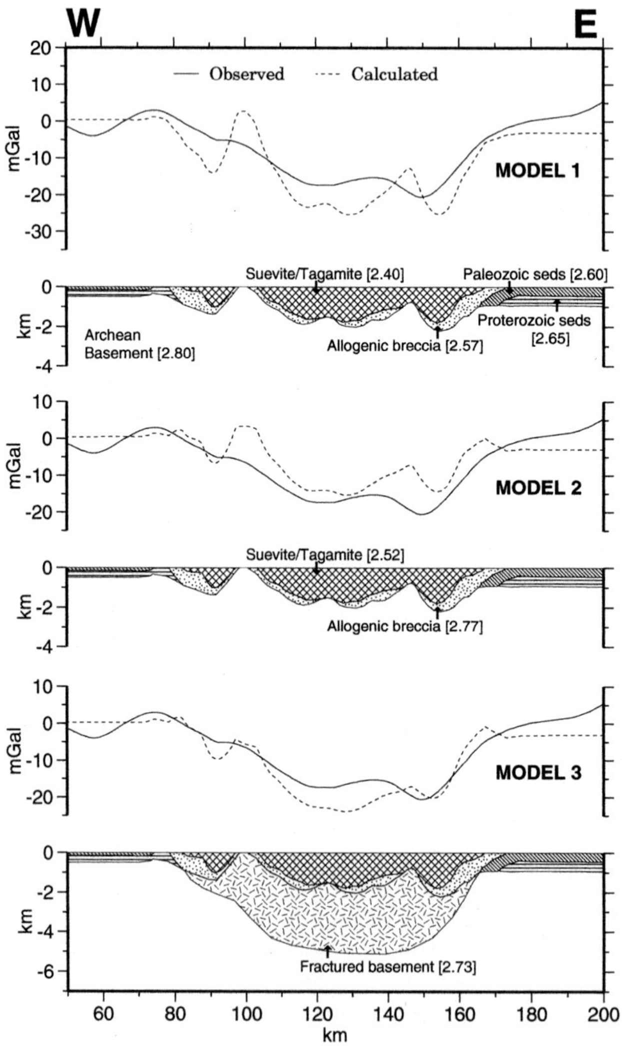

Fig. 7. Initial gravity models. Densities given in gcm^{-3}. Densities in model 2 are the same as model 1 unless labelled otherwise. Similarly for models 2 and 3. See Fig. 1 for profile location.

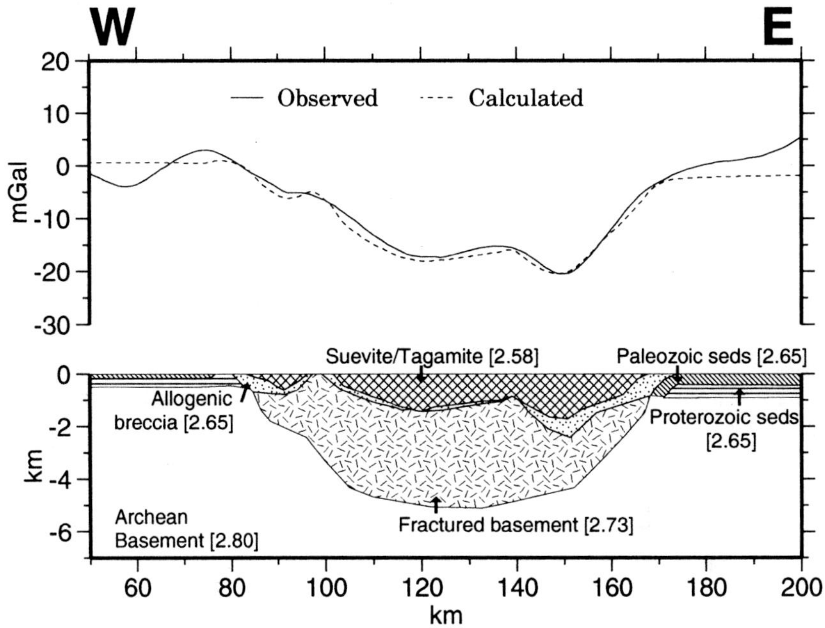

Fig. 8. Final gravity model. Densities are given in gcm^{-3}. See Fig. 1 for profile location.

geometry will tend to enhance this effect. For the western uplift, the model effectively has a 0.4 gcm^{-3} density contrast at the surface, which appears excessive. Model 2 is based on densities from IPEX (Table 1) and overall produces gravity amplitudes smaller than those observed. Again, the annular uplift effect is greater than observed. Model 3 is equivalent to model 2 but includes a volume representing fractured basement beneath the crater floor. A density contrast of -0.07 gcm^{-3} was used to match the observed low over the structure. The general fit is better than previous models, particularly as the contrast between the annular uplift and surrounding crater fill is now reduced.

Of the three models, the first is based on the most extensive petrophysical data collection (Table 4), with many more samples than IPEX. The overall density distribution, however, appears unrealistic. In this model, only the crater fill (above the true crater floor) contributes to the calculated anomaly; whereas samples from drillholes into the parautochthonous rocks of crater floors also show reduced densities, e.g., at Siljan, Sweden (Dyrelius 1988). Furthermore, gravity lows are observed over structures that are eroded to below this level, e.g., at Söderfjärden, Finland (Lauren et al. 1978), Carswell, Canada (Innes et al. 1964). Based on modelling of gravity data at several complex structures, Pilkington and Grieve (1992) showed that the structural or stratigraphic uplift (SU) is a better measure than the physical true crater floor for the depth extent of the fractured/brecciated zone that contributes most to the gravity anomalies. For Popigai, SU . 10 km, using the relationship $SU = 0.086D^{1.03}$ from Grieve and Pilkington (1996).

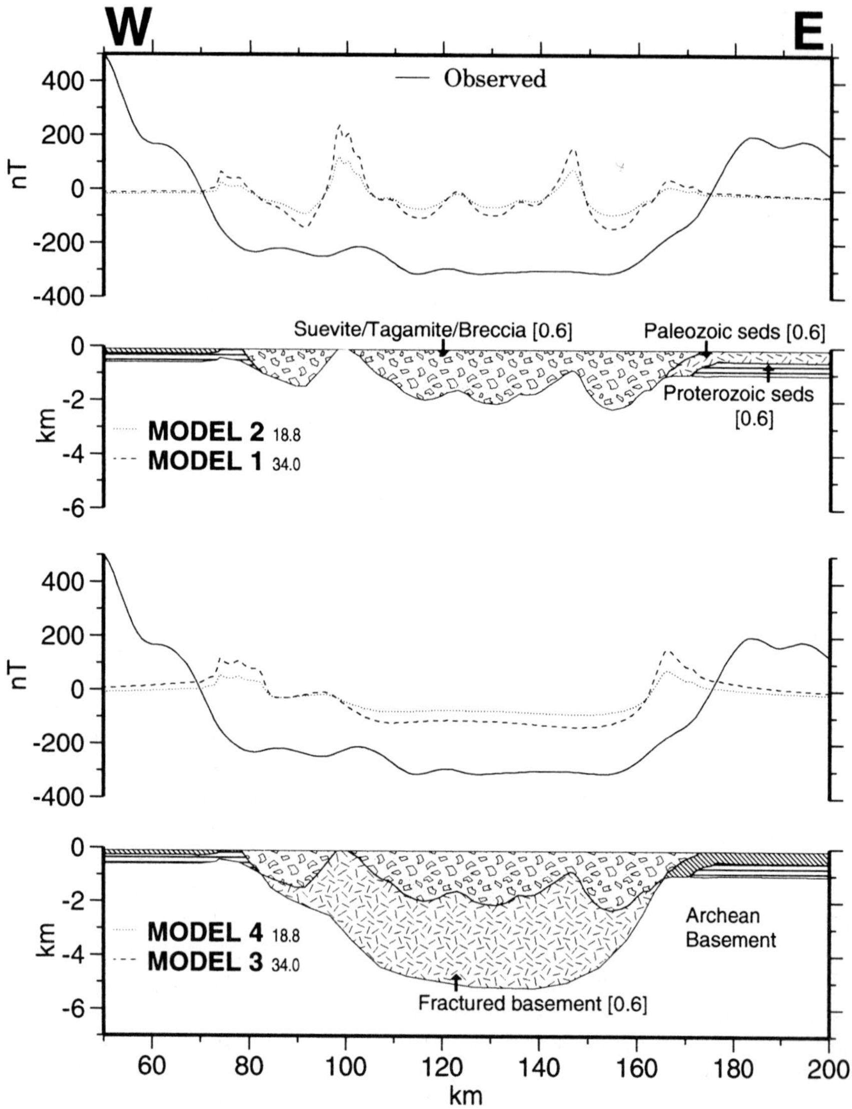

Fig. 9. Initial magnetic models. Susceptibilities are in units of 10^{-3} SI. In each of the two plots, basement susceptibilities are given next to model number. Susceptibilities for models 3, 4 are same as models 1, 2 unless labelled otherwise. See Fig. 1 for profile location.

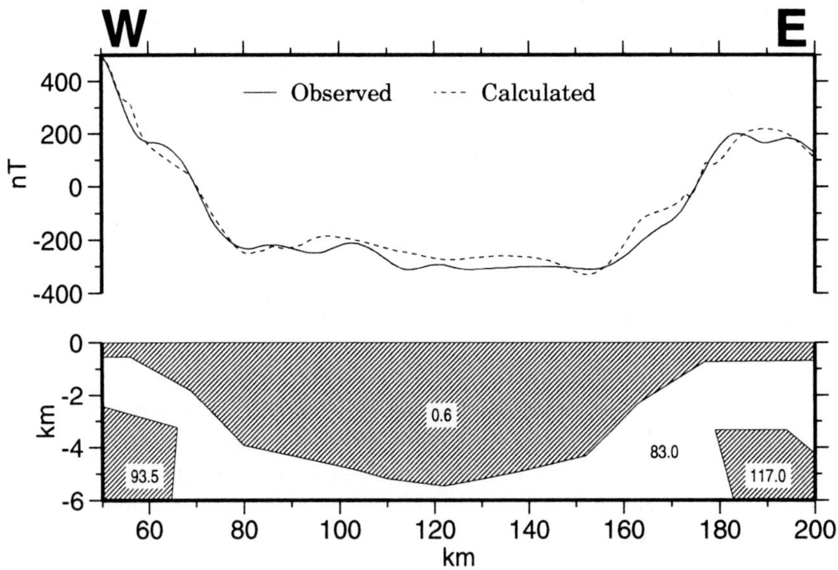

Fig. 10. Final magnetic model. Susceptibilities are in units of 10^{-3} SI. Since all crater-related units have the same susceptibility, they are grouped together. See Fig. 1 for profile location.

Furthermore, the large density contrast between crater fill and surrounding basement (-0.4 gcm^{-3}) in model 1 is much greater than values determined by sampling or modelling at other impact structures, where the average contrast between units of reduced density and the surrounding target rocks is 0.15 - 0.25 gcm^{-3} (Pilkington and Grieve 1992). For these reasons, we use model 3 as a starting point and adjust body geometries and densities by trial and error until an acceptable fit to the data is reached. Fig. 8 shows the final model, which retains most of the structural elements of the cross-section of Fig. 2 from Masaitis et al. (1980). The main adjustments to model 3 were the shifting (~7 km) of the eastern part of the annular uplift to coincide with the weak (~5 mGal) relative gravity high (cf. Fig. 3b) and reduction of the breccia density contrast to -0.15 gcm^{-3}. The final model now has a more realistic (monotonic) increase in density with depth.

Assigning magnetization values to an initial model is simpler than the gravity case at Popigai, since the various petrophysical studies give consistent results. Tagamites, suevites and breccias are distinguished by very low susceptibilities, most being $<1.0 \times 10^{-3}$ SI compared to Archean basement values of $>20.0 \times 10^{-3}$ SI. Remanent magnetizations were measured only on IPEX samples and indicate for tagamites, with Q . 2, that remanence will reinforce the induced magnetization levels. Remanence vectors in the suevites and the allogenic breccia are expected to be more variable and assumed to average to zero (cf. Beals et al. 1963). Hence, the total magnetic effect of the crater fill, consisting of an induced component in the suevites, breccias and tagamites, plus a remanent contribution to the latter, remains small compared to that for the basement. However, to account for this non-zero effect, we set the crater fill to 0.6×10^{-3} SI (cf. Tables 1 and 4). Two values for Archean basement were used: 34.0×10^{-3} SI, representing the upper end of the Masaitis et al. (1975) range, and 18.8×10^{-3} SI, the average of the Savrasov (1967) values. The Earth's magnetic field used in modelling has an intensity of 60500 nT, declination 5°W and inclination 85°N. No regional field was removed from the magnetic data prior to modelling.

Figure 9 shows the initial models (the same as Fig. 7 but with no breccia units) and their calculated effects for the two basement susceptibilities, with and without a fractured basement volume, as used in the gravity modelling. For the latter models (1 and 2, Fig. 9), the calculated fields mimic the shape of the basement/crater fill interface, in particular, causing highs over the annular uplift. The overall effect is a superposition of short wavelength anomalies on the smooth long wavelength (~100 km) magnetic low observed over the structure. Including a volume of reduced magnetization beneath the crater floor (models 3 and 4) produces a closer fit to the data but with an amplitude that is too small (by 100 - 150 nT), even when using the maximum measured basement susceptibility value of 34×10^{-3} SI (Table 3). The low caused by the model could be amplified by the presence of a remanent magnetization component in a direction opposing the present-day geomagnetic field. However, IPEX sample measurements indicate remanent components with steep inclinations parallelling the present field direction (Table 1, Fig. 6). Therefore, the model effect can only be amplified by increasing the susceptibility of the basement. The final model (Fig. 10) features such an increase (to 83×10^{-3} SI) plus adjustment to the geometry of the base of the fractured basement volume. The presence in the model of a zone of reduced magnetization beneath the crater floor is supported by drill core sampling at several Canadian craters, which show both induced and remanent magnetization levels well below those of target rocks (Scott et al. 1997). A major difference in the magnetic model (compared to the gravity case) is the presence of the additional low-susceptibility volume at the western edge of the crater. This is not thought to represent part of the structure itself but accounts for the NW-SE-trending linear magnetic low that is tangential to the western crater edge (cf. Fig. 4a). Additional, deeper units corresponding to changes in basement lithology/structure were also modelled to account for field variations exterior to the crater (Fig. 10).

8 Discussion

Modelling of profile gravity and magnetic data across the Popigai impact structure suggests the presence of a large volume of target material beneath the crater floor, characterized by reduced densities and magnetizations. Since the gravity and magnetic data used in modelling are regional in nature, we regard the final model as preliminary, and subject to refinement using higher resolution data. Similarly, the lack of positional and statistical information on the older petrophysical measurements does not allow hard constraints to be applied to the physical properties of the final models. However, the effects of the first-order structure of the crater have been accounted for by modelling. For the gravity case, the observed anomaly is a combination of the simple circular gravity low due to the fractured basement and shorter wavelength gravity effects resulting from the crater floor topography. The topographic variations lead to the ring-shaped gravity high that coincides with the mapped position of the annular basement uplift (Figs. 1 and 2). For the magnetic case, the anomaly over the crater is the sum of the magnetic effects of the crater fill and the fractured basement volume below. Both of these are essentially non-magnetic compared to the surrounding highly-magnetized Archean basement. Petrophysical measurements suggest that remanent magnetization effects are negligible and that magnetic anomalies related to internal crater morphology are not expected.

The geophysical signature of Popigai contrasts with most other structures of comparable size in lacking a central circular gravity high related to a central structural uplift or high-amplitude magnetic anomalies caused by thick melt/suevite deposits. Clearly, no significant density contrast exists between the central uplift and its surrounds. For this reason and because the only constraints on its form would be from morphometric relationships, we have not included a model body to represent a central uplift in the models. Central gravity highs are well developed at structures in mixed targets where uplifted crystalline rocks within the central uplift are juxtaposed with surrounding, less dense, sedimentary formations, e.g., Chicxulub, Mexico; Carswell, Canada. The sedimentary cover at Popigai is not thick enough to lead to this situation. Nonetheless, the existence of central highs at structures formed in mostly crystalline targets, such as at Manicougan, suggest target stratigraphy is not the only factor controlling the final density distribution.

The absence of any magnetic signature related to the extensive impact melt rocks at Popigai highlights the differences in the types of anomalies observed at various craters. High-amplitude, short-wavelength magnetic anomalies are observed at several impact structures with mapped impact melt rocks, such as Dellen and Mien in Sweden, and Morokweng in South Africa. Conversely, several impact sites with extensive melt sheets do not exhibit these kinds of anomalies, e.g., Clearwater, Canada and Lappajärvi, Finland. Chemical analyses of impact melt rocks from many of these structures indicate sufficient iron is present to produce magnetic anomalies (Grieve et al. 1977; Allen et al. 1982). Petrographic work also shows that hydrothermal alteration effects (which may produce or destroy magnetic carrier minerals) are present at sites, with and without associated magnetic anomalies. At Popigai, Masaitis et al. (1975) observed that little mineralogical alteration of ferrimagnetic minerals has occurred. Compared to unaffected target rocks, the Fe_2O_3 : FeO ratio of tagamites and suevites is reduced, which may partly explain the lack of magnetite and, hence, the absence of significant magnetic anomalies associated with these lithologies. In oxidized glasses from Popigai, iron occurs in secondary mica minerals and limonite. In the tagamites, the iron is found in pyrrhotite and minor ilmenite, but a significant portion is present in paramagnetic minerals, such as enstatite. The key to whether short-wavelength magnetic anomalies are present or not at a given impact structure, must therefore lie in the particular physical and chemical conditions associated with the impact melt rock formation, if present, and its subsequent alteration.

Acknowledgements

We thank J.D. Fairhead (GETECH Ltd.) for kindly providing gravity data and allowing publication of associated images. W.R. Roest provided access to topographic data and the modelling program GM2D. Fieldwork was supported by a National Geographical Society grant awarded to R.A.F.G. This work resulted from the International Popigai Expedition (IPEX) in July/August 1997. Petrophysical work was carried out in the Geological Survey of Finland paleomagnetic laboratory. Mike Thomas provided helpful comments on the text. Jean Pohl, Herbert Henkel, Seppo Elo, and Jüri Plado provided formal reviews of the paper. This is Geological Survey of Canada contribution 2000127.

References

Allen CC, Gooding JL, Keil K (1982) Hydrothermally altered impact melt rock and breccia: contributions to the soil of Mars. Journal of Geophysical Research 87: 10083-10101

Beals CS, Innes MJS, Rottenberg JA (1963) Fossil meteorite craters. In: Middlehurst BM, Kuiper GP (eds) The Moon, Meteorites and Comets. University of Chicago Press, Chicago, pp 235-284

Cisowski SM, Fuller M (1978) The effect of shock on the magnetism of terrestrial rocks, Journal of Geophysical Research 83: 3441-3458

Dyrelius D (1988) The gravity field of the Siljan ring structure. In: Boden A, Eriksson KG (eds) Deep Drilling in Crystalline Bedrock, Vol 1, The deep gas drilling in the Siljan impact structure, Sweden and astroblemes. Springer-Verlag, New York, pp 85-94

Grieve RAF, Pilkington M (1996) The signature of terrestrial impacts. Australian Geological Survey Organisation Journal 16: 399-420

Grieve RAF, Dence MR, Robertson PB (1977) Cratering processes: As interpreted from the occurrence of impact melts. In: Roddy DJ, Pepin RO, Merrill RB (eds) Impact and Explosion Cratering. Pergamon Press, New York, pp 791-814

Hargraves RB, Perkins WE (1969) Investigations of the effect of shock on natural remanent magnetism. Journal of Geophysical Research 74: 2576-2589

Hastings DA, Dunbar PK (1998) Development and assessment of the Global Land One-km Base Elevation Digital Elevation Model (GLOBE). International Society of Photogrammetry and Remote Sensing Archives, 32: 218-221.

Henkel H (1992) Geophysical aspects of meteorite impact craters in eroded shield environment, with special emphasis on electric resistivity. Tectonophysics 216: 63-90

Henkel H (1994) Standard diagrams of magnetic properties and density - A tool for understanding magnetic petrology. Journal of Applied Geophysics 32: 43-53

Hittelman A, Buhmann R, Meyers H (1991) USSR magnetic anomaly data. In: Taylor P (ed) IAGA Working Group V-9 Newsletter 3, Goddard Space Flight Center, Greenbelt

Innes MJS (1964) Recent advances in meteorite crater research at the Dominion Observatory, Ottawa, Canada. Meteoritics 2: 219-241

Korhonen JV, Säävuori H, Kivekäs L (1997) Petrophysics in the crustal model program of Finland. In: Autio S (ed) Current Research 1995-1996, Geological Survey of Finland Special Paper 23: 157-173

Lauren L, Lehtovaara J, Bostrom R, Tynni R (1978) On the geology of the circular depression at Söderfjärden, western Finland. Geological Survey of Finland Bulletin 297: 1-38

Masaitis VL (1994) Impactites from Popigai crater. In: Dressler BO, Grieve RAF, Sharpton VL (eds) Large Meteorite Impacts and Planetary Evolution. Geological Society of America Special Paper 293: 153-162

Masaitis VL (1998) Popigai crater: Origin and distribution of diamond-bearing impactites. Meteoritics & Planetary Science 33: 349-359

Masaitis VL, Mikhaylov MV, Selivanovskaya TV (1975) The Popigai meteorite crater (in Russian). Nauka Press, Moscow, 124 pp (English translation: NASA TT F-16900, 1976)

Masaitis VL, Danilin VN, Mashchak MS, Raikhlin AI, Selivanovskaya TV, Shadenkov EM (1980) The geology of astroblemes (in Russian). Nedra Press, Leningrad, 231 pp (English translation: NASA TM-77743, 1984)

Pilkington M, Grieve RAF (1992) The geophysical signature of terrestrial impact craters. Reviews of Geophysics 30: 161-181

Plado J, Pesonen LJ, Puura V (1999) Effect of erosion on gravity and magnetic signatures of complex impact structures: Geophysical modeling implications. In: Dressler BO, Sharpton VL (eds) Large Meteorite Impacts and Planetary Evolution II. Geological Society of America Special Paper 339: 229-240

Pohl J, Bleil U, Hornemann U (1975) Shock magnetization and demagnetization of basalt by transient stress up to 10 kbar. Journal of Geophysics 41: 23-41

Savrasov DI (1967) Physical features of rocks in Western Yakutiya (in Russian). In: Fotiadi EE (ed) Geological results of the geophysical investigation in Siberia and the Far East. Nauka Press, Novosibirsk, 514 pp

Scott RG, Pilkington M, Tanczyk EI (1997) Magnetic investigations of the West Hawk, Deep Bay and Clearwater impact structures, Canada. Meteoritics & Planetary Science 32: 293-308

U-Pb Analyses of Zircons from the Popigai Impact Structure, Russia: First Results

Richard Armstrong[1], Sergei Vishnevsky[2], and Christian Koeberl[3]

[1]Research School of Earth Sciences, Australian National University, Canberra 0200, A.C.T., Australia.
[2]Institute of Mineralogy and Petrology, 3 Universitetsky Prospekt, 630090 Novosibirsk-90, Russia.
[3]Institute of Geochemistry, University of Vienna, Althanstrasse 14, A-1090 Vienna, Austria. (christian.koeberl@univie.ac.at)

Abstract. Zircons were extracted from impact melt rock samples from the 35.7 Ma Popigai impact structure in Russia. U-Pb SHRIMP age dating of four zircons resulted in ages of 1982 Ma for two of the zircons and younger ages for the other two grains. Minimum age estimates of the younger grains can be derived from their 912 and 306 Ma ^{207}Pb / ^{206}Pb ages. No shock features are obvious in these grains; thus, the Pb loss could be recent. The older ages represent ages of zircon derived from the target lithology. No evidence for resetting of the isotopic system at 35 Ma in the zircons was found in this pilot study.

1 Introduction

It has been shown that it is possible to date large impact structures by using U-Pb zircon dating techniques. This may be achieved in two ways, *viz.* (1) by dating zircons that have crystallized from an impact melt (e.g., at Vredefort – Kamo et al. 1996; Morokweng – Koeberl et al. 1997a), or (2) by dating the event from the determination of lower intercept concordia ages from shock metamorphosed zircons, which have planar deformations or polycrystalline textures (e.g., Chicxulub – Krogh et al. 1993; Kamo and Krogh 1995). The present study is a first step in the examination of the U-Th-Pb isotopic system in zircons from the Popigai impact structure, which were separated from suevites and melt rocks, to determine if it is possible to constrain the impact age from such analyses.

2
Background: The Popigai Impact Structure

The Popigai structure (centered at 71°30'N and 111°00'E) is located on the Anabar Shield, Siberia, Russia. It is exposed within Archean crystalline rocks, with overlying Proterozoic to Mesozoic sedimentary sequences (e.g., Masaitis 1994, 1998; Vishnevsky and Montanari 1999), and is the largest Cenozoic impact structure known on Earth. For location maps, geological maps, and details about the geology of the Popigai structure, see the reviews of, for example, Masaitis (1998, 1999) and Vishnevsky and Montanari (1999). The structure comprises a circular basin with several ring features, but without a topographic outer rim. It is separated from the surrounding areas by a complex concentric zone of uplifted and overturned blocks and klippen of about 100 km in diameter. Slightly deformed rocks occur at a maximum distance of about 70 km from the crater center. The structure hosts a large variety of impactites, including suevites, impact melt rocks, and other impact breccias.

Archean basement rocks, which are exposed to the south of the structure, comprise the Verkhne-Anabar and the Khapchan series. These rocks are strongly deformed and belong mainly to amphibolite and granulite facies of regional metamorphism, with various gneisses being the dominant rock type. The average composition of impact melt rocks from Popigai is similar to that of the average composition of Anabar Shield gneisses (Vishnevsky and Montanari 1999). Granitoids of various compositions make up about 10 % of the gneiss volume. The sedimentary cover has a total thickness of about 1.6 km and ranges from quartz arenites, dolomites, and sandstones of Proterozoic age to Cambrian dolomites, marls, and limestones, to Permian sandstones. A Mesozoic cover is exposed at some locations to the northeast of the structure and includes Triassic volcano-sedimentary rocks (which are not found in any impact breccias), some Jurassic marine and continental formations, and Cretaceous siltstones, sandstones, coal-bearing sands, and clays, all of which feature prominently in suevitic breccias at the crater (e.g., Masaitis 1998; Vishnevsky and Montanari 1999).

Popigai is characterized by thick breccia and impact melt deposits of impressive appearance, which are described in detail by, e.g., Masaitis (1994) and Vishnevsky and Montanari (1999). The impact melt rock (local name "tagamite") forms up to 100-m-thick layers with an appearance that is similar to that of a volcanic lava flow. Masaitis (1994) quotes a total volume of melt rocks at Popigai of 1750 km^3. These melt rocks were found to contain evidence for an extraterrestrial component, and undigested as well as partly digested clasts within the melt rock contain shocked minerals, providing evidence for an impact origin. Impact-derived diamonds have been found in impact melt rocks within the crater (see, e.g., Koeberl et al. 1997b; Masaitis, 1998). Details on petrography and geochemistry of the impact melt rocks are given by Masaitis (1994) and Vishnevsky and Montanari (1999).

Bottomley et al. (1997) determined an Ar-Ar age of 35.7 ± 0.8 Ma for melt rocks from the Popigai structure, which is indistinguishable from the (biostatigraphic) age of the 85-km-diameter Chesapeake Bay structure (cf. Poag

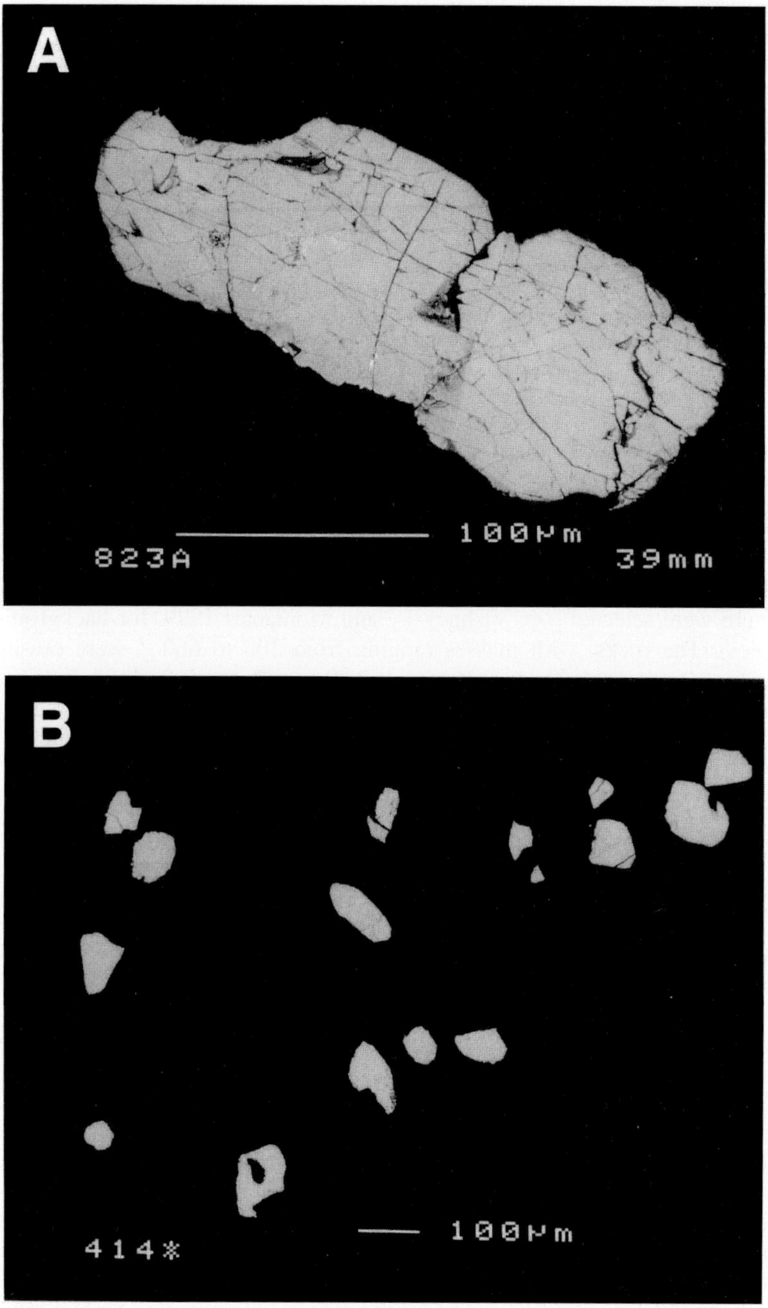

Fig. 1. Back-scattered electron images of zircons isolated from Popigai impact melt rocks. a) A fractured "group A" grain from sample 823; b) "Group B" grains from sample 414.

and Aubry 1995). In addition, there are two coeval impact ejecta layers with a near-global distribution: the North American tektite/microtektite layer, and a layer with clinopyroxene-bearing spherules that (at locations with both layers) occurs a few cm below the microtektite layer (for a review, see Montanari and Koeberl 2000). It is now commonly assumed (e.g., Glass and Koeberl 1999; Vonhof and Smit 1999) that the global Late Eocene microkrystite layer originated from the Popigai impact event. Confirmation of such a link has to be done by using isotope geochemical methods, as radiometric age determinations do not allow to resolve an age difference of 10 or 20 k.y. The results of Whitehead et al. (2000) seem to confirm the association of the lower of the two Late Eocene impactoclastic layers with the Popigai impact structure (the upper – microtektite – layer seems to be associated with the Chesapeake Bay impact structure), and also that two large impact events occurred at that time within a few tens of thousands of years.

3 Samples and Analytical Procedures

Seven samples of impact melt rocks ("tagamites") from the Popigai impact structure were selected (see Vishnevsky and Montanari 1999, for background on samples). The rocks, with masses ranging from 300 to 660 g, were crushed to <0.25 mm grain size, and a heavy (>2.89 gcm^{-3}) fraction was extracted in bromoform. Subsequently, the nonmagnetic fraction of the residue was extracted and cleaned of adhering rock remnants for two days at room temperature in a HF / H_2SO_4 solution (500 ml water + 200 ml conc. HF + 200 ml conc. H_2SO_4). The residues contained various transparent and translucent silicates, opaques, garnet, and zircon fragments. Optically, the zircons seem to represent various generations; they were divided into two groups. Group A are clear zircons with a variety of colors, sizes, and forms; some of them seem to show evidence of granulation and/or planar features, and are fractured (Fig. 1a). These were interpreted as pre-impact zircons. Group B are isometric grains or fragments thereof, without any traces of melting along the grain boundaries. These were assumed to represent post-impact zircons that may have crystallized from the impact melt. As these grains were thought to be most promising, four "group B" grains from one of the samples (No. 414) were used in the present pilot study (Fig. 1b).

Zircons were mounted in epoxy, polished to half their thickness to reveal any internal structure, and examined in Vienna under normal transmitted and reflected light and with a JEOL scanning electron microscope (SEM). At the Research School of Earth Sciences (RSES) the grains were mounted together with the RSES zircon standards SL13 and AS3 and studied by cathodoluminescence imaging before being analyzed by the SHRIMP ion probe. The SHRIMP data have been reduced in a manner similar to that described by Compston et al. (1992) and Williams and Claesson (1987). The U/Pb ratios in the unknowns were normalized to a $^{206*}Pb / ^{238}U$ value of 0.1859 (equivalent to an age of 1099.1 Ma; Paces and

Table 1. SHRIMP U-Th-Pb zircon analyses of sample 414B from the Popigai impact structure

Grain-Spot	U ppm	Th ppm	$^{232}Th/^{238}U$	% $^{206}Pb_c$	$^{206}Pb^*$ ppm	$^{207}Pb^*/^{206}Pb^*$	err	$^{207}Pb^*/^{235}U$	err	$^{206}Pb^*/^{238}U$	err	$^{206}Pb/^{238}U$ Age		$^{207}Pb/^{206}Pb$ Age		err corr	% Conc.
414B8.1	644	613	0.98	0.26	25.7	0.0525	±0.0010	0.335	±0.012	0.0463	±0.0014	291.9	± 8.7	306	± 43	0.849	5
414B13.1	160	100	0.65	0.07	49.1	0.1218	±0.0010	6.01	±0.19	0.358	±0.011	1,973	±52	1,982	± 15	0.966	0
414B7.1	597	357	0.62	0.06	181	0.1217	±0.0006	5.92	±0.18	0.352	±0.011	1,946	±51	1,982	± 8.3	0.988	2
414B3.1	86	60	0.72	-	3.95	0.069	±0.0039	0.516	±0.033	0.0539	±0.0018	338	±11	912	±110	0.510	63

Errors are 1 sigma; Pb_c and Pb^* indicate the common and radiogenic portions, respectively; common *Pb* corrected using measured ^{204}Pb; *% Conc.*, % concordant

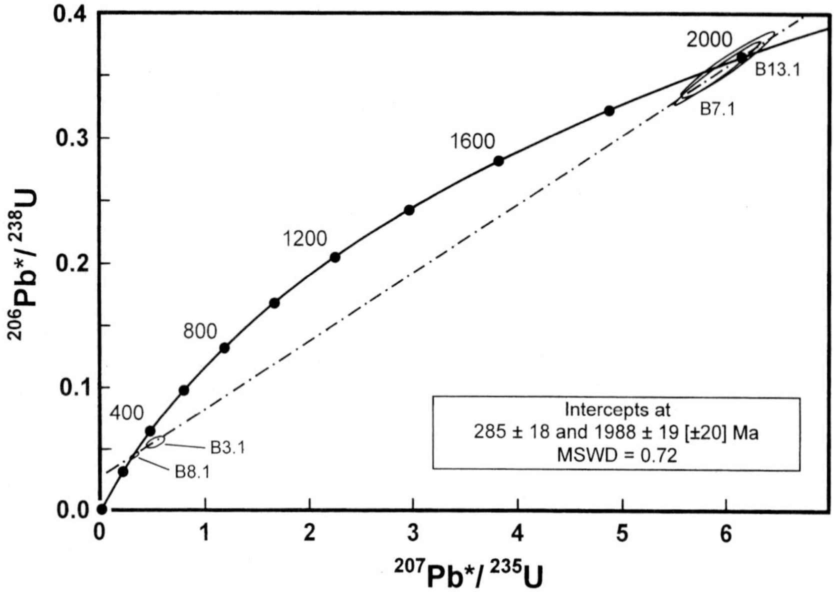

Fig. 2. Isotopic data for four zircon grains from impact melt rock sample 414B, plotted on a conventional Wetherill concordia diagram. No ages close to the Ar-Ar impact age of about 35 Ma are apparent.

Miller 1989) for AS3. The U and Th concentrations were determined relative to those measured in the SL13 standard. Correction for common Pb was made using the measured ^{204}Pb and the appropriate common Pb composition assuming the model of Cumming and Richards (1975). All statistical analyses and calculations were done using the Isoplot/Ex software (version 2.4; Ludwig 1999).

4 Results

Initial microscopic examination of all zircons from the seven melt rock samples revealed none of the shock-characteristic planar features that have been found for shocked zircons from other impact structures (e.g., Bohor et al. 1993; Kamo et al. 1996). Many of the zircons were, however, highly fractured (Fig. 1a); most are either anhedral or fragments of once larger grains (Fig. 1b). Cathodoluminescence imaging showed that the zircons were zoned, with either concentric or sector compositional zoning, or a combination of both styles of zoning. A number of crystals show cores and rims, suggesting complex geological histories.

The four "group B" zircons selected for these reconnaissance analyses were from sample 414 (see Vishnevsky and Montanari, 1999). The results for these grains are given in Table 1 and are plotted on a conventional Wetherill concordia plot in Fig. 2. The first important point to note is that all of the zircons are far too old to be related to the Cenozoic impact event. It is therefore concluded that these zircons could not have crystallized from impact melt and, hence, cannot be used to directly date the event. The four analyses lie on a discordia line (Fig. 2) with an upper intercept date calculated at 1988 ± 19 Ma and a lower intercept at 285 ± 18 Ma (uncertainties are 95 % error limits; MSWD = 0.7). Considering the distribution of the data points, this is essentially a two-point isochron, as is reflected in the low MSWD, which does in this case not reflect a statistical significance. The two zircons B7.1 and B13.1 (which are 98 % and 100 % concordant, respectively) that constrain the upper intercept, give identical ^{207}Pb / ^{206}Pb ages with a weighted mean of 1982 ± 14 Ma (95 % confidence limit). The other two analyses plot near the lower intercept of the discordia, which appears to be a simple mixing line, and implies no genetic significance. There is no simple relationship between apparent Pb-loss and U (and/or Th) content as grain B3.1 has by far the lowest U content of the grains analyzed. No Pb loss at 35 Ma is evident. There is no reason to link the younger grains to the 1982 Ma event. Minimum age estimates of these grains are given by the ^{207}Pb / ^{206}Pb ages of 912 and 306 Ma, respectively (Table 1). There is some similarity to zircons from the Sudbury structure, which, while giving concordant ages, were also unshocked and gave ages of the Archean target lithologies (Krogh et al. 1996). The 1982 Ma age of the two 414B grains from Popigai could well represent the Proterozoic crystalline basement.

We conclude that the “group B” zircons from the Popigai impact melt are not suitable to U-Pb dating of the impact event, as they have clearly not crystallized from the impact melt, and also have not suffered the unique shock-induced Pb-loss history required in order to obtain a realistic lower intercept age. These four zircons analyzed so far are inherited from the local basement.

Acknowledgments

This work has been supported in part by the Austrian FWF, project Y58-GEO (to C.K.). We are grateful to S. Kamo, J. Plado, W.U. Reimold, and M. Vaasjoki for critical comments that helped to improve this short note.

References

Bohor BF, Betterton WJ, Krogh TE (1993) Impact-shocked zircons: discovery of shock-induced textures reflecting increasing degrees of shock metamorphism. Earth and Planetary Science Letters 119: 419-424

Bottomley RJ, Grieve RAF, York D, Masaitis V (1997) The age of the Popigai impact event and its relations to events at the Eocene/Oligocene boundary. Nature 388: 365-368

Compston W, Williams IS, Kirschvink JL, Zhang Z, Ma G (1992) Zircon U-Pb age for the Early Cambrian time-scale. Journal of the Geological Society London 149: 171-184

Cumming GL, Richards JR (1975) Ore lead isotope ratios in a continuously changing Earth. Earth and Planetary Science Letters 28: 155-171

Glass BP, Koeberl C (1999) ODP Hole 689B spherules and Upper Eocene microtektite and clinopyroxene-bearing spherule strewn fields. Meteoritics & Planetary Science 34, 197-208

Kamo SL, Krogh TE (1995) Chicxulub crater source for shocked zircons from the Cretaceous-Tertiary boundary layer, Saskatchewan: Evidence from new U-Pb data. Geology 23: 281-284

Kamo SL, Reimold WU, Krogh TE, Colliston WP (1996) A 2.023 Ga age for the Vredefort impact event and a first report of shock metamorphosed zircons in pseudotachylite breccias and Granophyre. Earth and Planetary Science Letters 144: 369-387

Koeberl C, Armstrong RA, Reimold WU (1997a) Morokweng, South Africa: A large impact structure of Jurassic-Cretaceous boundary age. Geology 25: 731-734

Koeberl C, Masaitis VL, Shafranovsky GI, Gilmour I, Langenhorst F, Schrauder M (1997b) Diamonds from the Popigai impact structure, Russia. Geology 25: 967-970

Krogh TE, Kamo SL, Sharpton VL, Marin LE, Hildebrand AR (1993) U-Pb ages of single shocked zircons linking distal K/T ejecta to the Chicxulub crater. Nature 366: 731-734

Krogh TE, Kamo SL, Bohor BF (1996) Shock metamorphosed zircons with correlated U-Pb discordance and melt rock with concordant protolith ages indicate an impact origin for the Sudbury structure. In: Basu A, Hart S (eds) Earth Processes: Reading the Isotopic Code. American Geophysical Union, Geophysical Monograph 95: 343-353

Ludwig KR (1999) Isoplot/Ex (version 2.4): A geochronological toolkit for Microsoft Excel. Berkeley Geochronology Center Special Publication 1a, 46 pp

Masaitis VL (1994) Impactites from Popigai crater. In: Dressler BO, Grieve RAF, Sharpton VL (eds) Large Meteorite Impacts and Planetary Evolution. Geological Society of America, Special Paper 293: 153-162

Masaitis VL (1998) Popigai crater: Origin and distribution of diamond-bearing impactites. Meteoritics & Planetary Science 33: 349-359

Masaitis VL (1999) Impact structures of northeastern Eurasia: The territories of Russia and adjacent countries. Meteoritics & Planetary Science 34: 691-711

Montanari A, Koeberl C (2000) Impact Stratigraphy: The Italian Record. Lecture Notes in Earth Sciences, Vol. 93, Springer Verlag, Heidelberg, 364 pp

Paces JB, Miller JD (1989) Precise U-Pb ages of the Duluth Complex and related mafic intrusions, Northeastern Minnesota: Geochronological insights to physical, petrogenetic, palaeomagnetic and tectonomagmatic processes associated with the 1.1 Ga midcontinent rift system. Journal of Geophysical Research 98B: 13997-14013

Poag CW, Aubry M-P (1995) Upper Eocene impactites of the U.S. East Coast: Depositional origins, biostratigraphic framework, and correlation. Palaios 10: 16-43

Vishnevsky S, Montanari A (1999) Popigai impact structure (Arctic Siberia, Russia): Geology, petrology, geochemistry, and geochronology of glass-bearing impactites. In: Dressler BO, Sharpton VL (eds) Large meteorite impacts and planetary evolution II. Geological Society of America, Special Paper 339: 19-60

Vonhof HB, Smit J (1999) Late Eocene microkrystites and microtektites at Maud Rise (Ocean Drilling Project Hole 689B; Southern Ocean) suggest a global extension of the approximately 35.5 Ma Pacific impact ejecta strewn field. Meteoritics & Planetary Science 34: 747-755

Whitehead J, Spray JG, Grieve RAF, Papanastassiou DA, Ngo HH, Wasserburg GJ (2000) Late Eocene impact ejecta: geochemical and isotopic connections with the Popigai impact structure. Earth and Planetary Science Letters 181: 473-487

Williams IS, Claesson S (1987) Isotopic evidence for the Precambrian provenance and Caledonian metamorphism of high grade paragneisses from the Seve Nappes, Scandinavian Caledonides. II. Ion microprobe zircon U-Th-Pb. Contributions to Mineralogy and Petrology 97: 205-217

Impact-Generated Hydrothermal Systems: Data from Popigai, Kara, and Puchezh-Katunki Impact Structures

Mikhail V. Naumov

Karpinsky All-Russia Geological Research Institute (VSEGEI), Sredny Prospect 74, 199106 St. Petersburg, Russia. (mvn@mail.wplus.net)

Abstract. The basic and common features of impact-generated hydrothermal systems are considered, based on studies of the hydrothermal mineralization in large Russian impact structures - Popigai (diameter: 100 km), Kara (65 km), and Puchezh-Katunki (80 km). The dominant hydrothermal assemblages at all craters are clay minerals (smectites, chlorites, and mixed-layered smectites-chlorites), various zeolites, calcite, and pyrite; in addition, cristobalite, quartz, opal, anhydrite, gypsum, apophyllite, prehnite, epidote, andradite, ferrosalite, actinolite, and albite occur locally. In Puchezh-Katunki, the abundant hydrothermal mineralization within the central uplift area shows a distinct vertical distribution due to thermal gradients: an upper smectite-zeolite zone and a lower chlorite-anhydrite zone can be distinguished. In Kara and Popigai, where the hydrothermal alteration affected mainly the impact melt rocks filling the craters, a vertical zonation of clay minerals is only outlined, whereas the distribution of other minerals mainly results from different host lithologies. The mineral assemblages indicate post-impact hydrothermal alteration at temperatures of 50 - 350ºC, pH of 6 - 8, and Eh > -0.5. The Al^{IV} content in clay minerals and zeolites decreases from the base to the top of thick impact rock sequences, indicating the decrease of temperature and pH of the mineral-forming solutions. The composition of the hydrothermal mineralization in impact craters is determined by the target composition and by the occurrence of large amounts of shock-derived aluminosilicate materials, which result in alkaline properties and higher silica activities in the solutions.

1 Introduction

Every impact onto water-bearing planetary surfaces generates long-term hydrothermal activity in the resulting craters due to the transfer of kinetic energy of the impactor to the target. Several factors that control potential parameters of impact-induced hydrothermal activity can be distinguished (Naumov 1996): a) formation of a long-term (in large craters, many thousands of years - Crossey et al. 1994) thermal anomaly, causing convection of ground and meteoric water; b) creation of an extensive zone of fractured rocks, favorable to the circulation of solutions; c) shock deformation of mineral structures, which stimulates reactions between minerals and active fluids (Boslough and Cygan 1988). Consequently, hydrothermal alteration is a common mineral-forming process in impact structures.

The goal of this paper is to briefly show basic and common features for hydrothermal systems of different impact structures, based on the detailed study of secondary mineralization. At present, some evidence of impact-induced hydrothermal activity, mainly the occurrence of mineral associations including ore-forming parageneses, is known from 60 terrestrial craters (Fig. 1; Table 1). In addition, evidence for post-impact hydrothermal activity was obtained from fluid inclusion studies for: Charlevoix (Pagel and Poty 1975), Carswell (Pagel et al. 1985), Houghton (Bain and Kissin 1988), Siljan (Komor et al. 1988), Roter Kamm (Koeberl et al. 1989), Manson (Boer et al. 1996), Lockne (Sturkell et al. 1998), and Kärdla (Kirsimäe et al. 2002). In a series of craters, hydrothermal mineralization was studied in detail as a result of drilling projects (Stöffler et al. 1977; Stähle and Otteman 1977; Aldahan 1990; McCarville and Crossey 1996; Sturkell et al. 1998). Nevertheless, only few of them give the opportunity to reveal the 3D distribution of the hydrothermal mineralization and to construct a reliable model of the post-impact circulation system. In this respect, the three large Russian impact craters Popigai, Kara, and Puchezh-Katunki are some of the best available structures due to the detailed investigations, including drilling, which have been carried out at these locations. These craters are distinct in their structure, target rock compositions, amount and mode of distribution of impact melt (Table 2); thus, post-impact hydrothermal systems in various environments can be considered. Because of the similarity in hydrothermal mineral associations in the majority of terrestrial craters - smectites, chlorites, zeolites, calcite, anhydrite, epidote, pyrite, quartz (see Table 1), the above-mentioned three craters can be considered as examples for simulations of post-impact hydrothermal systems.

The present paper is based on results of studies of hydrothermal mineralization in the Popigai, Kara, and Puchezh-Katunki impact craters. In addition, data from some other Russian and Ukrainian craters (Boltysh, Jänisjärvi, Terny, Suavjärvi) were used as well. Special attention was paid to the spatial distribution of hydrothermal mineralization within the craters.

In the paper, the nomenclature of certain impact rock species is in accordance with the Russian classification adopted during the mapping of the Popigai and other Russian craters (Masaitis 1983, 1999). Massive impact melt rocks are considered as tagamites, and fragmental glass-bearing polymict breccias as suevites. Suevites are subdivided into varieties in accordance with the size of the fragments (ash, lapilli, agglomeratic, coarse fragmented) and the composition of the clastic material, or impact glass content (vitric, lithic, etc). Brecciated rocks

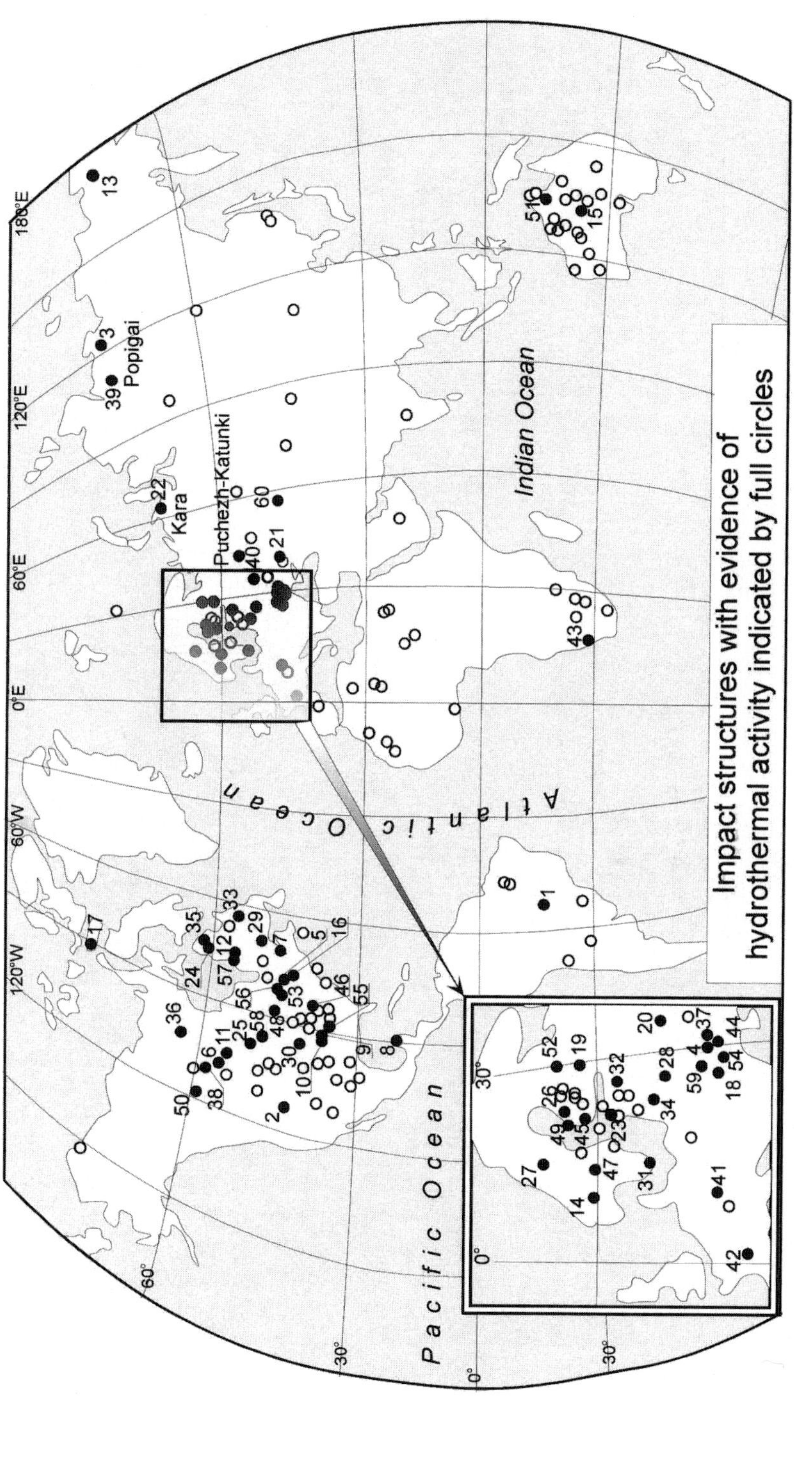

Fig. 1. Location of terrestrial impact craters where post-impact hydrothermal alteration has been found. Impact structures with evidence of hydrothermal activity are indicated by full circles, and other impact structures by open circles. Numbered impact structures correspond to those in Table 1.

Table 1. Post-impact hydrothermal mineralization from terrestrial impact craters

No. on Fig.1	Impact structure	Country	Diameter (km)	Target rocks	Post-impact hydrothermal minerals from different impact lithologies			References
					Lithic breccias	Suevites	Massive impact melt rocks	
1	Araguainha	Brazil	40	Cg	Qz Hem		Chl Qz Ca Fu	Engelhardt et al. 1992
2	Beaverhead	USA	60	St	Chl Qz Corr Ca Kfsp Sulf			Fiske et al. 1994
3	Beyenchime-Salaaty	Russia	8	Sc	Sulf Qz			Masaitis et al. 1980
4	Boltysh	Ukraine	24	Cg	Sm Chl Ep Ze Ca Grt		Sm Chl Ca Ze Stp Sulf	Masaitis et al. 1980; this paper
5	Brent	Canada	3.8	St+c	Chl Hem Ca Prh	Ze Ca Sulf Chl Br	Chl Sm Qz Hem Ze Act	Currie 1971a
6	Carswell	Canada	39	Sgd+St	Chl Ill Sulf Ca Hem Coff	Chl Ca Sulf Pb	Chl Qz Ze	Rondot 1983; Ruhlmann 1985
7	Charlevoix	Canada	54	Cgd+St+c	Ze Ca Prh Qz		Chl Sm Qz	Rondot 1983
8	Chicxulub	Mexico	180	St+c		Sm Anh Sulf Chl	Ep Qz Ab Kfsp Ca	Kring et al. 1991; Schuraytz et al. 1994
9	Crooked Creek	USA	7	Sc	Sulf Qz Br			Kiilsgaard et al. 1963
10	Decaturville	USA	6	St+c	Sulf			Offield and Pohn 1977
11	Deep Bay	Canada	13	Cgd		Chl		Dence et al. 1968
12	East Clearwater	Canada	26	Cgd	Sm Chl Ca Hem		Sm Qz Ca Sulf Hem	Allen et al. 1982; Palme et al. 1979
13	El'gygytgyn	Russia	18	V	Qz Ze			Ryabenko 1982
14	Gardnos	Norway	5	Cgd		Chl Stp		French et al. 1997
15	Gosses Bluff	Australia	24	St+c	Ze Kfsp			Milton et al. 1972
16	Holleford	Canada	2.3	C	Chl Ill			Beals 1960

Table 1. (cont.)

17	Houghton	Canada	24	St+c	Sulf Gps Br Ca Fu Hem Sm	Jar Ca Gps	Qz Ca Dl Ill	Metzler et al. 1988; Osinski et al. 2001
18	Ilyinets	Ukraine	4.5	Cgd	Chl Sm Ca Ze Ab Fu Sulf	Sm Ze Qz Hem Chl Kfsp	Ze Sm Chl Sulf Ca Fu Br	Valter and Ryabenko 1977
19	Jänisjärvi	Russia	14	Cgd		Chl Ca Ze	Qz Ca Chl Ep	Author's data
20	Kaluga	Russia	15	Cgd	Ca Chl Lm	Chl Ze Chc Act		Masaitis et al. 1980
21	Kamensk	Russia	25	St+c	Ca Sulf	Ca Sm Chl Ze		Movshovich and Milyavsky 1990
22	Kara	Russia	65	St+V	Ca Sulf Ze Apf Br	Sm Ca Sulf Ze Op	Sm Chl	Yushkin et al. 1976; Selivanovskaya et al. 1982; this paper
23	Kärdla	Estonia	4	Cgd+St	Chl Corr Hem Sulf Ill			Zhukov et al. 1987; Kirsimäe et al. 2002
24	Lac Couture	Canada	8	Cgd	Ep Chl Ser Hem			Beals et al. 1967
25	Lake Saint Martin	Canada	23	Cgd+Sc	Ca Ze Sm	Ca	Sm Chl Ca Qz	Allen et al. 1982; Reimold et al. 1990
26	Lappajärvi	Finland	23	Cg	Ze	Ze Sm Ca	Sm Chl Ca Chc Hem	Lehtinen 1976
27	Lockne	Sweden	7	Cgd+St	Ca Sulf Ze Qz			Sturkell et al. 1998
28	Logoisk	Belarus	15	Cgd+St	Ze	Ab Qz Chl	Chl Ill	Glazovskaya et al. 1991
29	Manicouagan	Canada	100	Cd	Ze Chl Hem		Sm Qz Chl Ze Hem	Floran et al. 1978; Currie 1972
30	Manson	USA	36	Cg+St+c	Chl Corr Qz Sm Ill Grt Act Prh Ep	Sm Corr Chl Ca Ze Qz Sulf		McCarville and Crossey 1996
31	Mien	Sweden	9	Cg	Chl Ep Ser		Sm Ze Qz	Allen et al. 1982; Stanfors 1969
32	Mishina Gora	Russia	2.5	Cgd+St	Chl Ca Kaol			Masaitis et al. 1980
33	Mistastin	Canada	28	Cd	Chl Ser Ze		Sm Ca Qz Chl Ze	Grieve 1975; Currie 1971b
34	Mizarai	Lithuania	5	Cgd	Chl Ep Ze		Sm Ca Qz Chl Ze	Motuza and Gailus 1980

Table 1. (cont.)

No. on Fig.1	Impact structure	Country	Diameter (km)	Target rocks	Post-impact hydrothermal minerals from different impact lithologies			References
					Lithic breccias	Suevites	Massive impact melt rocks	
35	New Quebec	Canada	3.4	Cg	Ep Hem			Currie 1965
36	Nicholson Lake	Canada	12.5	Cg	Ep Hem			Dence et al. 1968
37	Obolon	Ukraine	20	Cg+St	Ca Sd	Sm Chl		Valter and Ryabenko 1977
38	Pilot Like	Canada	6	Cg	Chl Hem Ser			Dence et al. 1968
39	Popigai	Russia	100	Cgd+St+c	Sm Ca Anh Chl Ze Prh Grt Gp	Sm Ca Chl Sulf Ze Qz Gp Ep Op	Sm Chl Qz Ze Ill Ca Sulf Op Act	This paper
40	Puchezh-Katunki	Russia	80	Cgd+St+c	Sm Ze Chl Anh Ca Sulf Apf Act Grt Ab Gp Qz Ep Prh	Sm Ze Ca Op	Sm Ze Ca	Naumov 1999
41	Ries	Germany	24	Cg+St+c	Sm Ca Chl Ze Anh Corr Qz	Sm Ze Ill		Engelhardt 1972; Stöffler et al. 1977; Newsom et al. 1986
42	Rochechouart	France	23	Cg	Chl Corr Ca Sulf	Chl Sm Sulf	Chl Qz	Lambert 1977; Reimold et al. 1987
43	Roter Kamm	Namibia	2.5	Cg	Qz Sulf Chl Ill	Ca		Koeberl et al. 1989; Reimold et al. 1994
44	Rotmistrovka	Ukraine	2.7	Cg	Ze Sm			Masaitis et al. 1980
45	Sääksjärvi	Finland	6	Cgd	Ze Chl Qz Chc		Chl	Papunen 1973
46	Serpent Mound	USA	6.4	Sc	Sulf			Heyl and Brock 1962
47	Siljan	Sweden	55	Cg+St+c	Sm Chl Ze Ep Ab Sulf Hem			Johansson 1984; Morad et al. 1989; Aldahan 1990
48	Slate Islands	Canada	30	C+Sc	Hem Chl	Sm Chl		Dressler and Sharpton 1997
49	Söderfjärden	Finland	5	Cgd	Ca Chl Ze	Ze	Chl Ze	Laurén et al. 1978
50	Steen River	Canada	25	Cg+St+c	Ze Qz Chl Ca	Ze Qz Chl Ca	Sm Ze	Carrigy 1968

Table 1. (cont.)

51	Strangways	Australia	26	S		Sm		Guppy et al. 1971
52	Suavjärvi	Russia	15	Cg	Sulf			Author's data
53	Sudbury	Canada	~250	Cgd		Chl Act Sulf Ca Sm Ep Sph		Roussel 1984; Ames and Gibson 1995
54	Terny	Ukraine	11	Cgd	Sm Anh Sulf Br Pb		Corr Chc Sulf Ca Stp Op	Valter et al. 1988; Nikolsky 1991
55	Wells Creek	USA	12	Sc	Sm Ca Dl Qz Sulf			Stearns et al. 1968
56	Wanapitei	Canada	7.5	Cg		Sm Ze	Op Sm Ze	Grieve and Ber 1994
57	West Clearwater	Canada	32	Cgd			Sm Hem	Allen et al. 1982
58	West Hawk Lake	Canada	2.4	Cd	Ca Sm Sulf Ill Corr At		Ze	Short 1970
59	Zapadnaya	Ukraine	3.2	Cg	Ca Sm Ill Qz Ze Chl Sulf	Sm Ze Sd Br Kfsp Sulf	Ze Anh Sulf	Tatarintsev et al. 1981; Valter et al. 1995
60	Zhamanshin	Kazakhstan	13	S+V		Sm Chl Ca		Author's data

Target rocks:

- *C*, crystalline rocks: *g*, of granitic composition; *gd*, of granodioritic composition; *d*, of dioritic composition.
- *S*, sedimentary rocks: *t*, terrigenous deposits; *c*, carbonate rocks; *t+c*, terrigenous and carbonate rocks.
- *V*, volcanogenic rocks.

Minerals: Ab, albite; *Act*, actinolite; *Anh*, anhydrite; *Apf*, apophyllite; *At*, antigorite; *Br*, baryte; *Ca*, calcite; *Chc*, chalcedony; *Chl*, chlorite; *Coff*, coffinite; *Corr*, corrensite; *Dl*, dolomite; *Ep*, epidote; *Fu*, fluorite; *Gp*, gypsum; *Grt*, garnet; *Hem*, hematite; *Ill*, Illite; *Jar*, jarozite; *Kaol*, kaolinite; *Kfsp*, potassium feldspar; *Lm*, limonite; *Op*, opal; *Qz*, quartz; *Pb*, pitchblende; *Prh*, prehnite; *Sd*, siderite; *Ser*, sericite; *Sm*, smectites; *Sph*, sphene; *Stp*, stilpnomelane; *Sulf*, sulfides; *Ze*, zeolites.

Crater diameters are from Grieve and Shoemaker (1994) and Masaitis (1999).

Table 2. Comparative characteristics of environments of impact-generated hydrothermal systems in Popigai, Kara, and Puchezh-Katunki impact structures

Crater	Popigai	Kara	Puchezh-Katunki
Coordinates	71°38'N; 111°11'E	69°10'N; 65°00'E	56°58'N; 43°43'E
Diameter (km)	100	65	42 (80 with terrace)
Age (Ma)	35.7 ± 0.2 [a]	70.3 ± 2.2 [a]	165 ± 3 [a]
Inner structure	Complex with peak ring of crystalline rock	Complex with central uplift of sedimentary rocks	Complex with central uplift of crystalline rocks
Target rocks	Crystalline + Sedimentary	Sedimentary	Crystalline + Sedimentary
Paleogeography	Continent	Shallow (~100 m) sea	Lake plain
Impact rocks	Suevites, tagamites, and lithic breccia	Suevites and lithic breccia	Lithic breccia with minor suevites
Volume of impact melt rocks (km^3)	2800 [d]	1700 [b]	3.2 [d]
Post-impact deposits (m)	Absent	<100 [b]	>450 [c]
Water/rock ratio	low	very high	high
Main area of detailed studies	Annular through and peak ring	Outer slope of annular depression	Central uplift

Data from:
[a] Masaitis 1999
[b] Mashchak 1990b
[c] Masaitis and Pevzner 1999
[d] Author's estimate.

and lithic breccias are classified as authigenic (parautochthonous) and allogenic (allochthonous) varieties. Allogenic lithic breccias are subdivided based on the size of fragments and clasts (coarse-grained or megabreccias, and fine-grained breccias or coptoclastites).

2 Samples and Analytical Methods

More than 2000 chips and core sections with hydrothermal mineralization were sampled from numerous outcrops and drill cores of the Popigai, Kara, and Puchezh-Katunki structures during field work and examination of impact rock collections at the All-Russia Geological Research Institute (VSEGEI), St. Petersburg. The locations of the samples are given below, where the hydrothermal mineralization in the impact structures is discussed. For all samples, thin sections were studied by optical microscopy. Preparation of mineral separates was performed by handpicking (for zeolitic minerals, calcite, and apophyllite); for sulfides, it involved magnetic separation and separation under heavy liquids followed by handpicking.

To identify phyllosilicates and zeolitic minerals, and to establish their composition and structural patterns, X-ray powder diffraction (XRD) and differential thermal analysis (DTA) were used. The chemical compositions of the minerals were determined by electron microprobe (EMP) analysis, X-ray fluorescence spectroscopy (XRF), and atomic absorption spectrometry (AAS). The trace element contents (Co, Ni, Cu, Pb, Zn, As, Ag, and Ti) in pyrite were determined by atomic emission spectroscopy (AES). All the analyses were performed by VSEGEI Labs, St. Petersburg. The analysts are cited below in Tables 3 - 10. The detailed analytical procedures for XRF, AES, and AAS methods, as used at the VSEGEI Laboratories, are given by Rudnik and Smyslov (1986).

EMP analyses were performed on whole rock polished thin sections coated by carbon. The instrument used was a Cameca Camebax with a PDP-11/23 computer system, allowing the determination of 9 elements (Si, Ti, Al, Fe, Mn, Mg, Ca, Na, and K). The minerals are analyzed at acceleration voltage of 15 kV, beam currents of 2×10^{-9} A, beam diameter of 2 μm, and counting for 75 s. For phyllosilicate analyses, the beam was defocused to a diameter of 5 μm. Analytical lines of $K\alpha$ radiation, natural standards and ZAF correction procedures were used. Precision is ±3 rel%, detection limits are 0.01 wt% for Na_2O, and 0.03 - 0.04 wt% for the other elements.

XRF analyses of zeolites, calcite, and apophyllite were carried out on powdered samples weighing from 0.2 to 3.5 g, with a SRM-2 instrument. Accuracy levels as obtained from duplicate sample analysis are (in rel%) 0.7 for SiO_2, 7.0 for TiO_2, 2.0 for Al_2O_3, 3.4 for Fe_2O_3, 6.0 for MnO, 2.2 for MgO, 1.5 for CaO, 9.3 for Na_2O, and 5.4 for K_2O. The concentrations of SrO and BaO were determined by AAS using nitrogen oxide and acetylene flames. The measurements were performed with a Perkin-Elmer 305 spectrophotometer. Limit of detection and precision are 0.002 % and 4 rel% for SrO, and 0.003 % and 10 rel% for BaO, respectively.

AES measurements were carried out on 10-mg pyrite separates (of 97 % purity) with the STE-1 diffraction spectrometer. The method of emission spectroscopy using the d.c. arc with coaly electrode and Tb as an external standard (Mannkopff and Friede 1975; Vellon 1976; Rudnik and Smyslov 1986) was employed. Accuracy (for contents of 10 - 100 ppm) is (in rel%) 20 for Ni, 15 - 24 for Co, 15 - 23 for Cu, 23 - 24 for Pb, 20 for Ag, 14 - 31 for Ti, and 15 - 19 for Zn.

For XRD analysis, the powdered mineral separates as well as oriented <0.002 mm clay fraction preparations were used. The samples of the clay minerals were taken from the center of the sawed core sections to avoid the possibility of drilling mud contamination. Diffraction patterns were recorded using a DRON 3M diffractometer by scanning at 0.5 - 1° min^{-1} over the range $1° < \Theta < 36°$, with Fe_2O_3-filtered Co$K\alpha$ radiation, voltage of 30 kV, and current of 15 mA. The diffraction patterns were calculated as averages from three measurements, the error of angle determination was $\Delta\Theta < 0.01°$. Identifications of clay minerals and zeolites were made by reference to standard files (Joint Committee on Powder Diffraction Standards 1974) and by comparison with diffractograms of standard minerals (Brindley and Brown 1980; Dritz and Kossovskaya 1990; Gottardi and Galli 1985).

DTA analyses of zeolites and smectites were performed with a Q-1500D instrument from powdered to <60 mm samples weighing 200 - 500 mg. The samples were analyzed in air, within the temperature range from 20 to 1100°C, and at a heating rate of 15 - 20°C min^{-1}.

3 Hydrothermal Mineralization in Impact Structures

3.1 Kara Impact Structure

The Kara impact structure, which is situated on the northwestern slope of the Pai-Khoi Ridge, is the most representative object to study post-impact hydrothermal alteration of a thick suevite sheet filling the crater. The total thickness of the suevites in the Kara structure exceeds 1 km; their volume is estimated at 1700 km^3 (Mashchak 1990b). According to Mashchak (1990b), the crater fill is subdivided into two units: (1) a lower unit (of 0.7 - 0.9 km thick), composed of megabreccias and coarse fragmented suevites, the latter form irregular bodies up to 40 - 60 m thick; (2) an upper unit, composed of lapilli-sized and agglomerate-sized suevite (up to 1 km thick) and fine-grained lithic breccias (coptoclastites, 0.2 - 0.3 km thick). Thin (up to 10 - 15 m thick) lenses of tagamites occur locally in the top of the lower unit. The suevites contribute 46 vol% of impact rocks and contain more than 95 vol% of impact melt. On the modern erosion surface the suevites are exposed below Pliocene-Quaternary deposits along river valleys in outer parts of the crater; they are also found in boreholes in the southern and southeastern parts of the crater (Fig. 2).

The formation of the crater on the shallow (about 100 m) shelf (Mashchak 1990a) has caused abundant impact-induced hydrothermal activity, which mainly affects the suevite deposits. Hydrothermal alteration of the suevites is displayed in the recrystallization of vitroclasts and in the development of mineral assemblages forming veins and vugs, as well as impregnated mineralization in the fine-grained matrix of the suevites. Locally, the secondary mineralization contributes up to 30 % to rock volume (in the valleys of the Sopcha-Yu and Kara Rivers in the southern part of the crater).

The number of common vein minerals is limited; it includes pyrite, calcite, and analcime (Fig. 2). Calcite is the most abundant; two generations of calcite are distinguished: (1) an earlier generation, represented by yellowish or white, transparent rhombohedral or prismatic crystals, which form, together with analcime and pyrite, veins up to 10 cm thick and vugs in the matrix of suevite and allogenic breccia; (2) a later generation, represented by greasy and milky cryptograined calcite, which is developed in central parts of analcime vugs, or forms separate vugs. The total content of chemical admixtures (Fe, Mn, Mg) in calcite was determined to be less than 1.5 wt% (Table 3), but increases somewhat

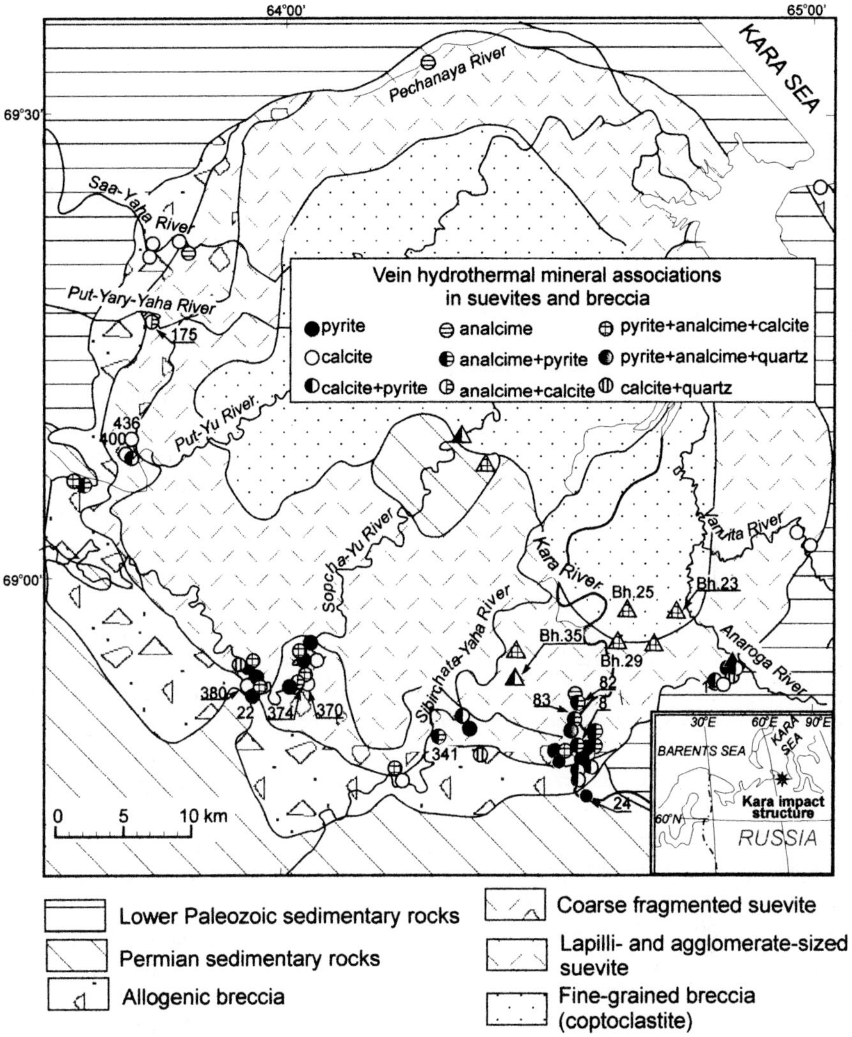

Fig. 2. The distribution of the hydrothermal mineralization in the Kara impact structure. The general geology follows Mashchak (1990b). Occurrences of hydrothermal minerals in outcrops are shown by circles, and in drill cores, by triangles. Numbered outcrops or boreholes indicate locations of samples described in the text and tables.

from the earlier to the later generation. Pyrite occurs in salbands of analcime-calcite veins; it forms also vugs and disseminated mineralization in the matrix of the suevites, mainly in the coarse fragmented suevite. Locally (in the Kara River valley), pyrite contributes 30 vol% to the matrix of the suevites. Pyrite occurrences in different areas are characterized by the strong preference of the cubic habit of crystals, by low total content (less than 0.5 %) of trace elements

Table 3. Chemical composition[a] (in wt%) of hydrothermal calcite from the Kara impact structure

Core/outcrop [b]	370*	23	400	380	370	29	35
Depth (m)		198				160	228
FeO	0.24	0.02	0.04	0.06	0.65	-	0.40
MnO	0.13	0.51	0.30	0.30	0.36	1.27	0.52
MgO	0.01	0.02	0.05	0.23	0.02	-	0.02
CaO	56.99	55.38	56.30	55.95	54.67	55.32	56.57
CO_2	41.56	44.59	44.26	42.32	44.30	44.00	41.46
Total	98.92	100.52	100.86	98.81	100.01	100.59	98.98

[a] XRF data, analysed by B.A. Tsimoshenko, VSEGEI, operation conditions are given in the section 2

[b] Numbers of samples correspond to numbers of outcrops and boreholes, shown in Fig. 2.

Short description of the samples: *370**, analcime-pyrite-calcite zoned vein with yellowish rhombohedral calcite; *23/198*, druses of fine-crystalline scalenohedral calcite in a suevite; *400*, calcite vein of 5 cm thick in a brecciated coaly shale (coarse fragment in a suevite); *380*, coarse crystalline calcite vein to 10 cm thick, with minor pyrite and mordenite; *370*, later calcite from analcime-calcite vugs in the suevite matrix; *29/160*, milky calcite vugs in a suevite; *35/228*, later calcite cementing suevite.

(Cu, Pb, Zn, Ti, Ni, Co), and by very high Ni/Co ratios (3.5 - 10), which are similar to that of pre-impact pyrite from sedimentary target rocks (Table 4). Analcime is characterized by insignificant variations in composition (common abundance in silica, depletion in Na, and addition of Mg in structural formula) and in crystal structure (lattice parameter 13.698 Å < a < 13.720 Å, Table 5). Thus, apart from the identical series of main hydrothermal minerals, their chemical compositions and structural features, which reflect chemical and PT-conditions of mineral formation, are characterized by constancy within the entire suevite sheet.

Other secondary minerals (chabazite, laumontite, stilbite, mordenite, quartz, opal) were identified by optical microscopy and XRD analysis (Table 6) in rare cases. In addition, barite and apophyllite were observed locally (Yushkin et al. 1976).

As inferred from DTA (Selivanovskaya et al. 1982) and XRD data (Fig. 3), all impact glasses from suevites are to some extent transformed into aggregates of dioctahedral and trioctahedral smectites with a predominance of saponite; also, an admixture of chlorite and willemseite (a nickel-bearing phyllosilicate) is commonly found. In rare cases, mixed-layered chlorite-smectite with the smectite content of <20 rel% predominates in altered glasses. The d_{001} values on the powder diffraction patterns show that saponites from altered glasses of the lower part of the suevite sheet (samples 436/3, 175, and K-8o on Fig. 3) are characterized by the prevalence of Ca and Mg among interlayer cations (14 Å < d_{001} < 15 Å), whereas in saponites from the upper part (samples from core of borehole 23 from Fig. 3), Na and K play a significant role (12 Å < d_{001} < 13 Å). In addition, chlorites and mixed-layered chlorite-smectite minerals are abundant only in altered glass from the lower part of the suevite blanket. The vesicles in glass are filled by Ca-montmorillonite, and by calcite, pyrite, and opal, and, in some cases, by zeolites.

Thus, from the distribution of secondary clay minerals replacing impact glass fragments, a vertical zonation of hydrothermal alteration is weakly outlined.

Based on interrelations between secondary minerals in zoned veins and vugs, the order of crystallization of hydrothermal minerals in the suevites is as follows: saponite (mixed-layered saponite-chlorite) pyrite calcite, laumontite analcime calcite, mordenite, chabazite pyrite, calcite. The appearance of pyrite- and calcite-bearing veins in clastic dikes, which formed during isostatic relaxation of the crater after lithification of a friable matrix of the suevites (Mashchak 1990b), indicates that the post-impact hydrothermal activity exists during the entire early

Table 4. Some trace elements concentrations [a] (in ppm) in hydrothermal pyrite from the Kara and the Puchezh-Katunki impact structures

Core/outcrop	Depth (m)	Ni	Co	Cu	Pb	Zn	As	Ti	Ag
Kara impact structure									
370*		32	-	69	68	170	200	370	n.d.[b]
23	188	27	-	14	54	150	-	240	n.d.
25	231	37	-	56	39	140	-	320	n.d.
23	377	71	21	57	43	130	100	910	n.d.
1		34	5	21	-	170	-	280	n.d.
83		43	7	39	-	140	100	460	n.d.
24		320	46	530	190	150	-	190	n.d.
22		58	21	22	-	130	-	1900	n.d.
Puchezh-Katunki impact structure									
VDW	2567	180	160	290	9	-	n.d.	n.d.	-
VDW	2850	13	43	17	7	-	n.d.	n.d.	-
33	421	70	30	260	36	110	n.d.	n.d.	2
38	401	81	26	35	19	-	n.d.	n.d.	-
50	314	170	46	30	5	-	n.d.	n.d.	-
64	268	62	100	26	6	-	n.d.	n.d.	-
VDW	4252	190	590	410	75	17	n.d.	n.d.	-
Limit of detection		0.5	3	0.5	5	10	100	5	1

[a] Emission spectroscopy data, analysed by V.I. Latikainen, VSEGEI, instruments and operation conditions are described in section 2

[b] n.d., not determined

Numbers of samples correspond to numbers of outcrops and boreholes, which are shown in Fig. 2 (for Kara) and Fig. 8 (for Puchezh-Katunki).

Short description of the samples:

The Kara structure: *23/188* and *370**, analcime-pyrite-calcite zonal veins in a suevite; *25/231*, analcime-pyrite vein in a suevite; *23/377, 83*, and *1*, pyrite disseminated in the fine-clastic matrix of a suevite; *24*, pre-impact pyrite concretions from a calcareous sandstone outside of the crater; *22*, pre-impact pyrite from a large block of calcareous shale within the allogenic megabreccia.

The Puchezh-Katunki structure: *VDW/2567*, pyrite disseminated in a recrystallized and brecciated biotite-amphibole gneiss; *VDW/2850*, pyrite-anhydrite vein in a brecciated gneiss; *33/421*, pyrite-calcite-analcime vein in a brecciated and shocked gneiss; *38/401*, pyrite from the cement of a redeposited suevite; *50/314*, pyrite-calcite vein within polymict allogenic breccia; *64/268*, pyrite disseminated in an intensely recrystallized and brecciated amphibole gneiss; *VDW/4252*, pre-impact pyrite (disseminated mineralization) from Archean gneiss.

Table 5. Chemical compositions [a] (in wt %) and lattice constants [b] of analcime from the suevites of the Kara impact structure

Core/outcrop	23	25	341	29	82	374
Depth (m)	398	231		214		
SiO_2	57.24	56.96	58.00	56.72	58.67	56.45
Al_2O_3	22.40	22.33	22.14	22.31	21.68	22.88
MgO	0.61	0.68	0.79	0.71	0.62	0.60
Na_2O	12.87	12.82	12.46	13.14	12.52	12.95
Sum	93.11	93.09	93.39	92.88	93.49	92.92
Formulae based on 96 O atoms						
Si	32.81	32.78	33.06	32.67	33.39	32.48
Al	15.13	15.14	14.88	15.14	14.53	15.46
Mg	0.52	0.58	0.67	0.61	0.53	0.51
Na	14.30	14.30	13.77	14.67	13.81	14.44
Lattice constants						
a (Å)	13.698	13.704	13.720	n.d.[c]	13.700	13.705
V (Å^3)	2570.2	2573.0	2582.7	n.d.	2571.2	2574.4

[a] EMP data, analysed by O. Yakovleva, VSEGEI, instruments and operation conditions are described in section 2
[b] Lattice constants as determined by Yu.S. Dyakonov, VSEGEI, using DRON-3M diffractometer, Ni-filtered Co*Kα* radiation, internal standard NaCl, scanning speed 0.5°min^{-1}. *a* values are calculated from 14 diffraction patterns (*h k l*=2 2 0; 4 2 2; 4 3 1; 5 2 1; 4 4 0; 6 4 0; 7 2 1; 6 3 3; 6 5 1; 7 3 2; 9 3 2; 7 6 3; 7 7 2; 10 1 1); precision is ±0.002 Å
[c] n.d. – not determined

Numbers of samples correspond to numbers of outcrops and boreholes, which are shown in Fig.2

Short description of the samples: *23/398*, analcime-pyrite veins in a calcareous shale; *25/231*, *29/214*, and *374*, analcime-pyrite veins in a suevite; *341*, analcime-calcite vug in the matrix of a coarse fragmented suevite; *82*, black analcime with coaly microinclusions from vugs in a suevite

Fig. 3. (opposite) XRD powder patterns, obtained by using a DRON 3M diffractometer, by scanning at 1°min^{-1} over the range $1° < \Theta < 36°$, Fe-filtered Co*Kα* radiation, for altered impact glasses from the suevites of the Kara impact structure. Analyst, O.Yu. Dubic, VSEGEI. Identifications of clay minerals were made by references to standard files (Joint Committee on Powder Diffraction Standards (JCPDS 1974). Fractions of altered impact glass (about 0.5 g) were separated by handpicking from suevite samples presented by M.S. Mashchak (VSEGEI). Samples *23/137*, *23/188*, and *23/214* are from borehole 23 (at the depths of 137, 188, and 214 m, respectively), Anaroga River; samples *175*, *8o*, and *436/3* are from outcrops at Put-Yary-Yaha, Put-Yu, and Kara Rivers, respectively. Locations of the samples are shown on Fig. 2. Samples from the 23 core represent the upper part of the suevite blanket (so-called lapilli-sized suevites), and other samples represent its lower part (coarse fragmented suevites). Peak labeling indicates the diffraction patterns of smectite (sm), chlorite (chl), mixed-layered smectite-chlorite (sm-chl), willemseite (will), calcite (ca), pyrite (pyr), feldspar (fsp), quartz (qz), and stilbite (stb). Willemseite, $(Ni,Mg)_6Si_8O_{20}(OH)_4$, was identified in samples 23/137 and 23/214 from the presence of a 9.45 Å line; other characteristic lines of willemseite (with d = 4.57 Å, 3.14 Å, 2.52 Å, etc.) are overlapping with lines from the smectite patterns.

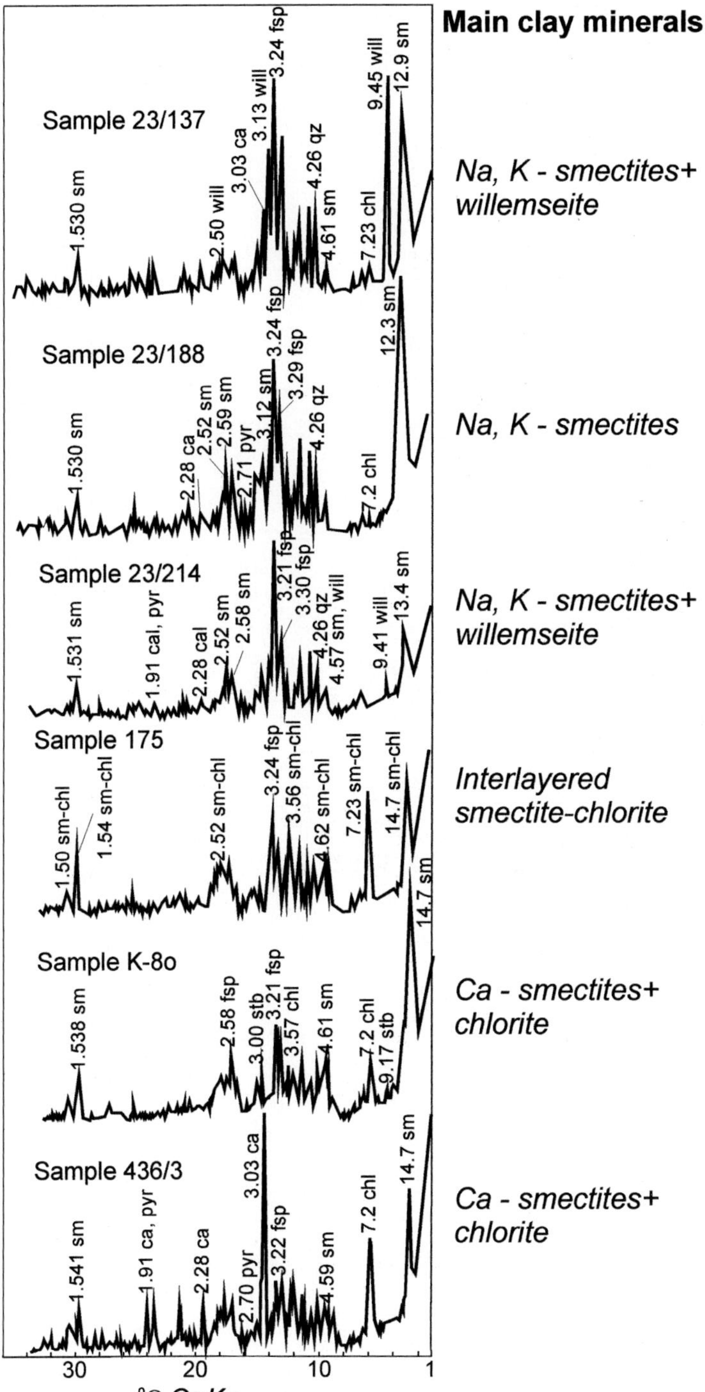
Main clay minerals
Sample 23/137
1.530 sm
2.50 will
3.03 ca
3.13 will
3.24 fsp
4.26 qz
4.61 sm
7.23 chl
9.45 will
12.9 sm
Na, K - smectites+ willemseite
Sample 23/188
1.530 sm
2.28 ca
2.52 sm
2.59 sm
2.71 pyr
3.12 sm
3.24 fsp
3.29 fsp
4.26 qz
7.2 chl
12.3 sm
Na, K - smectites
Sample 23/214
1.531 sm
1.91 cal, pyr
2.28 cal
2.52 sm
2.58 sm
3.21 fsp
3.30 fsp
4.26 qz
4.57 sm, will
9.41 will
13.4 sm
Na, K - smectites+ willemseite
Sample 175
1.50 sm-chl
1.54 sm-chl
2.52 sm-chl
3.24 fsp
3.56 sm-chl
4.62 sm-chl
7.23 sm-chl
14.7 sm-chl
Interlayered smectite-chlorite
Sample K-8o
1.538 sm
2.58 fsp
3.00 stb
3.21 fsp
3.57 chl
4.61 sm
7.2 chl
9.17 stb
14.7 sm
Ca - smectites+ chlorite
Sample 436/3
1.541 sm
1.91 ca, pyr
2.28 ca
2.70 pyr
3.03 ca
3.22 fsp
4.59 sm
7.2 chl
14.7 sm
Ca - smectites+ chlorite
30
20
10
1
°Θ CoKα

Table 6. XRD powder patterns of zeolites from the Kara and Popigai impact structures

Impact crater	Kara								Popigai					
Mineral	Analcime		Mordenite		Chabazite		Stilbite		Heulandite		Chabazite		Stilbite	
Sample	374		KR-8a		370*		KR-8		6854/241		6854/557		6854/557	
	I	d (Å)	I	d (Å)	I	d (Å)	I	d (Å)	I	d (Å)	I	d (Å)	I	d (Å)
Most intense peaks in X-ray diffraction pattern; I, normalized relative intensity	6	6.186	35	13.6	55	9.390	100	9.17	80	8.86	55	9.31	90	9.20
	90	5.598	100	9.10	20	6.912	16	5.32	15	7.85	25	6.89	10	5.42
	20	4.848	40	6.58	15	5.590	35	4.68	30	6.64	35	5.60	50	4.70
	10	3.662	35	6.39	25	5.006	15	4.58	20	5.89	30	5.00	35	4.32
	100	3.427	15	6.07	70	4.333	60	4.29	20	5.23	12	4.67	100	4.11
	15	3.027	50	5.77	20	3.879	100	4.06	55	5.07	90	4.31	22	3.74
	60	2.925	70	4.53	35	3.587	15	3.75	30	4.62	20	3.87	12	3.58
	20	2.688	18	4.13	16	3.431	25	3.42	22	4.33	35	3.58	35	3.45
	18	2.501	85	4.00	14	3.243	30	3.20	12	4.18	16	3.45	50	3.24
	12	2.223	20	3.84	18	3.185	90	3.05	70	3.95	100	2.93	100	3.07
	16	1.900	20	3.56	100	2.936	30	3.00	40	3.88	30	2.89	55	2.82
	20	1.740	85	4.48	32	2.893	40	2.78	25	3.40	25	2.84	15	2.72
	10	1.865	100	3.40	16	2.608	15	2.60	15	3.31	15	2.68	18	2.60
	10	1.687	100	3.22	16	2.510	14	2.56	22	3.16	14	2.49	10	2.29
	10	1.414	40	2.90	10	2.109	10	2.51	70	2.96	10	2.09	20	2.07
	10	1.357	20	2.69	12	1.806	12	2.37	30	2.79	18	1.79	10	1.85
	8	1.593	22	2.56	10	1.712	14	2.04	24	2.72	10	1.72	12	1.69
	6	1.496	25	2.52			8	1.90	10	2.42	8	1.90		

Difractograms were obtained using a DRON-3M diffractometer, by scanning at $1^{\circ}min^{-1}$ over the range $4^{\circ} < 2\Theta < 36^{\circ}$, Fe_2O_3-filtered Co$K\alpha$ radiation, (for samples 374 and KR-8a, at $1^{\circ}min^{-1}$, Ni-filtered Co$K\alpha$ radiation, scanning speed $0.5^{\circ}min^{-1}$), analysed by Yu.S. Dyakonov, VSEGEI
Numbers of samples from the Kara crater correspond to the outcrop numbers, which are shown in Fig. 2
Short description of the samples: *374*, analcime-pyrite vein in a suevite; *370**, analcime-pyrite-calcite zonal vein in a suevite; *KR-8*, zeolite vugs in the fine-clastic matrix of a coarse fragmented suevite
For description of the samples from the Popigai crater, see Table 9

modification stage of the crater. However, data on the rate of compaction of high-temperature pyroclastic rocks (Sheridan and Ragan 1976) indicate that the lithification of the suevites can take less than a few hundred years.

The most intensive hydrothermal mineralization occurs in the southern part of the crater (valleys of the Kara and Sopcha-Yu Rivers). This is probably due to the abundance of coal-bearing shales and limestones of Early Paleozoic age among lithic fragments in suevites from these locations. These coaly and carbonate lithologies probably provide geochemical conditions controlling the precipitation of pyrite and associated minerals. In contrast, in the northern part of the crater, lithoclasts belong exclusively to Permian terrigenous rocks. There is no correlation between the intensity of hydrothermal alteration and the content of impact melt fragments in suevites; in contrast, the suevites that are most enriched in glass from the Peschanaya River area are only weakly altered.

3.2 Popigai Impact Structure

The Popigai structure in Northern Siberia (Masaitis et al. 1980, 1998, 1999, and references therein) is characterized by its good preservation and exposure, and detailed investigations that were made through deep drilling and from abundant outcrops of diverse impact lithologies. The main elements of the inner structure of the Popigai crater are: (1) a central depression of 40 km in diameter filled mainly by suevites and fine-grained lithic breccia (coptoclastites); (2) a peak ring of disturbed crystalline target rocks of 45 km in diameter; (3) an annular trough of 15 - 20 km in width, filled by coarse allogenic breccia and impact melt rocks (suevites and tagamites); (4) an outer zone of deformed crystalline and sedimentary target rocks.

The impact lithologies are subdivided into three main units that vary considerably in thickness and show gradual transitions, from bottom to top in the vertical section: (1) a lower unit, which contains polymict megabreccia composed of fragments of sedimentary and crystalline target rocks, and crystalline megabreccia composed of fragments of shocked crystalline target rocks; (2) a middle unit, which contains a complex blanket of impactites, including thick (up to 600 m) tagamite sheets and overlying vitric and crystal-vitric suevites, which are composed mainly of black glass fragments (vitroclasts) and fragments of shocked crystalline rocks and their minerals; (3) an upper unit containing vitric-lithic suevites, which are composed mainly of greenish-gray vitroclasts and fragments of sedimentary rocks and coptoclastites; thin and shallow tagamite bodies also occur. The total volume of impact rocks has been suggested to exceed 5000 km^3, and impact melt rocks contribute up to 2800 km^3. In contrast to the Kara crater, the bulk of the impact melt is contained in massive impact melt rocks (tagamites). Shocked and brecciated crystalline basement (authigenic breccia) is exposed below the impactites and impact breccia in the peak ring in the NW sector of the crater (Majachika Upland), as well as in some other areas.

The study and sampling of the hydrothermal mineralization was carried out on some well-exposed outcrops (Sogdoku Upland, Chordu, Kysym, Usun-Taas-Kystyga) or on drill core samples (Balagan-Yuregue area, Majachika Upland,

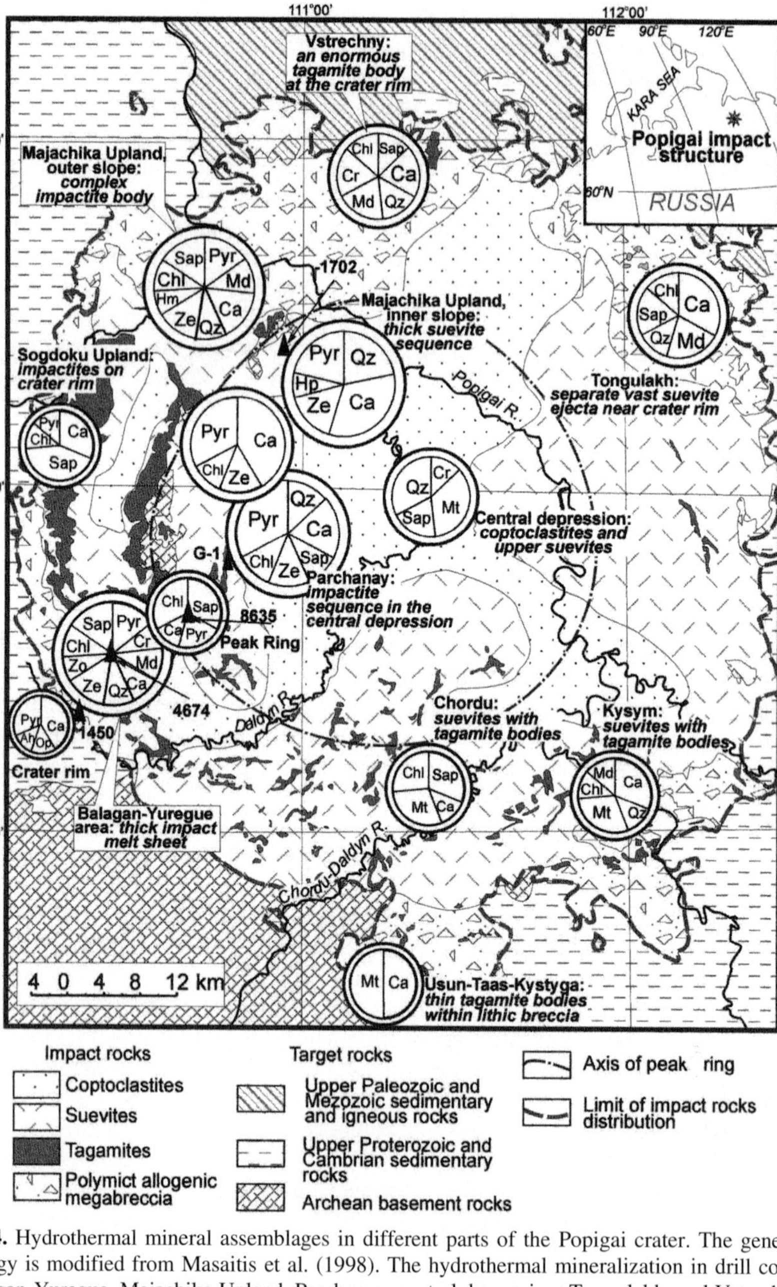

Fig. 4. Hydrothermal mineral assemblages in different parts of the Popigai crater. The general geology is modified from Masaitis et al. (1998). The hydrothermal mineralization in drill cores (Balagan-Yuregue, Majachika Upland, Parchanay, central depression, Tongulakh, and Vstrechny areas), in outcrops (Sogdoku Upland, Chordu, Kysym, and Usun-Taas-Kystyga areas), and both in G-1 drill core (see Fig. 5) and in outcrops (Parchanay area) was studied. The size of circles shows the relative abundance of the hydrothermal mineralization. Numbered triangles are the drill cores shown in Fig. 5.

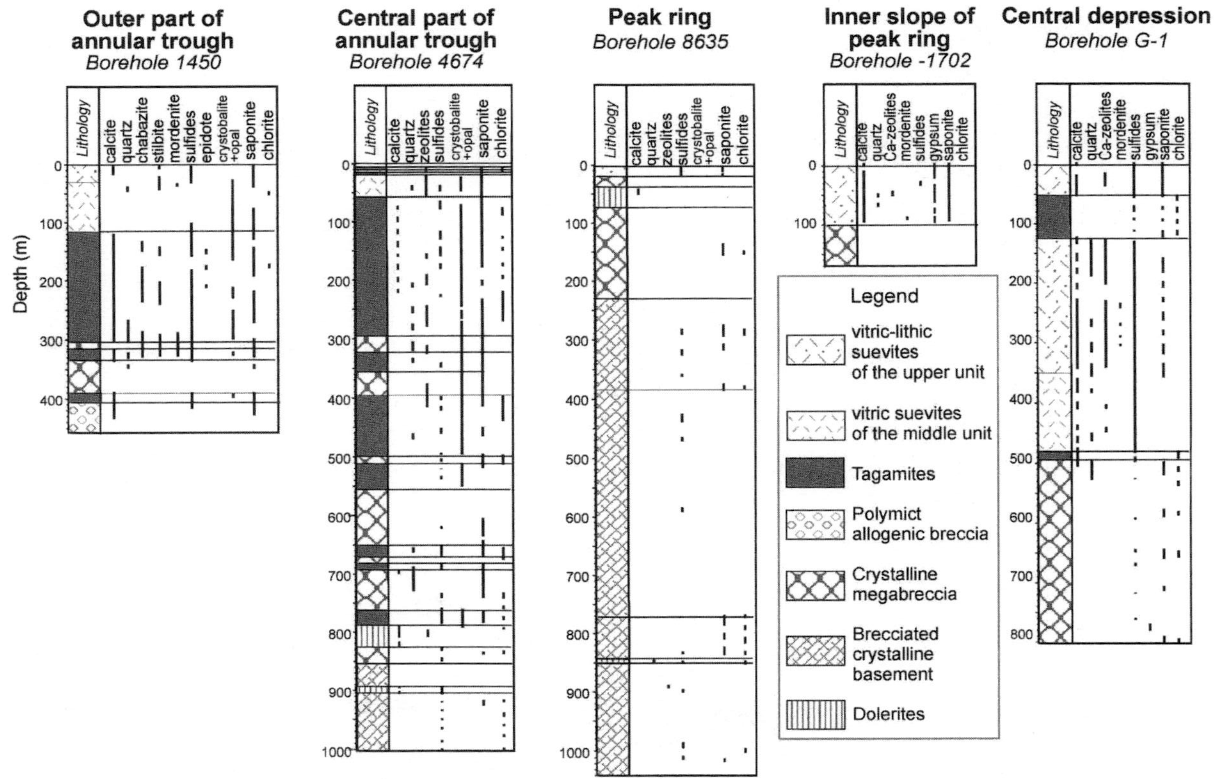

Fig. 5. Distribution of main hydrothermal minerals in selected drill cores that show the main structural elements of the Popigai impact structure. Dashed lines indicate sporadic occurrence of a mineral. Locations of drill cores are shown in Fig. 4.

Table 7. Representative electron microprobe analyses [a] of post-impact smectites from different impact lithologies of the Popigai impact structure

Host rock	Coptoclastite		Vitric-lithic suevite from the upper unit		Crystal-vitric and vitric suevite in the middle unit	
Core (outcrop) [a]	97	5002	-1702	99-06	1016	0924
Depth (m)		114	77		139	189
Mineral [b]	Nontronite	Al-Fe beidellite	Montmoril-lonite	Fe-Mg saponite		Mg-Fe saponite
wt%						
SiO_2	42.99	48.18	50.42	43.96	44.25	38.28
TiO_2	0.26	1.27	1.27	1.25	1.37	1.62
Al_2O_3	6.63	16.42	17.10	14.45	11.41	13.37
FeO*	25.69	9.94	7.76	22.21	20.58	22.07
MnO	0.33	0.16	0.16	0.28	0.64	0.19
MgO	0.48	2.12	4.20	8.77	10.54	12.83
CaO	3.01	1.69	2.54	1.54	1.20	1.94
Na_2O	-	0.08	0.11	-	-	-
K_2O	-	0.62	0.47	0.63	0.14	0.44
Total	79.39	80.40	83.30	93.09	90.13	90.74
Formula based on 22 O atoms						
Si	7.17	7.36	7.38	6.48	6.73	5.9
Ti	0.03	0.15	0.14	0.14	0.16	0.19
Al^{IV}	0.83	0.64	0.62	1.52	1.27	2.10
Al^{VI}	0.47	2.32	2.33	0.99	0.77	0.32
Fe^{3+}	3.00	1.14	0.85	-	-	0.48
Fe^{2+}	0.33	-	-	2.74	2.62	2.13
Mn	0.05	0.02	0.02	0.03	0.10	0.03
Mg	0.12	0.48	0.92	1.93	2.39	2.95
Ca	0.64	0.28	0.40	0.24	0.2	0.32
Na	-	0.02	0.03	-		-
K	-	0.12	0.09	0.12	0.03	0.09
n	2	3	3	2	2	3

Analyses by L.S. Bogomolnaya, VSEGEI, analytical conditions are described in the section 2; *n* is the number of repeated analyses

[a] Locations of samples: *97*, Kysym-Yuregue Creek (the southeastern part of the Popigai crater); *5002/114*, borehole 5002, Balagan-Yuregue area; *-1702/77*, borehole -1702, the southeastern slope of the Majachika Upland; *99-06*, Parchanay Creek; *1016/139*, borehole 1016, Majachika Upland; *0924/189*, the northwestern slope of the Majachika Upland; *4674/275, 4674/489, 4290/150, 6854/200*, and *6854/557*, from the thick tagamite sheet, Balagan-Yuregue area, boreholes 4674, 4290, and 6854; *31*, lens-like tagamite body of 10 m thickness within suevite deposits, Chordu-Daldyn River.

[b] Mineral names are in accordance with the classification of Dritz and Kossovskaya (1990).

* Total iron as FeO.

Table 7. (cont.)

Tagamite					Megabreccia	
6854		4674	4290	4674	31	6854
200		489	150	275		557
Fe-Al montmorillonite	Al-Fe beidellite	Al-Fe montmorillonite	Fe-Mg saponite	Mg-Fe saponite		
54.93	56.32	60.08	43.62	43.68	46.26	44.93
-	0.46	-	-	0.07	0.60	-
5.89	19.55	14.19	5.72	7.88	9.55	8.59
20.47	3.24	7.55	22.89	21.43	18.36	18.81
0.58	-	-	0.40	0.09	-	-
4.85	3.97	5.23	7.04	12.11	15.6	12.71
2.47	0.40	3.40	2.52	2.11	0.87	1.82
-	0.60	-	0.95	0.31	-	-
0.17	3.30	0.48	-	0.39	0.35	0.35
89.36	87.84	90.94	83.14	88.07	91.59	87.21
7.86	7.70	7.99	7.33	6.86	6.78	6.95
0.05	0.05	-	-	0.01	0.07	-
0.14	0.30	0.01	0.67	1.14	1.22	1.05
0.85	2.85	2.21	0.70	0.32	0.43	0.52
2.20	0.37	0.75	-	-	0.32	-
0.00	-	-	3.22	2.81	1.77	2.43
0.07	-	-	0.06	0.01	-	-
1.03	0.81	1.04	1.76	2.83	3.41	2.93
0.38	0.06	0.48	0.45	0.35	0.14	0.30
-	0.16	-	0.31	0.09	-	-
0.03	0.58	0.08	-	0.08	0.07	0.07
3	3	4	3	3	3	3

Parchanay, central depression, Tongulakh, Vstrechny) (Fig. 4). These areas contain all impact lithologies and represent all structural elements of the crater; their general geology is given by Masaitis et al. (1998, 1999) and by Mashchak and Selivanovskaya (1988).

In comparison with the Kara structure, post-impact hydrothermal alteration of Popigai rock is rare. Nevertheless, secondary minerals are present in all impact lithologies from different parts of the crater (Fig. 4). They are developed mainly in tagamites, suevites, and impact glasses. Within allogenic breccia and, especially, in parautochthonous rocks of the peak ring and crater rim, the hydrothermal mineralization is less common.

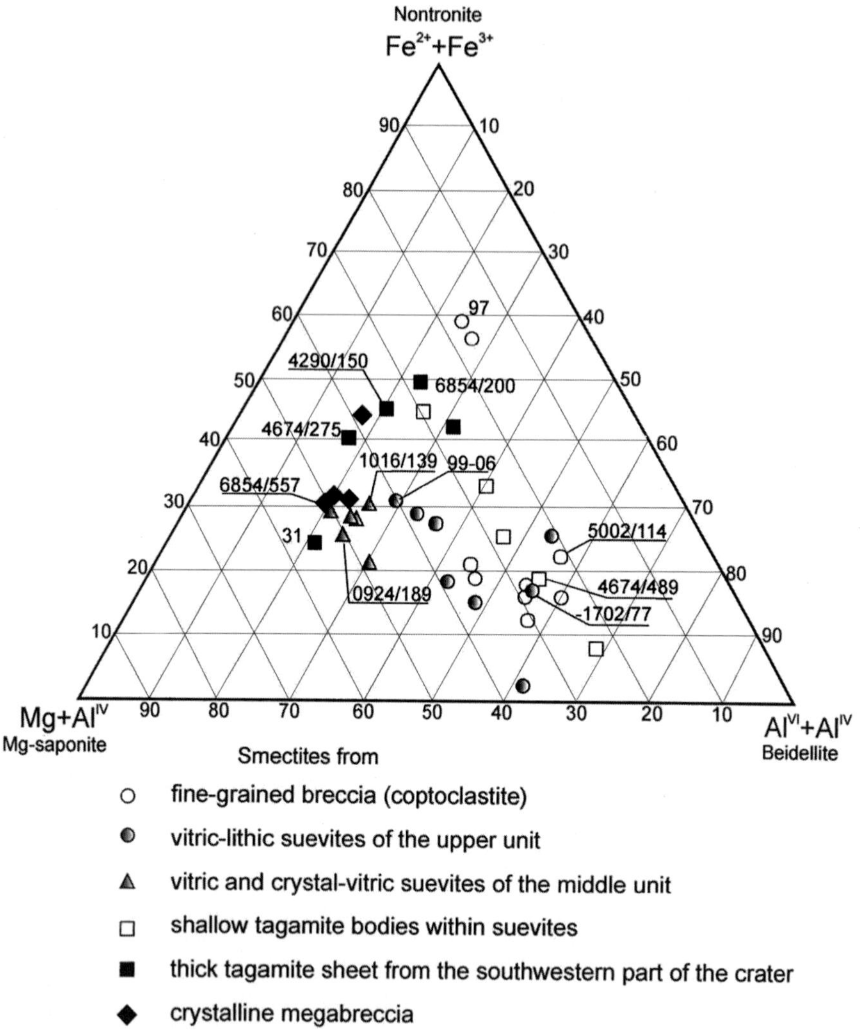

Fig. 6. Ternary $Fe^{2+}Fe^{3+}$-$MgAl^{IV}$-$Al^{VI}Al^{IV}$ diagram for secondary smectites, which replace impact glass and fill cavities in different impact lithologies from the Popigai crater (results of 36 electron microprobe analyses [EMP] are plotted). The samples are from drill cores in the Balagan-Yuregue area and the Majachika Upland, and from the southeastern part of the crater (Chordu and Kysym areas). Normative compositions of smectites were calculated on the basis of 28 O atoms. Representative EMP analyses of smectites (indicated by numbers on the diagram) are given in Table 7. The data show a general trend for variations in the clay mineral composition from high-alumina varieties (beidellite) to Mg-Fe varieties (saponite) from top to bottom in a generalized impact rock sequence (from coptoclastites to lithic suevites, to vitric and crystal-vitric suevites, to tagamites, and to crystalline megabreccia).

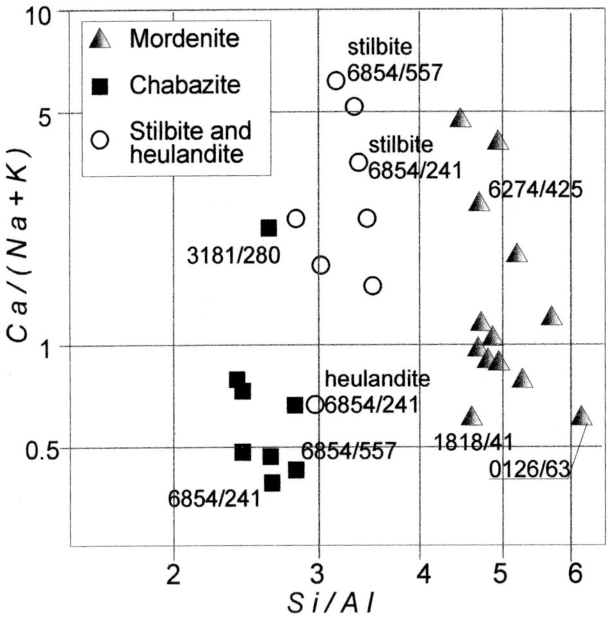

Fig. 7. Plot of Ca / (Na + K) vs. Si / Al ratios of hydrothermal zeolites from the Popigai crater (results of 29 electron microprobe [EMP] analyses are shown). Zeolites were sampled from drill cores in the Balagan-Yuregue area and Majachika Upland. The mineral compositions were calculated on the basis of 72 O atoms for mordenite, stilbite, and heulandite, and of 24 O atoms for chabazite. Representative EMP analyses of zeolites (indicated by numbers on the diagram) are given in Table 8. The data show that all zeolitic minerals vary strongly in the Ca / (Na + K) ratio and deviate from the stoichiometric Si/Al ratio, that is equal of 5.0 for mordenite, 2.0 for chabazite, and 3.0 for heulandite and stilbite. This indicates that local variation of the composition of hot-water solutions is an important factor for zeolic mineral distribution.

The main secondary minerals are similar to those found at the Kara crater: hydrous phyllosilicates locally replace the matrix of impact glass, but, much more frequently, occupy pores and fissures. Calcite, zeolites (predominantly mordenite), pyrite, quartz, cristobalite, and opal are common hydrothermal minerals in vugs and amygdules. From interrelations between minerals infilling cavities, the following generalized order of the mineral formation is established: silica modifications (quartz, cristobalite, opal), hydrous phyllosilicates (smectites and chlorites), zeolites, calcite, and pyrite.

The spatial differentiation of hydrothermal minerals is more clear than at the Kara structure. Figure 4 shows the distribution of mineral assemblages across the whole structure, and Fig. 5 indicates the occurrences of hydrothermal minerals in drill cores, which were selected as representatives for the main structural elements of the crater (annular trough, peak ring, central depression). In both cases the dependence of both the mineral types and of the intensity of hydrothermal

Table 8. Representative electron microprobe analyses of post-impact zeolites from the Popigai crater

Mineral	Mordenite			Stilbite		Heulandite	Chabazite		
Core	0126	6274	1818	6854	6854	6854	6854	6854	3181
Depth (m)	63	425	41	557	241	241	241	557	280
wt%									
SiO_2	70.58	66.35	66.67	58.09	59.71	57.46	52.30	54.27	53.82
Al_2O_3	9.84	12.15	12.34	14.68	14.88	16.21	16.73	16.14	18.37
FeO	-	-	0.25	-	-	-	-	0.40	-
MgO	-	-	0.11	-	-	-	-	-	-
CaO	4.22	5.37	3.24	8.06	7.39	5.53	4.16	4.07	7.61
Na_2O	3.69	0.35	2.78	0.87	1.16	4.09	4.22	2.60	2.01
K_2O	0.15	1.18	0.29	-	-	0.69	2.66	4.36	-
Total	88.48	85.40	85.61	82.09	83.13	83.99	80.07	81.44	81.82
Si/Al	6.08	4.63	4.59	3.18	3.41	3.01	2.65	2.85	2.49
Ca/Na	0.61	2.65	0.60	6.10	3.51	0.67	0.38	0.41	2.09

Analysed by O.A. Yakovleva, VSEGEI; instruments and operation conditions are described in section 2

Host rocks for zeolite samples: *0126/63*, crystal-vitric suevite, eastern slope of the Majachika Upland; *6274/425*, cataclased within megabreccia of crystalline rocks underlying the thick tagamite sheet, Balagan-Yuregue area; *1818/41*, tagamite from a shallow body within the suevite sequence, Majachika Upland; *6854/241*, tagamite with 40% of lithic clasts from the lower part of the thick impact melt sheet, Balagan-Yuregue area; *6854/557* and *3181/280*, megabreccia of crystalline rocks, Balagan-Yuregue area

alteration on the composition of host rocks is obvious. Cristobalite and mordenite, together with saponite and mixed-layered phyllosilicates, dominate among secondary phases in impact melt rocks, whereas in underlying crystalline megabreccia and in the uppermost suevite filling the central depression calcite, quartz, pyrite, and low-silica zeolites are prevailing. Thus, the vertical zonation of hydrothermal mineral formation displayed in cross-sections is mainly caused by the distribution of impact lithologies in the rock sequence.

Nevertheless, the distribution of hydrous phyllosilicates, which replace impact glass in different impact lithologies, indicates the presence of an approximately vertical zonation, similar to that at Kara. In coptoclastites and suevites of the upper unit, dioctahedral smectites (beidellite or montmorillonite) dominate, whereas in crystal-vitric and vitric suevites of the middle unit, Ca-saponites ($14.7\ \text{Å} < d_{001} < 15.3\ \text{Å}$), mixed-layered chlorite-saponites, and Fe-chlorites are developed (Table 7). These data indicate an increase of the Fe and Mg contents and a decrease of the Si, Al, Na, and K contents in clay minerals downward in the vertical section from the upper unit to the middle unit (Fig. 6). However, in tagamites, which form thick sheets in the southwestern sector of the crater (Balagan-Yuregue area), both trioctahedral and dioctahedral (both montmorillonite and beidellite) smectites occur often together. This indicates that local compositional variations of solutions, which reflect a local inhomogeneity of the target, are more significant than temperature gradients within a thick impact rock sequence.

The diversity of conditions of hydrothermal mineral-forming processes within Popigai crater is manifested also in the wide range of zeolite compositions. Two zeolitic associations are distinguished: (1) mordenite associated with cristobalite, and, in rare cases, with minor chabazite, occurs in cavities of shallow tagamite and suevite bodies, as well as in cataclasites within the polymict megabreccia; (2) stilbite, Na-chabazite, and, more rarely, heulandite, together with calcite, form zonal veins and geodes within the deep-seated breccia of crystalline rocks and impact melt sheets (predominantly in their near-bottom parts enriched in rock clasts), and in vitric-lithic suevites of the upper unit. A zonation of zeolite formation is maintained in veins, from margins to center, as following: stilbite, chabazite, and calcite.

Representative EMP analyses of the zeolitic minerals are presented in Fig. 7 and Table 8. All zeolitic species vary to a great extent in their Si/Al and Ca/Na ratios. Although the chabazite analyzed is unusually high in silica, its identification is confirmed by XRD powder patterns (see Table 6).

Thus, the hydrothermal mineralization in the Popigai structure is different from the Kara crater in being less intense and in having a greater variety in the composition of minerals due to the variety of impact lithologies. On the other hand, some common features for the hydrothermal mineralization in the Popigai crater can be noted. In particular, the wide-spread occurrence of minerals such as cristobalite and mordenite, which replace only amorphous materials, is typical; this is probably due to the much more extensive shock transformation of the target rocks in this crater.

3.3 Puchezh-Katunki Impact Structure

The Puchezh-Katunki crater is located in the center of the Russian plain. The structure is buried beneath Mesozoic-Cenozoic deposits; however, a large amount of core drilling has been carried out there (Masaitis and Pevzner 1999). This makes Puchezh-Katunki one of the best objects to study post-impact hydrothermal alteration.

The bulk of the impact rocks at the Puchezh-Katunki crater is represented by lithic breccias. Impact melt rocks (predominantly suevites) occur mainly on the slopes of the central uplift (Fig. 8) and contribute only 0.2 % to the total volume of impact rocks. The wide-spread hydrothermal mineralization is confined to the area of the central uplift; outside, pyrite, gypsum, calcite, and zeolites are disseminated locally in the matrix of the suevites and polymict breccia. The central uplift with a diameter of 8 - 10 km is composed mainly of blocks of shocked and brecciated Archean gneisses and amphibolites (authigenic breccia) and is partly overlain (on the slopes of the uplift and within the central pit up to 525 m in depth) by thin layers of polymict allogenic breccia, suevites, redeposited suevites, and clayey sediments of the crater lake.

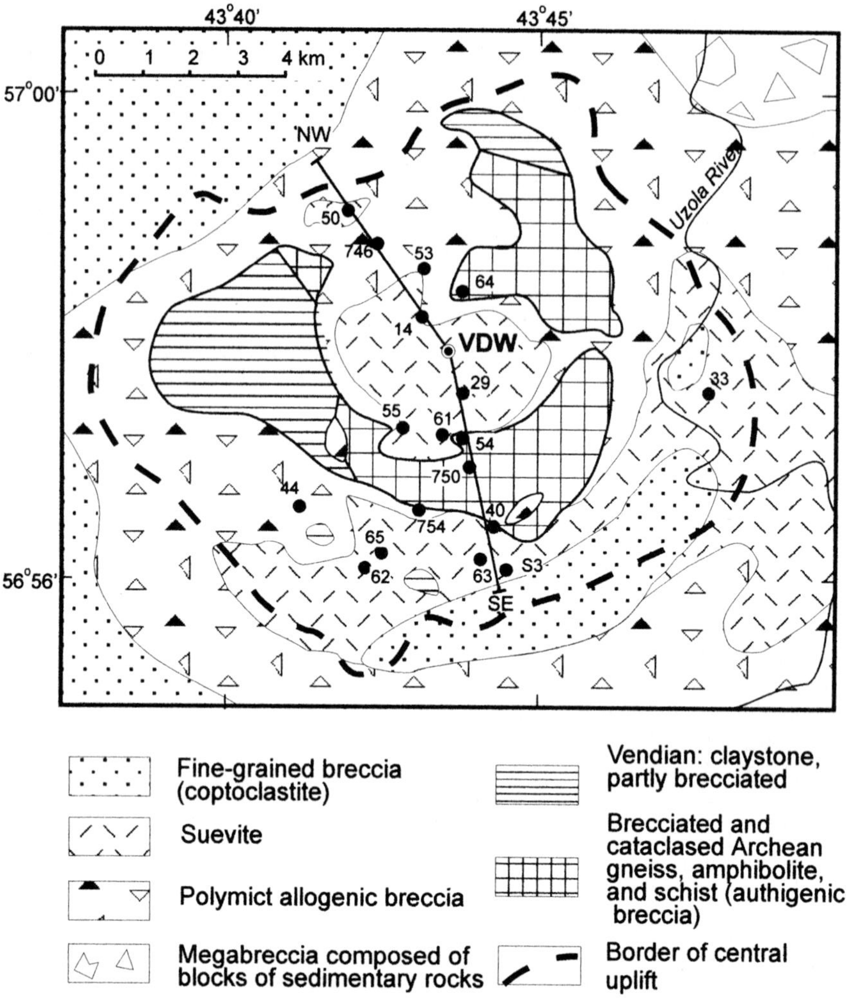

Fig. 8. General geology of the central uplift of the Puchezh-Katunki impact structure showing locations of drill cores (indicated by numbered filled circles) referred to Tables 9 and 10, and on Figs. 9, 11, and 12. The location of the cross-section (Fig. 11) is shown by NW-SE solid line. VDW is the Vorotilovo Deep Well.

Due to drilling, secondary mineralization can be traced in the Vorotilovo Deep Well (VDW) to 5.3 km depth, as well as laterally in the area of the central uplift. The degree of hydrothermal alteration of rocks depends on both their permeability and degree of preceding thermal recrystallization; correspondingly, it decreases downward through the section as well as outwards from the crater center.

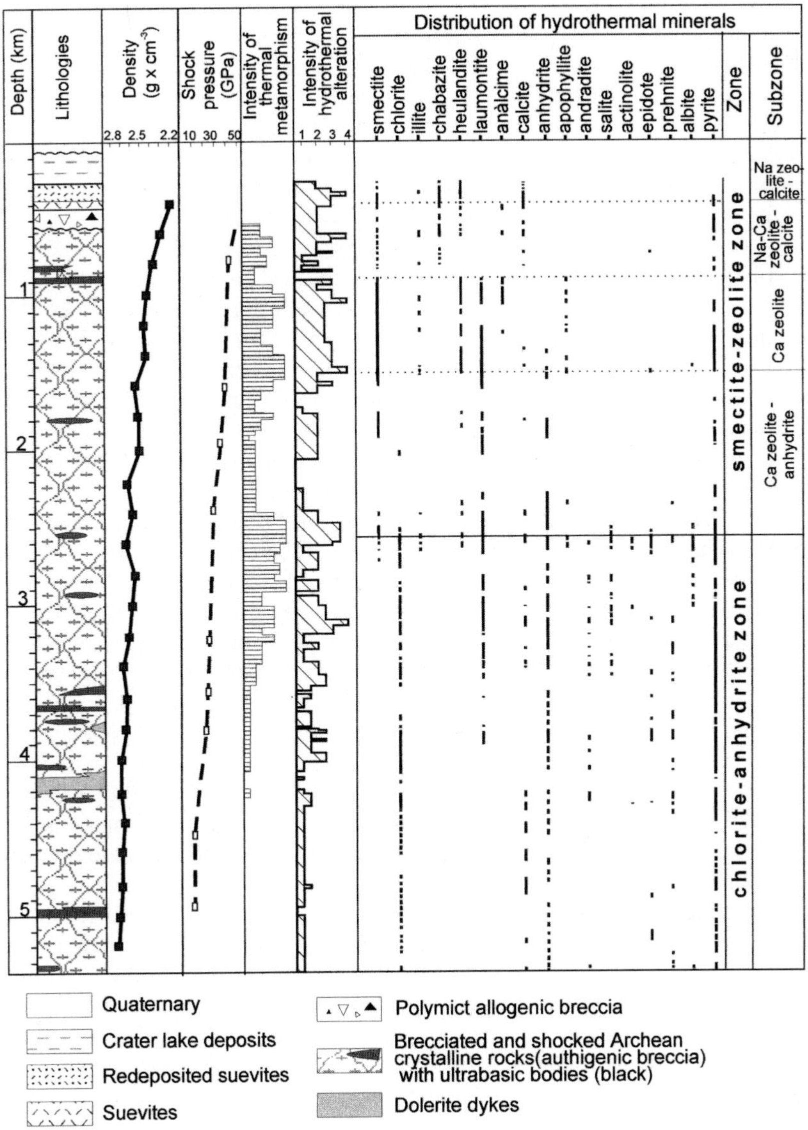

Fig. 9. Distribution of main hydrothermal minerals in the Vorotilovo Deep Well section. Data on shock pressure and thermal metamorphism are after Masaitis and Pevzner (1999). The intensity of thermal metamorphism is shown in arbitrary units by rectangular areas for 50-m-depth intervals. The intensity of hydrothermal alteration is in arbitrary units, reflecting the content of secondary minerals calculated for 15-m depth intervals from thin section study: *1*, <0.5 %; *2*, 0.5 - 2 %; *3*, 2 - 4 %; *4*, >4 %.

The list of hydrothermal minerals in the Puchezh-Katunki structure includes smectites, chlorite, illite, calcite, pyrite, various zeolites, anhydrite, gypsum, apophyllite, quartz, opal, prehnite, epidote, andradite, ferrosalite, actinolite, and albite. Their representative chemical analyses are given elsewhere (Naumov 1993, 1999). The spatial distribution of the hydrothermal mineralization is characterized by a vertical zonation. It is best studied in the VDW core, which was drilled in the crater center (Masaitis and Pevzner 1999). Two alteration zones, named after the main minerals that occur in these zones, can be distinguished (Fig. 9): the upper smectite-zeolite zone and the lower chlorite-anhydrite zone.

The *smectite-zeolite zone* contains thin layers of redeposited suevites, suevites, and polymict allogenic breccia, plus the shocked and brecciated target rocks, to 2550 - 2600 m depth. The compositions of the main minerals of the zone - both smectites (Table 9) and, especially, zeolites (Table 10) - vary with depth. For smectites, this variation appears in substitution of dioctahedral Al-montmorillonite by trioctahedral Mg-Fe varieties (saponites) downward in the section, the content of tetrahedral aluminia increasing with depth in these saponites (Table 9). At depths of 2500 - 2600 m, the saponites are gradually replaced by chlorites (Fig. 10).

Among zeolites, heulandite, chabazite, analcime, stilbite, laumontite, clinoptilolite, erionite, philippsite, harmotome were identified by EMP, XRD, and DTA analyses (Naumov 1993). Both a vertical zonation and some lateral inhomogeneity in the distribution of zeolites are noted (Fig. 11). On the whole, the Ca/Na and Al/Si ratios in zeolites increase downwards in the section (Fig. 12), so that some zeolite subzones can be distinguished, from top to bottom (Fig. 9):

(a) the *Na zeolites-calcite subzone* contains redeposited suevites, and, locally, underlying suevite layer. Zeolites are represented by Na-heulandite, Na-chabazite, and, in some locations, erionite and clinoptilolite. They, together with calcite, form the cement of the redeposited suevites and suevites within the central pit and the disseminated mineralization in these rocks within the annular depression. Among the clay minerals, which replace impact glass, montmorillonite dominates over saponite (Fig. 10).

(b) the *Na-Ca zeolites – calcite subzone* contains suevites, polymict allogenic breccia, and shocked and brecciated basement rocks to 550 - 850 m depth. In this subzone, zeolite-bearing assemblages vary laterally, indicating significant chemical variability of the mineral-forming solutions. Common are heulandite-calcite, analcime-calcite, chabazite-analcime-(heulandite)-calcite, stilbite-analcime, apophyllite-analcime-heulandite-gypsum, apophyllite-erionite-heulandite associations (Fig. 11). On the whole, the lateral distribution of zeolites consists of both an alternation of areas with high-silica zeolites (heulandite, erionite) and low-silica zeolites (chabazite, analcime) (Fig. 11), and an increase of the Si and Na contents in chabazite (Table 10), a zeolite occupying only a thin layer (no more 100 m thick) at the top of the authigenic breccia, outward from the crater center (Naumov 1993). Clay minerals are mainly represented by saponite within this subzone.

(c) the *Ca zeolites subzone* is penetrated only by the VDW core at a depth interval of 850 - 1800 m. Laumontite and Ca-heulandite, with minor apophyllite, chabazite, scolecite, and quartz occur there.

Table 9. Electron microprobe analyses of hydrothermal clay minerals from the Puchezh-Katunki impact structure

Mineral	Chlorite		Mg-Fe saponite					Montmorillonite	
Core	VDW	VDW	VDW	VDW	VDW	64	55	44	65
Depth (m)	3240	2578	2596	2470	1404	270	244	275	170
wt%									
SiO_2	27.37	29.17	35.71	38.80	40.34	41.09	39.57	61.04	40.46
TiO_2	1.03	-	-	-	-	0.17	-	0.31	0.88
Al_2O_3	17.13	17.19	11.31	11.46	7.58	5.28	6.31	14.56	17.47
FeO*	28.16	24.83	18.40	19.20	23.50	17.35	22.74	8.18	18.72
MnO	0.21	0.24	-	-	-	0.25	0.29	0.28	0.08
MgO	11.95	17.74	15.49	14.14	12.94	14.58	11.98	6.69	4.94
CaO	0.47	0.29	1.08	1.61	3.40	1.49	1.96	2.35	1.75
Na_2O	-	0.39	0.32	0.51	-	1.50	0.53	0.82	1.04
K_2O	0.19	-	1.67	1.23	-	-	0.06	0.21	0.25
Total	86.50	90.11	83.98	86.95	87.76	81.71	83.44	94.44	85.59
	Formula based on 28 O atoms		Formula based on 22 O atoms						
Si	5.95	5.96	5.93	6.19	6.60	6.89	6.70	7.96	6.10
Ti	0.17	-	-	-	-	0.02	-	0.03	0.10
Al^{IV}	2.05	2.04	2.07	1.81	1.40	1.04	1.26	0.04	1.90
Al^{VI}	2.33	2.10	0.14	0.34	0.06	-	-	2.20	1.20
Fe^{3+}	-	-	1.09	0.52	0.16	0.10	0.52	0.56	2.00
Fe^{2+}	5.12	4.24	0.94	1.79	2.63	2.38	2.45	-	0.36
Mn	0.04	0.04	-	-	-	0.03	0.05	0.03	0.01
Mg	3.87	5.43	3.83	3.35	3.15	3.64	3.02	1.30	1.11
Ca	0.11	0.06	0.19	0.27	0.60	0.27	0.36	0.33	0.28
Na	-	0.15	0.11	0.16	-	0.49	0.17	0.21	0.30
K	0.13	-	0.35	0.25	-	-	0.01	0.05	0.04
n	2	2	2	2	2	3	3	6	3

Analysed by O.A. Yakovleva, VSEGEI; operating conditions are described in section 2
n is the number of duplicate analyses

Samples of altered rocks: *VDW/3240,* Brecciated biotite-garnet-amphibole gneiss, *VDW/2578* and *VDW/2596,* intensely recrystallized and brecciated biotite-amphibole gneisses, *VDW/2596* and *VDW/2470*, pyroxene-amphibole schist; *VDW-1404*, shocked and recrystallized biotite-amphibole gneiss, *55/244* and *64/270*, intensely recrystallized, shocked, and brecciated amphibole gneiss, *44/275*, vesicular tagamite; *65/170*, impact glass from suevite

* Total iron as FeO

Table 10. Representative chemical analyses of zeolites and apophyllite from different alteration subzones of the Puchezh-Katunki impact structure

Subzone	Na zeolite – calcite			Na-Ca zeolites - calcite					
Mineral [a]	Heulandite [b]	Chabazite [b]	Clinoptilolite [b]	Heulandite [c]	Heulandite [b]	Chabazite [a]			Erionite [c]
Core [d]	VDW	VDW	50	54	S3	VDW	61	63	S3
Depth (m)	334	337	325	297	424	551	295	315	425
wt%									
SiO_2	60.06	53.87	62.21	59.07	60.36	50.93	48.42	51.36	54.37
TiO_2	0.03	-	-	-	0.06	-	-	-	0.02
Al_2O_3	14.86	14.23	13.05	15.39	15.87	18.52	18.68	15.95	16.53
FeO	-	0.10	-	0.12	0.88	-	-	0.11	0.81
MgO	-	0.05	-	0.26	0.14	-	-	0.19	-
CaO	3.43	4.32	2.57	4.90	5.05	7.30	6.26	4.69	3.96
Na_2O	2.42	8.55	3.43	3.11	1.71	2.83	3.64	6.35	2.78
K_2O	0.29	0.75	0.99	0.55	0.13	0.20	0.37	0.73	2.42
SrO	1.54	n.d.	n.d.	1.65	n.d.	0.32	0.33	0.15	0.13
BaO	-	n.d.	n.d.	0.19	n.d.	0.44	-	-	0.44
H_2O^+	n.d.	n.d.	n.d.	15.66	n.d.	19.83	21.93	20.73	18.04
Total	82.60	81.87	82.25	100.90	84.20	100.37	99.63	100.26	99.93
	Formula on the basis of								
	72(O)	24(O)	72(O)			24(O)			72(O)
Si	28.32	8.83	29.05	27.36	27.69	8.40	8.28	8.62	26.47
Al	8.24	2.75	7.18	8.40	8.58	3.60	3.76	3.16	9.40
Ca	1.73	0.76	1.28	2.43	2.50	1.31	1.15	0.84	2.06
Na	2.21	2.72	3.10	2.79	1.57	0.91	1.21	2.07	3.57
K	0.18	0.16	0.59	0.33	0.10	0.03	0.08	0.16	1.50
H_2O				24.18		10.84	12.51	11.60	29.27

[a] Structural parameters and optical constants of the zeolites and apophyllite were given earlier (Naumov 1993)
[b] electron microprobe data, analysed by L.S. Bogomolnaya, VSEGEI; analytical procedures are described in section 2
[c] X-ray fluorescence spectroscopy data, analysed by A.B. Safro and I.G. Ginzburg, VSEGEI; Ba and Sr were determined by atomic absorption spectrometry, analysed by V.N. Tarasova, VSEGEI
[d] Location of boreholes is shown in Figs. 8 and 11

Short description of the samples: *VDW/334* and *WDW/337,* zeolites forming the cement matrix of a redeposited suevites; *50/325*, zeolite from the cement of polymict allogenic breccia; *62/181*, zeolite-calcite geodes in a suevite; *54/294, 54/297, 61/295, 63/315*, and *33/421*, zeolite-calcite veins in a brecciated and shocked Archean gneisses; *S3/424, S3/425, S3/559*, zeolite-apophyllite veinlets in a brecciated and shocked Archean gneisses; *VDW/1238*, apophyllite vein in a brecciated gneiss, *VDW/1408, VDW/1467, VDW/2596,* zeolite veins and geodes in shocked and thermally-recrystallized gneisses; *VDW/2410*, laumontite amygdules in a cataclasite.

Table 10. (cont.)

Na-Ca zeolites - calcite				Ca zeolites			Ca zeolites - anhydrite	
Apophyllite ^{c}	Analcime ^{c}	Stilbite ^{c}	Philippsite ^{c}	Laumontite ^{c}	Heulandite ^{c}	Apophyllite ^{b}	Laumontite ^{b}	Heulandite ^{b}
S3	33	54	62	VDW	VDW	VDW	VDW	VDW
559	421	294	181	1467	1408	1238	2410	2596
wt%								
53.01	57.13	56.36	57.99	52.26	55.15	53.18	56.73	58.10
0.02	0.02	-	-	-	-	-	-	-
0.73	23.08	15.96	14.11	21.69	17.25	-	23.86	16.74
0.23	0.25	-	-	-	-	-	-	-
-	-	-	-	-	-	-	0.23	-
24.98	-	8.25	1.96	11.82	6.94	25.31	11.99	8.57
0.28	11.86	0.61	3.05	0.52	1.66	-	-	0.59
4.83	0.01	0.02	1.96	0.16	0.21	4.74	-	-
n.d.	0.01	n.d.	0.44	0.09	1.97	n.d.	n.d.	n.d.
n.d.	-	n.d.	0.50	0.70	0.70	n.d.	n.d.	n.d.
13.99	6.88	17.36	15.77	12.78	16.26	n.d.	n.d.	n.d.
98.07	99.22	98.56	96.60	100.02	100.14	83.23	92.58	84.00
Formula on the basis of								
21(O,F,OH^{-})	96(O)	72(O)	32(O)	48(O)	72(O)	21(O,F,OH^{-})	48(O)	72(O)
7.88	32.87	26.99	12.54	16.09	26.28	7.99	16.16	26.90
0.13	15.65	9.02	3.59	7.89	9.68	-	8.01	9.13
3.99	0.00	4.23	0.45	3.85	3.54	4.07	3.66	4.25
0.07	13.24	0.57	1.28	0.28	1.53	-	0.00	0.53
0.92	0.01	0.01	0.54	0.08	0.13	0.91	0.00	0.00
6.93	13.60	27.73	11.38	13.02	25.83			

(d) the *Ca zeolites – anhydrite subzone* (the depth interval 1800 - 2550 m) is distinguished by the appearance of anhydrite, together with laumontite and pyrite, as a common secondary mineral.

The *chlorite-anhydrite zone* contains brecciated and shocked basement rocks at depths of more than 2550 m. Common are chlorite, anhydrite, and pyrite, and locally (mainly in the depth interval 2900 - 3500 m), relatively high-temperature Ca-Fe-silicates (andradite, ferrosalite, epidote, prehnite, actinolite), and, in some cases, quartz also occurs. Zeolites are observed down to 3854 m and are represented by Ca-varieties (laumontite with minor heulandite) only in this zone.

In contrast to Kara, a strong variability in the composition and crystal habits of pyrite, which is a ubiquitous mineral in all impact lithologies, is observed. The Ni/Co ratio ranges from 0.27 to 3.7 (Table 4); it varies considerably from the Ni/Co ratio typical for pre-impact pyrite from Archean target rocks. Within the chlorite-anhydrite zone, tablets or prismatic crystals of pyrite are common, whereas in the smectite-zeolite zone, cuboctahedral, pentagon-dodecahedral, cubic and pyritohedral habits occur. Nevertheless, the $\delta^{34}S$ values of pyrite from the lower alteration zone are similar to that of pre-impact pyrite from crystalline target rocks (2.4 - 2.7 ‰ CDT and 2.8 ‰ CDT, respectively) (Naumov 1999).

In the periphery of the central uplift any evidence of a vertical zonation in the distribution of the minerals is less apparent. Some combinations of minerals from different zones (for example - zeolites, smectites, and epidote, actinolite) occur. Besides, the usual vertical sequence of clay minerals is locally inverted (chlorites are substituted by smectites downwards in the section), whereas the spatial distribution of zeolites is similar to one in the center of the crater (Fig. 13).

The generalized order of mineral crystallization in both smectite-zeolite zone (saponite/montmorillonite, laumontite/chabazite-calcite, heulandite, analcime-apophyllite, calcite, Fe-montmorillonite, calcite) and chlorite-anhydrite zone (andradite-salite-epidote-chlorite-calcite, anhydrite, laumontite) confirms the uniform decrease of the mineral formation temperature during the hydrothermal activity phase.

4 Discussion

The kinetic energy of an impact event and the paleogeographic characteristics of the target are the most important parameters for the development of a post-impact hydrothermal system. Thus, the most extensive hydrothermal alteration takes

Fig. 10. (opposite) X-ray powder diffraction patterns using a DRON 3M diffractometer by scanning at $1^{\circ}min^{-1}$ over the range $1° < \Theta < 32°$, Fe-filtered Co$K\alpha$ radiation, for <0.002 mm fractions of hydrothermally altered rocks from Vorotilovo Deep Well (VDW). A sample number indicates a core depth. Analysed by O.Yu. Dubic, VSEGEI. Identifications of clay minerals were made by references to standard files (Joint Committee on Powder Diffraction Standards (JCPDS 1974). Short description of the samples (all these contain secondary clay minerals, which both replace the matrix and fill cavities and fractures, from thin section study): *VDW-330*, altered impact glass from a redeposited suevite; *VDW-402*, altered impact glass from a suevite; *VDW-609*, intensely brecciated and shocked amphibolite; *VDW-789, VDW-1479, VDW-2568, VDW-2578*, intensely recrystallized shocked biotite-amphibole gneisses (so-called coptoblastolithes, after Masaitis and Mashchak 1996); *VDW-962*, tagamite; *VDW-1315*, brecciated and recrystallized amphibolite; *VDW-1797*, brecciated garnet-biotite-amphibole gneiss; *VDW-2485*, quartzitic gneiss; *VDW-3093*, brecciated amphibolite. Peak labeling indicates diffraction patterns of smectite (sm), chlorite (chl), and illite (il). The admixture of finely dispersed quartz and feldspar occurs, as indicated by some peaks in the Θ interval from 12° to 18°. Inferred from the d_{06} value, a clay mineral from the VDW-330 sample is a dioctahedral smectite (Ca-montmorillonite with b_o = 8.98 Å). In the VDW-402 sample, both dioctahedral (with b_o = 8.92 Å) and trioctahedral (saponite with b_o = 9.18 Å) smectites occur. In other samples, saponite with 9.20 Å < b_o < 9.22 Å is only smectite mineral. Below 2485 m depth, chlorite patterns appear on diffractograms. The occurrence of chlorite is confirmed by the displacement of the (001) pattern from 14.4 - 14.7 Å to 13.8 Å by heating to 600°C, and by low intensity of a peak with d = 17.8 Å when the sample was analyzed after solvation with ethylene glycol.

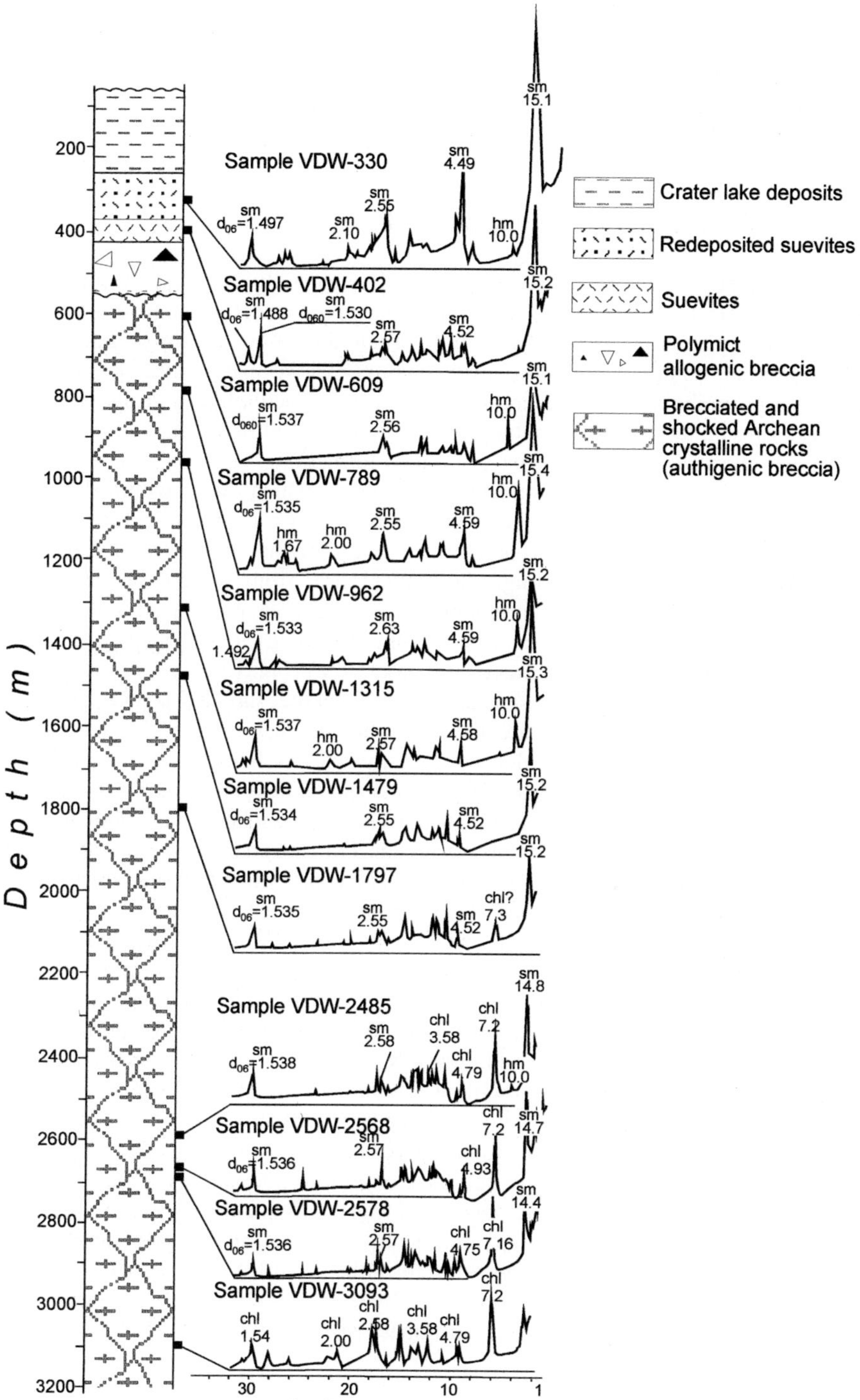
Depth (m)
200
400
600
800
1000
1200
1400
1600
1800
2000
2200
2400
2600
2800
3000
3200
Sample VDW-330
Sample VDW-402
Sample VDW-609
Sample VDW-789
Sample VDW-962
Sample VDW-1315
Sample VDW-1479
Sample VDW-1797
Sample VDW-2485
Sample VDW-2568
Sample VDW-2578
Sample VDW-3093
30
20
10
1
°Θ CoKα
Crater lake deposits
Redeposited suevites
Suevites
Polymict allogenic breccia
Brecciated and shocked Archean crystalline rocks (authigenic breccia)

place in craters that formed in a shelf environment or in intra-continent shallow basins (Kara, Puchezh-Katunki, etc.). If a crater formed in continental conditions, the intensity of alteration is much less, although the volume of the circulation cell can be estimated to exceed 1000 km^3 (the total volume of impact rocks where the hydrothermal mineralization occurs) as at Popigai.

4.1. Main Features of Hydrothermal Minerals

The dominant hydrothermal assemblage in all impact structures studied is: clay minerals (smectites, chlorites, and mixed-layered smectites-chorites)–various zeolites–calcite–pyrite, indicating hydrothermal alteration at low temperatures (less than 300 - 350°C); in addition, cristobalite, quartz, opal, anhydrite, gypsum, apophyllite, prehnite, epidote, andradite, ferrosalite, actinolite, and albite occur locally. The minerals with variable composition (clay minerals and zeolites) are the most informative for conditions of mineral formation.

The compositions of post-impact clay minerals, which occur within different impact lithologies from some craters, are given in Tables 11 and 12. Three impact rock units, each of them being characterized by a specific number and distribution of clay minerals, can be distinguished; these are, from bottom upward in a generalized rock sequence: (1) authigenic breccia; (2) coherent impact melt sheets, together with associated vitric and crystal-vitric suevites; (3) a blanket of lithic suevites and fine-grained lithic breccia (coptoclastite) (Fig. 14).

Fe-Mg chlorites and saponites with high Al^{IV} content (about 2.0 and 1.5 per 8 [Al + Si] atoms, respectively) are the predominant clays within the authigenic breccia. The Fe/Mg ratio in chlorites and smectites is like that of target rocks (Fig. 14a). Within coherent impact melt sheets (tagamites) and coarse fragmented vitric and crystal-vitric suevites, which are closely associated with tagamites, a diversity of the clay minerals is observed; the Popigai crater is an example (see Table 7). Both trioctahedral and dioctahedral smectites with a very variable Fe/Mg ratio (ranging from 0.2 to 2.5), and also chlorites (with Al^{IV} content 0.4 - 1.3 per 8 [Al + Si] atoms) occur, but saponite, with an Al^{IV} content ranging usually from 0.7 to 1.3, and mixed-layer smectite-chlorites are common. Clay compositions do not correlate with target compositions of craters (Fig. 14b). From a ternary Al-Fe-Mg diagram of clay mineral compositions, two different trends of hydrothermal clay formation are distinguished: (1) addition of Mg with invariable Fe and (2) addition of Fe with invariable Mg. The first trend is characteristic mainly of thick tagamite sheets and the second one of both tagamite and suevite bodies. Clay minerals, which replace impact glass or occur in vesicles and cavities of suevites and coptoclastites, are mainly represented by montmorillonite and, in some cases, by Al-Fe beidellite, but saponite occurs locally, frequently together with a dioctahedral smectite; the Kara crater is an example (analyses 20 and 21 in Table 12). No alteration trends appear in this unit (Fig. 14c).

Thus, Fe-Mg clay minerals are replaced by high-alumina varieties upward in a generalized vertical section through an impact rock sequence. The increase of the

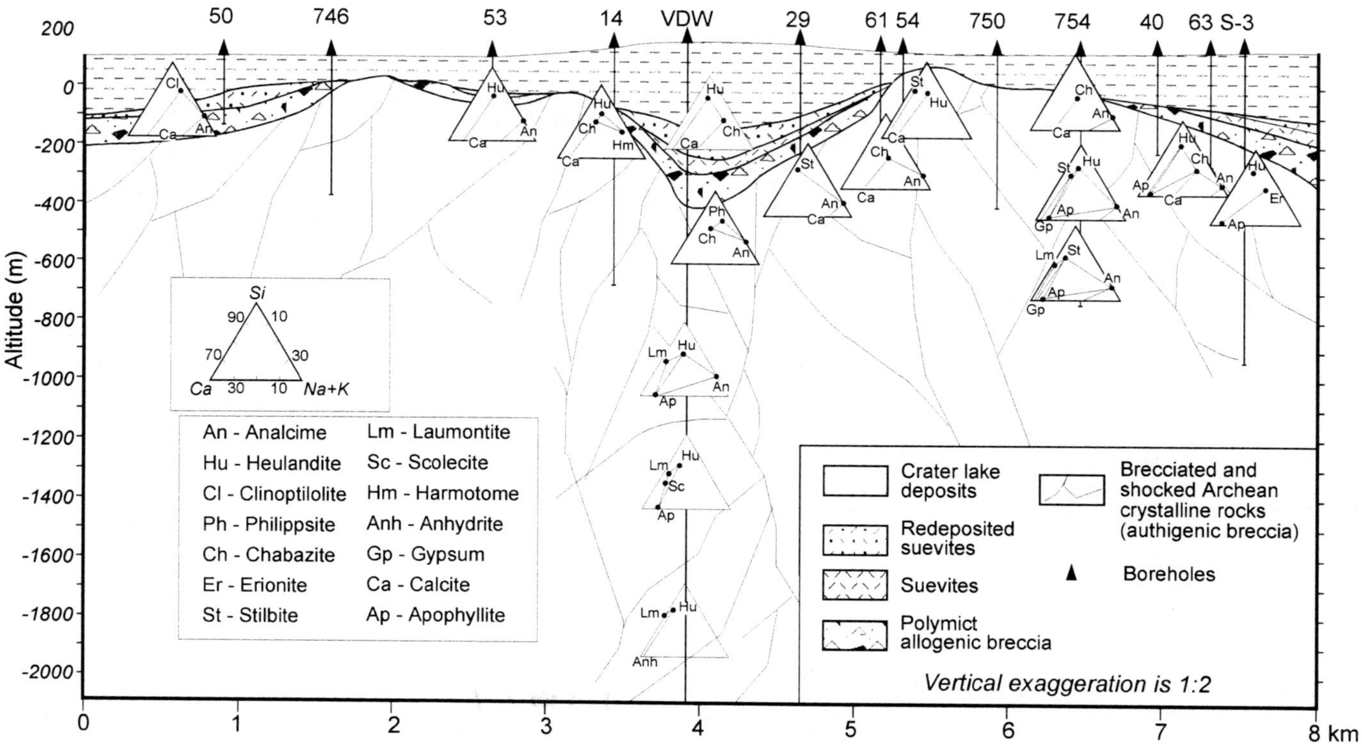

Fig. 11. Distribution of zeolite-bearing mineral associations shown in Ca-(Na+K)-Si ternary diagrams, in a cross-section through the central uplift of the Puchezh-Katunki impact structure. Vertical scale is exaggerated 2 times. Compositions of minerals are calculated from electron microprobe and X-ray fluorescence spectroscopy data. Both a vertical and lateral distribution of zeolitic mineral associations is indicated (see text). VDW is the Vorotilovo Deep Well.

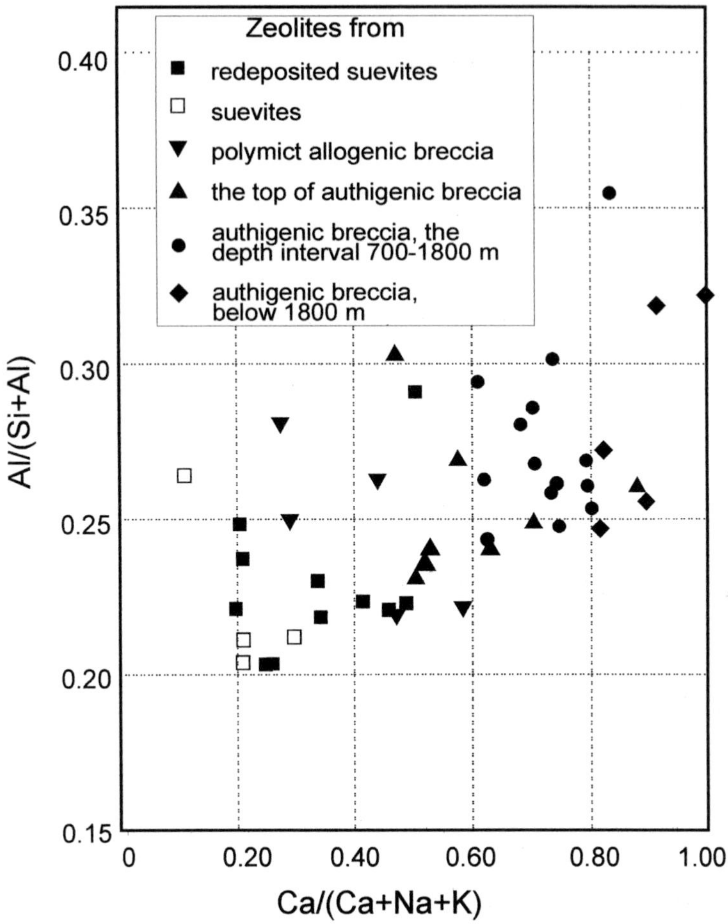

Fig. 12. Plot of Ca / (Ca+Na+K) vs. Al / (Si+Al) ratios of hydrothermal Ca-Na zeolites from different parts of a generalized vertical section through the central uplift of the Puchezh-Katunki impact crater (results of 48 X-ray fluorescence spectroscopy and electron microprobe analyses are used). Representative chemical data of zeolites are given in Table 10. The data show a general increase of Ca and, to a lesser content, Al contents in zeolite compositions downward in the vertical section.

Si content is the main trend of the compositional variation of both chlorites and smectites in this direction.

Apart from the spatial distribution of clay minerals within craters, all EMP analyses of post-impact chlorite fall into a restricted compositional range, with formula proportions of $Si > 5.9$ (cations per 28 atoms) and values $0.48 < Fe/(Fe+Mg) < 0.58$, which are equivalent to diabantite. The enrichment in silica and higher Fe content is characteristic of rarer Fe-Mg minerals of variable composition. In particular, all known occurrences of post-impact garnet in Manson

(McCarville and Crossey 1996), Puchezh-Katunki (Naumov 1999), and Boltysh are represented by andradite. Post-impact epidote is more ferrous than pre-impact epidote from target rocks, as indicated by EMP analyses at Siljan (Aldahan 1990) and Puchezh-Katunki (Naumov 1999).

Despite of the strong variability of clay minerals, the target composition is a factor, which has a significant influence on the original compositions. This is best expressed by occurrences of clay minerals with unusual compositions reflecting the compositional features of the targets [e.g., high-aluminia trioctahedral

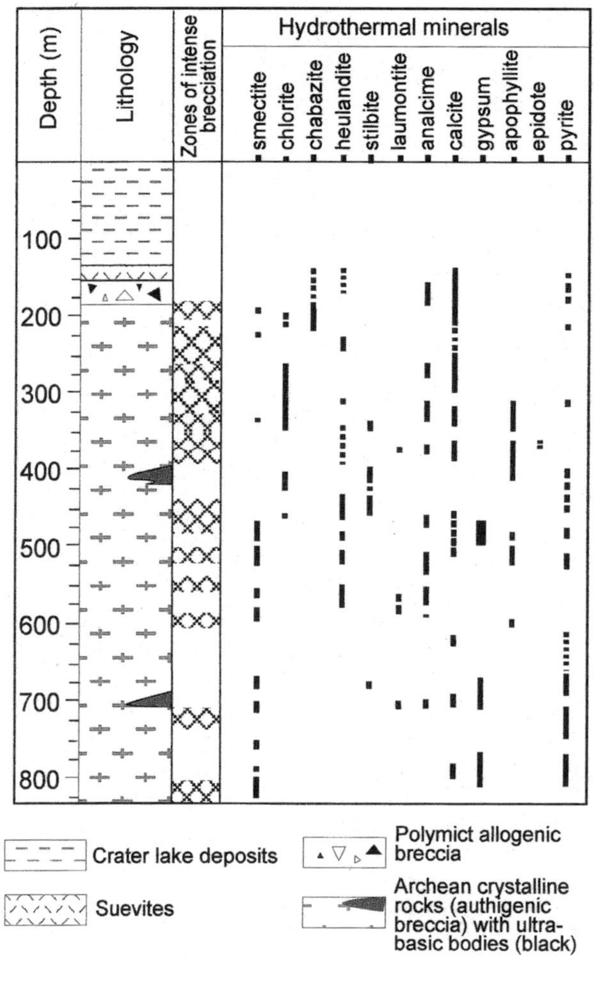

Fig. 13. Distribution of hydrothermal minerals in a vertical section of the southern part of the central uplift of the Puchezh-Katunki impact structure (drill cores 40, 63, 754). Dashed lines indicate the sporadic occurrence of a mineral. Compared with the Vorotilovo Deep Well (VDW), an inverse distribution of clay minerals is found. The distribution of zeolites and associated minerals is similar to that in the VDW section.

Table 11. Chemical analyses of hydrothermal chlorites from some impact structures

Impact structure	Siljan [a]	Puchezh-Katunki (5) [b]	Charlevoix [c]	Jänisjärvi (2) [b]	Brent [b] (2)	Sääksjärvi [c]	Rochechouart [c]	Popigai (4) [b]	Ries [c]
wt%									
SiO_2	28.20	28.92	27.20	33.91	33.09	34.60	28.00	32.33	33.00
TiO_2	0.10	0.32	-	0.05	-	-	0.06	-	1.60
Al_2O_3	16.32	17.92	16.90	12.95	12.88	11.40	14.20	15.00	9.00
FeO	28.11	25.59	30.40	29.38	27.16	23.80	23.40	28.98	19.90
MnO	1.32	0.20	-	-	0.55	-	-	0.29	-
MgO	13.44	15.92	12.10	14.84	13.11	14.20	11.90	11.98	9.20
CaO	0.05	0.42	0.09	0.30	1.36	1.00	0.26	1.13	0.27
Na_2O	-	0.24	0.07	0.22	0.90	-	-	-	2.40
K_2O	-	0.06	0.07	-	0.16	0.51	0.29	0.08	0.14
Total	87.54	89.59	86.83	91.65	89.21	85.51	78.11	89.79	75.51
	Formula based on 28 O atoms								
Si	6.12	5.95	5.99	7.02	7.01	7.62	6.77	6.90	8.00
Ti	0.02	0.05	-	0.01	-	-	0.01	-	0.29
Al^{IV}	1.88	2.05	2.01	0.98	0.99	0.38	1.23	1.10	-
Al^{VI}	2.29	2.30	2.37	2.18	2.23	2.58	2.82	2.68	2.57
Fe	5.09	4.41	5.59	5.09	4.81	4.38	4.73	5.18	4.03
Mn	0.24	0.03	-	-	0.10	-	-	0.05	-
Mg	4.34	4.88	3.97	4.57	4.14	4.66	4.29	3.81	3.32
Ca	0.01	0.09	0.02	0.07	0.31	0.24	0.07	0.26	0.07
Na	-	0.10	0.03	0.09	0.37	-	-	-	1.12
K	-	0.02	0.02	-	0.04	0.14	0.09	0.02	0.04

Electron microprobe chemical data from:
[a] Aldahan (1990)
[b] Author's data (analysed by O.A. Yakovleva, VSEGEI; analytical conditions are described in section 2; the number of duplicate analyses is given in parentheses)
[c] Allen et al. (1982)

Location of the samples: *Puchezh-Katunki*, chlorites from VDW core, the depth interval 2578-3620 m; *Jänisjärvi*, from impact melt rocks, Iso-Selkäsaari Island; *Brent*, borehole 1-59, sample B1-59-2787,5 (sample provided by R.A.F. Grieve); *Popigai*, from borehole 1124 (Majachika Upland), the depth 187 m.

Host rock for the samples analyzed: *Siljan* and *Puchezh-Katunki*, brecciated crystalline basement rocks (authigenic breccia); *Charlevoix*, pseudotachylite; *Jänisjärvi, Brent*, and *Sääksjärvi*, coherent impact melt sheets; *Rochechouart* and *Popigai,* suevites associated with coherent impact melt sheets; *Ries*, suevite.

smectites in Jänisjärvi and Wanapitei, or high-Fe clays (so-called "hisingerite") in Terny (Nikolsky 1991)].

Post-impact zeolites also exhibit a well-defined spatial distribution. Although there is a general dependence of the zeolitic assemblages in impact craters on the

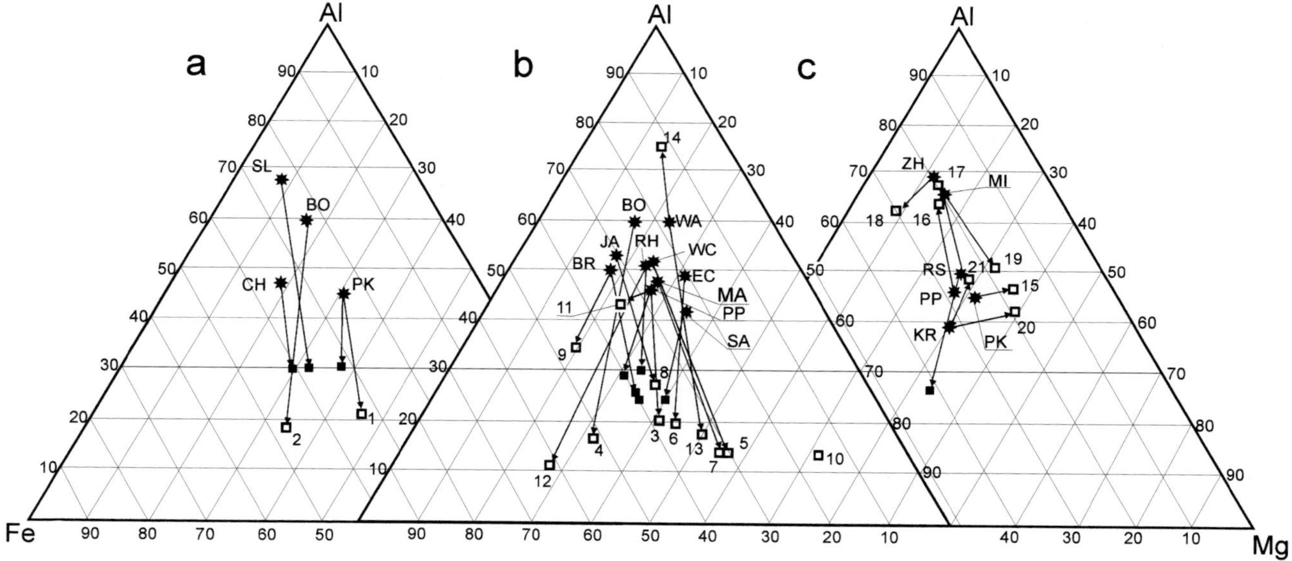

Fig. 14. Composition of post-impact clay minerals from brecciated and shocked crystalline basement rocks (authigenic breccia). (a) From coherent impact melt sheets and associated suevites, (b) from suevite-coptoclastite ejecta blankets, (c) based on averages of electron microprobe analyses (Tables 11 and 12). Filled squares indicate chlorites, and open squares smectites. Asterisks indicate average target compositions and, consequently, impact melt compositions. Numbered open squares correspond to numbers of analyses in Table 12. The points showing clay composition and average target composition from an impact structure are connected by arrow-headed solid lines, indicating a general trend of chemical alteration during clay mineral formation.

Data on average target compositions are from: Masaitis et al. (1980) for Boltysh (BO), Janisjärvi (JA), Kara (KR), and Popigai (PP); Grieve (1978) for Brent (BR); Rondot (1983) for Charlevoix (CH); Allen et al. (1982) for East Clearwater (EC), Ries, and West Clearwater (WC); Floran et al. (1978) for Manicouagan (MA); Masaitis and Pevzner (1999) for Puchezh-Katunki (PK); Lambert (1977) for Rochechouart (RH); Papunen (1973) for Sääksjärvi (SA); Aldahan (1990) for Siljan (SL); Grieve and Ber (1994) for Wanapitei (WA). For Zhamanshin, data from Florensky and Dabizha (1980) were recalculated. For Nicholson Lake, no data on average target composition exist.

Table 12. Chemical analyses (in wt%) of hydrothermal smectites and mixed-layered smectite-chlorites from some impact structures

Host rocks	Authigenic breccia		Coherent impact melt sheet (tagamite)						
Structure	Puchezh-Katunki [a]	Boltysh [a]	Popigai [a]	Boltysh [a]	West Clearwater [b]	East Clearwater [b]	Manicouagan [b]	Jänisjärvi [a]	Brent [a]
No. in Fig. 14	1	2	3	4	5	6	7	8	9
n	12	3	8	16	10	9	6	3	11
SiO_2	39.10	38.41	44.94	44.59	39.10	40.10	47.60	37.25	41.67
TiO_2	-	0.11	0.20	0.25	0.15	0.52	-	-	-
Al_2O_3	8.39	7.40	7.95	5.25	5.00	7.60	6.10	11.79	16.35
FeO*	20.24	25.83	20.02	22.10	13.00	18.40	17.50	25.79	25.21
MnO	0.11	0.07	0.13	0.27	-	-	-	0.33	0.06
MgO	13.83	10.45	11.78	5.52	14.00	12.5	17.3	12.00	6.90
CaO	1.91	1.88	1.74	1.40	0.72	3.4	1.90	1.50	0.78
Na_2O	0.57	1.04	0.32	0.76	1.10	0.99	0.17	-	-
K_2O	0.59	0.19	0.23	0.84	0.16	0.45	0.35	-	0.17
Total	84.74	85.31	87.48	81.01	73.20	84.00	90.90	88.66	89.35
	Formula based on 22 O atoms								
Si	6.44	6.48	7.01	7.32	7.11	6.73	7.04	6.00	6.71
Ti	-	0.01	0.02	0.03	0.02	0.07	-	-	-
Al^{IV}	1.56	1.47	0.99	0.68	0.89	1.27	0.96	2.00	1.29
Al^{VI}	0.07	0.18	0.47	0.13	0.18	0.23	0.10	0.24	1.38
Fe^{3+}	0.52	0.46	-	1.97	-	-	-	1.26	-
Fe^{2+}	2.01	2.95	2.61	0.91	1.98	2.58	2.16	1.58	3.32
Mg	3.39	2.63	2.74	1.38	3.80	3.12	3.81	2.88	1.29
Ca	0.34	0.34	0.30	0.25	0.14	0.61	0.30	0.26	0.15
Na	0.18	0.34	0.10	0.24	0.38	0.32	0.05	-	0.00
K	0.12	0.04	0.05	0.17	0.04	0.10	0.07	-	0.02

[a] Electron microprobe chemical data (analysed by O.A. Yakovleva and L.S. Bogomolnaya (analyses of smectites from Popigai), VSEGEI); analytical conditions are described in section 2; *n* is the number of duplicate analyses.
[b] Data from Allen et al (1982)
[c] Data from Grieve and Ber (1994)

Samples analyzed: *1*, Average composition of smectites from brecciated and shocked crystalline basement rocks (see Table 9); *2*, 10331 borehole drilled in the center of the Boltysh crater, the depth interval 723-731m (the samples provided by A.I. Raikhlin); *3*, average composition of smectites from the thick tagamite sheet, Balagan-Yuregue area (see Table 7); *4*, 11475 borehole (see Fig. 15), the depth interval 628-830 m (the samples provided by A.N. Danilin); *8*, Impact melt rock from Iso-Selkäsaari Island, Jänisjärvi Lake, sample Ja-4; *9*, Brent, borehole 1-59, sample B1-59-2787.5 (a sample provided by R.A.F. Grieve); *10*, impact melt breccia, sample

Table 12 (cont.).

Coarse fragmented suevite and associated thin tagamite bodies					Suevite and fine-grained lithic breccia (coptoclastite)						
Nichol-son Lake [a]	Popi-gai [a]	Roche-chou art [b]	Wana-pitei [a]	Wana-pitei [c]	Puchezh-Katunki [a]	Popi-gai [a]	Ries [b]	Zha-man-shin [a]	Mien [b]	Kara [a]	
10	11	12	13	14	15	16	17	18	19	20	21
2	7	2	2		7	15	5	3	10	3	10
46.35	40.54	42.32	48.76	55.10	59.70	54.26	50.30	59.51	58.70	39.53	46.64
-	1.12	0.06	-	0.30	0.41	1.43	0.39	1.06	0.32	1.52	0.50
5.40	12.82	12.78	6.62	22.50	13.21	19.66	16.20	15.18	14.00	15.5	11.66
9.87	19.25	10.55	15.28	4.90	9.16	9.67	5.27	10.32	7.50	11.14	8.72
-	0.33	0.23	-	-	0.25	0.08	-	0.21	-	0.04	0.02
21.56	12.05	18.68	14.14	3.00	7.88	3.55	2.38	1.41	6.30	10.6	4.89
1.14	1.39	1.08	2.30	0.20	1.74	2.13	2.18	1.65	1.80	1.91	2.96
-	0.06	-	-	0.80	0.49	0.08	0.05	0.52	0.19	0.81	0.85
0.35	1.18	0.64	0.53	1.40	0.40	0.64	0.35	1.95	0.71	0.57	0.20
84.67	88.74	86.34	87.63	88.20	93.24	91.5	77.12	91.81	89.5	81.76	76.44
Formulae based on 22 O atoms											
7.08	6.28	6.39	7.35	7.47	7.89	7.26	7.76	7.98	7.92	6.58	7.56
-	0.13	0.01	-	0.03	0.04	0.14	0.05	0.21	0.03	0.19	0.06
0.92	1.72	1.61	0.65	0.53	0.11	0.74	0.24	0.02	0.08	1.42	0.44
0.05	0.61	0.66	0.53	3.06	1.94	2.36	2.64	2.38	2.15	1.62	1.79
0.44	-	0.38	-	0.50	0.75	0.97	-	0.90	0.76	0.18	0.99
0.60	2.49	0.74	1.93	-	-	-	0.72	0.21	-	1.37	0.09
4.91	2.78	4.2	3.18	0.61	1.55	0.71	0.56	0.28	1.27	2.63	1.18
0.19	0.23	0.18	0.37	0.03	0.25	0.30	0.36	0.24	0.26	0.34	0.51
0.06	0.02	-	-	0.21	0.12	0.02	0.02	0.14	0.05	0.26	0.27
0.10	0.23	0.12	0.24	0.30	0.07	0.11	0.06	0.32	0.12	0.12	0.04

IDN-32-65; *11*, average composition of smectites from the suevite deposits in the northwestern part of the crater (boreholes 0924, 1016, 1124,5, and 2518, Majachika Upland); *13*, vesicular impact melt rock, sample MSWX-135,7 5 (sample provided by R.A.F. Grieve); *15*, average composition of smectites replacing impact glass in suevites, boreholes 62 (175-186 m depth), and 65 (165-170 m depth) (see Fig. 8); *16*, average composition of smectites replacing impact glass fragments in suevites and coptoclastites (see Table 7); *18*, smectite filling vesicles in impact glass from a suevite, borehole 101 (228 m depth); *20* and *21*, smectites (trioctahedral variety with unusually high Al content and montmorillonite, respectively), which replace impact glass in suevites, borehole 25, the depth interval 145-154 m (samples provided by M.S. Mashchak)

*Total iron as FeO

SiO_2 content in the target (Naumov 1993), the spatial distribution of calcic and alkali zeolites within craters is the most important factor (Fig. 15). Within authigenic and allogenic lithic breccia, Ca-zeolites (mainly laumontite, stilbite, and chabazite, in some cases also thomsonite) and analcime are mainly developed, whereas in impact melt rocks high-silica and alkali zeolites (heulandite, analcime, mordenite, philippsite, Na-chabazite, and erionite) are predominant (Table 13; Fig. 15). This distribution is analogous, in general, to the vertical zonation described from Puchezh-Katunki. In addition, the enrichment in silica is characteristic for post-impact zeolites. This is shown both by the predominance of high-silica zeolites in impact structures and by a higher Si content in zeolites compared to their stoichiometric compositions [e.g., the Si/Al ratio in analcime is 2.10 - 2.14 for Puchezh-Katunki (Table 10), 2.15 - 2.28 for Kara (Table 6), and even 2.3 - 2.8 for Ries (Stähle and Otteman 1977)].

Table 13. List of zeolitic minerals known from impact structures

No. on Fig.1	Impact Structure	Impact and post-impact lithologies: Authigenic and allogenic breccia	Impact melt rocks	Crater lake deposits	References
4	Boltysh		Cp [a]	Cp [b]	Valter and Bobonich 1979; Author's data
7	Charlevoix	Lm [a] St Ch			Rondot 1983
15	Gosses Bluff	Hu St Ch			Milton et al. 1972
19	Jänisjärvi [a]		Ch		Author's data
21	Kara	An [a] (Lm [a])	An [a] (Ch [b] Md [b] St [b])		Yushkin et al. 1976; this paper
26	Lappajärvi [b]		Hu (St Ch Md Er)		Lehtinen 1976
27	Lockne	Lm			Sturkell et al. 1998
29	Manicouagan	Tm An			Currie 1972
30	Manson [a]	Lm An	An		McCarville and Crossey 1996
33	Mistastin	An Tm	St		Currie 1971b
39	Popigai [a]	St Ch Hu	Md		This paper
40	Puchezh-Katunki [a]	Hu Ch An Lm Er St	Hu An Ch Ph (Hm Cp)	Ch Hu (Er)	Naumov 1993, 1999
41	Ries [a]	An	An Er Cp Ph (Ch St)	An Cp	Stähle and Otteman 1977
47	Siljan [a]	Lm, St, Sl			Morad et al. 1989; Aldahan 1990
56	Wanapitei [a]		Ch Md		Author's data
58	West Hawk	An	Gm Zm		Short 1970
59	Zapadnaya [a]	An	An Er Hu		Valter et al. 1995

[a] Zeolites were identified by X-ray powder diffraction, differential thermal, and electron microprobe analyses
[b] Zeolites were identified by X-ray powder diffraction measurements

Abbreviations of zeolites: *An*, analcime; *Ch*, chabazite; *Er*, erionite; *Gm*, gmelinite; *Hu*, heulandite; *Hm*, harmotome; *Cp*, clinoptilolite; *Lm*, laumontite; *Md*, mordenite; *Ph*, philippsite; *Sl*, stellerite; *St*, stilbite; *Tm*, thomsonite; *Wa*, wallsite; *Zm*, zismondine; zeolites, which occur in rare cases, are given in parentheses

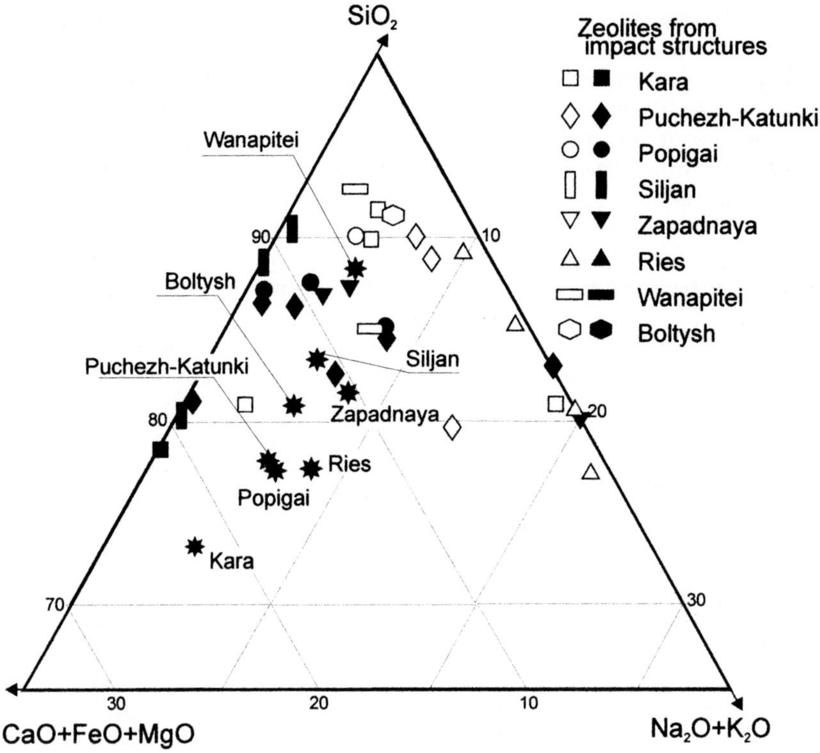

Fig. 15. Compositions of zeolites from impact structures plotted in a ternary SiO_2 - ($CaO + FeO + MgO$) - ($Na_2O + K_2O$) diagram, based on chemical analyses of zeolites. For Kara, Popigai, and Puchezh-Katunki, the average compositions for zeolitic minerals (Tables 4, 8, and 10) are given; for Kara, data of Yushkin et al. (1976) are also used. For Ries, data from Stähle and Otteman (1977), for Siljan, from Morad et al. (1989), for Zapadnaya, from Valter et al. (1995). For Boltysh and Wanapitei, averages from three repeat electron microprobe (EMP) analyses (in wt%) made by L.S. Bogomolnaya, VSEGEI; operation conditions are given in section 2. EMP results: Boltysh, clinoptilolite from a tagamite, borehole 11475 (588 m depth): SiO_2 61.73, TiO_2 0.29, Al_2O_3 12.75, MgO 0.39, CaO 3.09, Na_2O 1.81, K_2O 0.82 (in wt%). Wanapitei, zeolites from a vesicular impact melt rock, sample MSWX-135.7 (the sample provided by R.A.F. Grieve): mordenite: SiO_2 63.81, Al_2O_3 9.53, CaO 3.28, Na_2O 0.76, K_2O 1.19; chabazite: SiO_2 53,50, Al_2O_3 16.75, FeO 0.09, CaO 4.90, Na_2O 1.22, K_2O 4.56 (in wt%). Zeolites from different craters are marked by different symbols. Zeolites from authigenic and allogenic breccias are shown by filled symbols, and from impact melt rocks, by open symbols. Asterisks are average target compositions and, consequently, impact melt compositions. Data on average target compositions are from Masaitis et al. (1980) for Boltysh, Kara, and Popigai; Masaitis and Pevzner (1999) for Puchezh-Katunki; Allen et al. (1982) for Ries; Aldahan (1990) for Siljan; Grieve and Ber (1994) for Wanapitei; Gurov et al. (2000) for Zapadnaya.

Thus, three main features of the hydrothermal mineralization are: (1) the spatial distribution between different impact lithologies, which have a definite location within craters; a decrease in the Al^{IV} content from the base to the top of thick impact rock sequences is the main alteration trend of the composition of both clay minerals and zeolites, indicating a decrease in temperature and pH of the mineral-forming solutions in this direction; (2) some common features (e.g., enrichment in silica of hydrothermal minerals, higher Fe content in Mg-Fe minerals of variable composition) of the chemical composition (including the similarity of mineral assemblages from different craters); (3) an influence of the target composition on post-impact mineral associations.

4.2 Distribution of the Hydrothermal Mineralization in Impact Structures

The post-impact thermal anomaly in an impact crater (Masaitis and Naumov 1993) determines the formation of a hot-water circulation system in the central uplift area, as well as local convective cells in impact melt bodies, that fill annular depressions. Correspondingly, *central* and *peripheral* areas of the post-impact hydrothermal alteration can be distinguished, which differ not only in their location within craters, but also in the distribution of hydrothermal minerals.

The central area includes authigenic breccia (the central uplift of shocked basement rocks) and overlying suevites and lithic breccias, and crater lake deposits in the central parts of a crater. Both a vertical and a less expressed lateral zonation of the hydrothermal mineralization are main characteristics of this area; the Puchezh-Katunki structure is an example. Similar features of the vertical zonation of hydrothermal alteration have also been found in other craters where deep drilling in central uplifts was carried out: Ries (Stöffler et al. 1977; Stähle and Otteman 1977), Boltysh, Siljan Ring (Aldahan 1990), Manson (McCarville and Crossey 1996), and Zapadnaya (Tatarintsev et al. 1981).

Inferred from the vertical sections (Fig. 16), this zonation is expressed mainly in smectite-chlorite transition; the general absence of corrensite indicates both high temperature gradient and low water-rock ratio. This zonation is not caused by the composition of the target, nor by shock and post-shock thermal metamorphism, but the post-impact thermal field of an impact crater is the main factor controlling mineral formation. The temperature of crystallization of hydrothermal minerals decreases upward as well as outwards from the crater center. This conclusion is confirmed by a study of homogenization temperatures of post-impact fluid inclusions in the Siljan Ring impact structure (Komor et al. 1988). The zones of alteration belong to moderate-temperature (chlorite-anhydrite zone) and low-temperature (smectite-zeolite zone) facies of the impact-generated hydrothermal formation. Based on data from modern geothermal systems (Kristmannsdottir 1979; Slovtsov and Moskaleva 1989; Rychagov et al. 1993, 1994; Robinson and Santana de Zamora 1999), temperature intervals of mineral crystallization in these zones are estimated to be 180 - 350°C and 50 -200°C, respectively.

The peripheral area includes impactites and impact breccia infilling the annular depression. At Popigai and Kara, the hydrothermal mineralization in the peripheral area embraces the entire annular depression, where a number of low-temperature minerals, similar to the upper zone of the central area, occur: smectites, chlorites, zeolites, calcite, pyrite. The basic feature of the distribution of

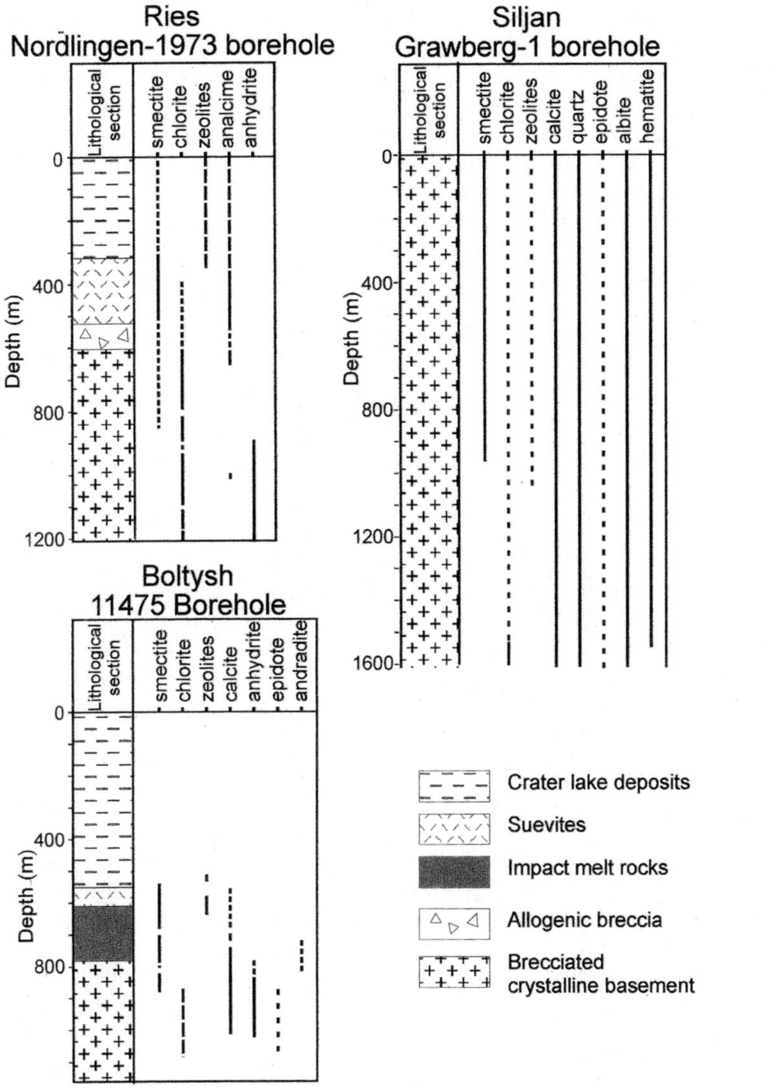

Fig. 16. General scheme for the distribution of hydrothermal minerals in deep drillholes from the Ries, Boltysh, and Siljan impact structures. For Ries, data from Stöffler et al (1977), for Siljan, from Aldahan (1990), and for Boltysh, author's data using optical microscopic study of thin sections and electron microprobe analyses of selected minerals. Dashed lines indicate sporadic occurrence of a mineral.

the hydrothermal mineralization there are the weak appearance or absence of any zonation and the constancy of chemical compositions and crystal structures of secondary minerals. The thick suevite layer in the Kara impact structure is a typical example: identical series of main hydrothermal minerals and similarity of their properties are observed in the whole layer. Nevertheless, in thick bodies of different impact lithologies, some spatial differentiation of secondary minerals can be observed; in this case, it is caused mainly by the inhomogeneity of the host rocks. At the same time, a vertical zonation of hydrous phyllosilicates within thick impact rock sequences is outlined in same cases, for example, in Kara and Slate Islands (Dressler and Sharpton 1997), which indicate the existence of thermal gradients during the initial stage of the hydrothermal activity.

4.3 Characteristics of Post-Impact Hydrothermal Solutions

The composition of hydrothermal mineralization indicates that

(a) superficial, meteoric and ground water (and, possibly, products of dehydration and degassing of minerals under shock) are the sources for hot-water solutions. This is confirmed by (1) post-impact fluid inclusion data from Charlevoix (Pagel and Poty 1975), Carswell (Pagel et al. 1985), Houghton (Bain and Kissin 1988), Siljan (Komor et al. 1988), Roter Kamm (Koeberl et al. 1989), Manson (Boer et al. 1996), Lockne (Sturkell et al. 1998), and Kärdla (Kirsimäe et al. 2002), which indicate low gas contents or a pure aqueous composition of the post-impact fluids, as well as their low salinity (0 - 13 % $NaCl_{eq}$) - with the exception of Roter Kamm and Lockne; (2) measurements of isotopic composition of oxygen and hydrogen from some hydrothermal minerals (Komor and Valley 1990; Davies et al. 1990);

(b) shocked target rocks are sources for the mineral components of the solutions. No evidence of any deep heat and mass addition in the post-impact hydrothermal activity is found;

(c) filtration occurs in liquid state only (mineral associations typical for vapor-dominated hot-water circulation systems are absent in impact craters);

(d) high rates of filtration (10^{-4} - 10^{-3} ms^{-1}) are assumed by analogy with a modern Pauzhetka hydrothermal system (Rychagov et al. 1993); this is indicated by intense development of high-aluminia smectites, as well as other evidence, e.g. crystallization of laumontite without calcite at deep levels of the hydrothermal column;

(e) the general succession of mineral crystallization in all environments (Table 14) confirms the uniform regressive course of the hydrothermal process.

From the composition of minerals (wide-spread development of Fe-chlorites, smectites with a higher Al^{IV}/Si ratio, and zeolites as the main hydrothermal minerals), it can be inferred that the chemical parameters of the mineral formation vary within a narrow interval of pH and correspond to weakly alkaline and near-neutral environments ($6 < pH < 8$). It is a consequence of the composition of host rocks, which consist mainly of shock-derived aluminosilicates and fusion glasses, the intense leaching of this material providing the above-mentioned properties of the solutions. Decrease of Al^{IV} content in phyllosilicates and zeolites progresses upward in a section, although in some craters local variations are possible. The common presence of sulfides in hydrothermal cells, together with rare occurrence

Table 14. Generalized succession of hydrothermal minerals crystallization

Successive stages	Metasomatic+veins	Veins (replacing and filling) and vugs	Veins (filling)	Latest veinlets
Impactites filling crater depression	actinolite←biotite← chlorite←smectites	crystobalite/ opal/ quartz	various zeolites anhydrite sulfides	calcite sulfides
Lower zone in central uplift	andradite salite epidote chlorite calcite	chlorite anhydrite	laumontite	
Upper zone in central uplift	saponite/ montmorillonite laumontite/ (chabazite+calcite)	calcite heulandite calcite analcime apophyllite	calcite Fe-montmorillonite	calcite

of iron oxides, indicates that the solutions maintained Eh values of >-0.5 V (for neutral environment) during the course of the hydrothermal process.

4.4 General Scheme of an Impact-Generated Hydrothermal System

Taking into consideration the decisive force of the temperature field on parameters of hydrothermal process, the proposed scheme of an impact-induced hydrothermal system is based on both petrological analysis of mineral associations in large craters, and the thermal history of impactites and shocked rocks (Newsom 1980; Masaitis and Naumov 1993; Crossey et al. 1994).

The regressive hydrothermal process in large impact structures can be subdivided into three successive stages (Fig. 17): 1) an initial stage, when the isotherms preserve their original configuration (the temperature decreased from top downward); 2) a main stage after an inversion of the thermal field; the thermal gradients during this stage range from about 30°C km^{-1} in the central part of the hydrothermal cell to 100°C km^{-1} at its periphery; 3) a final stage, during which thermal gradients are less than 10-30°C km^{-1} and hot-water circulation takes place in the near-surface area only. The vertical zonation of the mineral distribution that imitates a deep geothermal source of the thermal energy, results from the thermal evolution of an impact structure. The substitution of chlorites by smectites downward the section, which is found locally in peripheral parts of the central uplift in Puchezh-Katunki, is evidence for mineral formation during the initial stage. The vertical zonation of the distribution of the hydrothermal minerals in the central uplift form during the main stage when the circulation system incorporates the maximum amount of rocks. At last, zeolite-calcite cement of redeposited suevites and lithic breccia, as well as latest calcite veins, form during the final stage.

Impactites filling the true crater may create local thermal anomalies with gradients that are insignificant for providing zonations in the hydrothermal alteration. Nevertheless, thick impact melt rock sequences in large impact

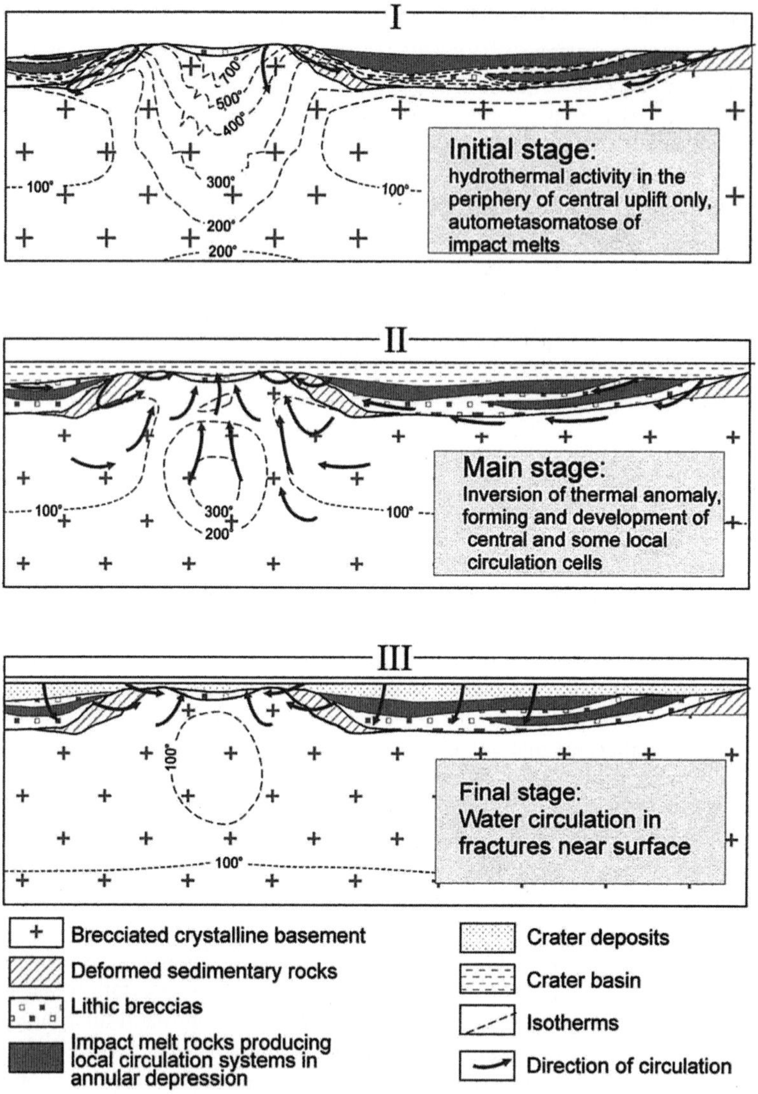

Fig. 17. A general scheme for impact-generated hydrothermal circulation in a complex impact crater. Three stages of regressive hydrothermal process are shown (see text).

structures can cause thermal gradients, which can exist during the early modification stage, as indicated by spreading of the secondary mineralization at Kara.

The following general scheme for the hydrothermal alteration of impact rocks is proposed. Superficial aqueous fluids infiltrate hot rocks of the central uplift and impactites filling the crater, reacting with shock-derived aluminosilicates and impact glass and forming Fe-Mg clay minerals. As a result of the selective removal of silica, alumina, and alkalis, plus under neutral environment also of Ca, the circulating fluids are weakly alkaline. The dissolution of finely dispersed silicates provides a constant super-saturation of solutions by silica, creating favorable conditions for the formation of Fe-smectites and zeolites – the main impact-generated hydrothermal minerals. If carbonate rocks and evaporites are significant in the target, the fluids have higher activities of Ca, Mg, Fe, S and CO_2, and high-calcium and calcium-ferriferous mineral associations are also formed. The newly formed mineral associations also show a vertical and horizontal distribution due to the thermal gradient in the authigenic breccia zone. Silica, Al, K, and Na are removed, but accumulate partly in the upper zone of the hydrothermal cell during the precipitation of zeolites.

During ascent in the thermogradient field, the hydrothermal solutions become more acidic as a result of both the absorption of OH^- ions, as rock-forming silicates are replaced by smectites and chlorites, and because of the increase of the CO_2 solubility due to the decrease of the temperature. This is an important factor against the decrease of pressure, which can cause dissolution of hydrocarbonate complexes that lead to boiling of solutions near the surface. The extent of rock transformation during hydrothermal processes is determined by the amount of fluids in the system.

The proposed scheme indicates a general trend of the impact-generated hydrothermal circulation process. Quantification of this scheme could give an estimate of the convective heat and mass transfer in impact-generated hydrothermal systems.

5 Conclusions

As a result of any impact event, a near-surface high-gradient zone of hot and high-permeability rocks is created. The parameters of this zone (gradients of T, P, permeability, etc.) allow hot-water circulation for several thousands of years or more. Hydrothermal circulation systems in impact structures are controlled by two factors: kinetic energy of the impact and properties of the target (mainly the availability of water).

Any impact-generated hydrothermal activity is expressed mainly in the formation of mineral assemblages, their distribution being controlled by characteristics of the thermal anomaly in a crater. The predominance of shock-disordered aluminosilicates and impact glasses in the host rocks determines that hydrothermal solutions are weakly alkaline and have higher activities of silica. These factors create favorable conditions for the formation of Fe-smectites and zeolites – the main minerals present in the hydrothermal column. Thus, mineralogical and geochemical features of impact-generated hydrothermal processes are governed by a combination of both chemical composition of target and physical properties of the host rocks as a result of shock metamorphism.

Products of impact-generated hydrothermal processes are similar to hydrothermal mineralization in volcanic areas, as well as in modern geothermal systems, because thermodynamic and hydrodynamic conditions are similar. Some distinctive features of the impact-induced hydrothermal mineralization can exist, though. For example, defects in the crystal structures of minerals, enrichment of zeolites and phyllosilicates by silica, low gas content in fluid inclusions, etc. Nevertheless, the geological environment of the mineral occurrences is the decisive factor for distinguishing of impact-generated hydrothermal formation from other hydrothermal systems.

Acknowledgements

Drs. V.L. Masaitis, M.S. Mashchak, and A.I. Raikhlin from VSEGEI, St. Petersburg, who provide important data (including lots of samples) on Popigai, Kara, and Puchezh-Katunki craters, are greatly appreciated. The geologists from Middle-Volga Geological Prospecting Enterprise (Nizhny Novgorod), FGSU "Nedra" (Yaroslavl), and Polar Geological Prospecting Enterprise (Khatanga) are acknowledged for the possibility to study the cores from Puchezh-Katunki and Popigai impact structures. I am especially grateful to Prof. Dr. Christian Koeberl and Dr. Kalle Kirsimäe, whose thorough and patient reviews of the original manuscript, criticism and comments led to very considerable improvements not only of the language and style, but also the substance of this paper.

References

Aldahan AA (1990) Alteration and mass transfer in cataclasites and mylonites in 6.6 km of granitic crust at the Siljan impact structure, central Sweden. Contributions to Mineralogy and Petrology 105: 662-676

Allen CC, Gooding JL, Keil K (1982) Hydrothermally altered impact melt rock and breccia: Contributions to the soil of Mars. Journal of Geophysical Research 87: 10,083-10,101

Ames DE, Gibson HL (1995) Controls on and geological setting of regional hydrothermal alteration within the Onaping Formation, footwall to the Errington and Vermilion base metal deposits, Sudbury structure, Ontario. Geological Society of Canada Current Research 1995-E. Ottawa, pp 161-174

Bain JG, Kissin SA (1988) A preliminary study of fluid inclusions in shock-metamorphosed sediments at the Haughton impact structure, Devon Island, Canada [abs]. Lunar and Planetary Institute Contribution 665. Houston, p D-7

Beals CS (1960) A probable meteorite crater of Precambrian age at Holleford, Ontario. Dominion Observatory Ottawa Publications 24: 117-142

Beals CS, Dence MR, Cohen AJ (1967) Evidence for the impact origin of Lac Couture. Dominion Observatory Ottawa Publications 31: 409-426

Boer RH, Reimold WU, Koeberl C, Kesler SE (1996) Fluid inclusion studies on drill core samples from the Manson impact crater: Evidence for post-impact hydrothermal activity. In:

Koeberl C, Anderson RR (eds) The Manson Impact Structure, Iowa: Anatomy of an impact crater. Boulder, CO, Geological Society of America Special Paper 302: 377-382

Boslough MB, Cygan RT (1988) Shock-enhanced dissolution of silicate minerals and chemical weathering on planetary systems. Lunar and Planetary Science 18: 443-453

Brindley GW, Brown G (eds) (1980) Crystal structures of clay minerals and their X-ray identification. Publications by Mineralogical Society of England, 500 pp

Carrigy MA (1968) Evidence of shock metamorphism in rocks from the Steen River structure, Alberta. In: French BM, Short NM (eds) Shock metamorphism of natural materials. Mono Book Corporation, Baltimore, pp 357-368

Crossey LJ, Kudo AM, McCarville P (1994) Post-impact hydrothermal systems: Manson impact structure, Manson, Iowa [abs]. Lunar and Planetary Sciences XXV. Lunar and Planetary Institute, Houston, pp 299-300

Currie KL (1965) The geology of the New Quebec crater. Canadian Journal of Earth Sciences 2: 141-160

Currie KL (1971a) A study of potash fenitization around the Brent crater, Ontario - a paleozoic alkaline complex. Canadian Journal of Earth Sciences 8: 481-497

Currie KL (1971b) Geology of the resurgent cryptoexplosion caldera at Mistastin Lake, Labrador. Geological Survey of Canada Bulletin 207: 1-62

Currie KL (1972) Geology and petrology of the Manicouagan resurgent caldera, Quebec. Geological Survey of Canada Bulletin 198: 1-154

Davies JF, Leroux MV, Whitehead RE, Goodfellow WD (1990) Oxygen-isotope composition and temperature of fluids involved in deposition of Proterozoic cedex deposits, Sudbury Basin, Ontario. Canadian Journal of Earth Sciences 27: 1299- 1303

Dence MR, Innes MJS, Robertson PB (1968) Recent geological and geophysical studies of Canadian craters. In: French BM and Short NM (eds) Shock metamorphism of natural materials. Mono Book Corporation, Baltimore, pp 339-362

Dressler BO, Sharpton VL (1997) Breccia formation at a complex impact crater, Slate Island, Lake Superior, Ontario, Canada. Tectonophysics 275: 285-312

Dritz VA, Kossowskaya AG (1990) Clay minerals: smectites, mixed-layer silicates. Nauka Press, Moscow, Russia, 214 pp

Engelhardt W (1972) Shock produced rock glasses from the Ries crater. Contributions to Mineralogy and Petrology 36: 265-292

Engelhardt W, Matthai SC, Walzebuck J (1992) Araguainha impact crater, Brazil, I: The interior part of the uplift. Meteoritics 27: 442-457

Fiske PS, Hargraves RB, Onstott TC, Koeberl C, Hougen SB (1994). Pseudotachylites of the Beaverhead impact structure: Geochemical, geochronological, petrographic, and field investigations. In: Dressler B, Grieve RAF, Sharpton VL (eds) Large Meteorite Impacts and Planetary Evolution. Boulder, CO, Geological Society of America Special Paper 293: 163-176

Floran RJ, Grieve RAF, Phinney WC, Warner JL, Simonds CH, Blanchard DP, Dence MR (1978) Manicouagan impact melt, Quebec. I: Stratigraphy, petrology, and chemistry. Journal of Geophysical Research 83: 2737-2759

Florensky PV, Dabizha AI (1980) Zhamanshin meteorite crater (in Russian). Nauka Press, Moscow, Russia, 126 pp

French BM, Koeberl C, Gilmour I, Shirey SB, Dons JA, Naterstad J (1997) The Gardnos impact structure, Norway: petrology and geochemistry of target rocks and impactites. Geochimica et Cosmochimica Acta 61: 873-904

Glazovskaya LI, Gromov EI, Parfenova OV, Ilkevich GI (1991) Logoisk astrobleme (in Russian). Nauka Press, Moscow, Russia, 135 pp

Gottardi D, Galli G (1985) Natural zeolites. Springer Verlag, Berlin-Heidelberg, 410 pp

Grieve RAF (1975) Petrology and chemistry of impact melt at Mistastin Lake crater, Labrador. Geological Society of America Bulletin 86: 1617-1629

Grieve RAF (1978) The melt rocks at Brent crater, Ontario, Canada. Lunar and Planetary Science 9: 2579-2608

Grieve RAF, Ber TJ (1994) Shocked lithologies at the Wanapitei impact structure, Ontario, Canada. Meteoritics 29: 621-631

Grieve RAF, Shoemaker EM (1994) The record of the past impacts on the Earth. In: Gehrels T (ed) Hazards due to comets and asteroids. Arizona university Press, Tuscon, pp 417-462

Guppy DJ, Brett R, Milton DJ (1971) Liverpool and Strangways craters, Northern Territory: two structures of probable impact origin. Journal of Geophysical Research 76: 5387-5393

Gurov EP, Gurova EP, Pavljuk VI, Sokur TM (2000) Zapadnaya astrobleme: Anatomy of an diamond-bearing impact structure (in Russian). Geologichny Zhournal 2: 29-38

Heyl AV, Brock MR (1962) Zink occurrences in the Serpent Mound structure, southern Ohio. US Geological Survey Professional Paper 450D: 95-97

Johansson A (1984) Geochemical studies on the Boda Pb-Zn deposits in the Siljan astrobleme, central Sweden. Geologiska Föreningen i Stockholm Forhandlingar 106: 15-25

Joint Committee on Powder Diffraction Standards, Selected powder diffraction data for minerals, search manual, Pub. M-1-23. Swarthmore, Pa, 1974

Kiilsgaard TH, Heyl AV, Brock MR (1963) The Crooked Creek disturbance, Missouri. US Geological Survey Professional Paper 450E: 14-20

Kirsimäe K, Suuroja S, Kirs J, Kärki A, Polikarpus M, Puura V, Suuroja K (2002) Hornblende alteration and fluid inclusions in Kärdla impact crater, Estonia – an indication for the post-impact hydrothermal activities. Meteoritics & Planetary Sciences (accepted)

Koeberl C, Fredriksson K, Götzinger M, Reimold WU (1989) Anomalous quartz from the Roter Kamm impact crater, Namibia: Evidence for post-impact activity? Geochimica et Cosmochimica Acta 53: 2113-2118

Komor SC, Valley JW (1990) Deep drilling at the Siljan Ring impact structure: oxygen-isotope geochemistry of granite. Contributions to Mineralogy and Petrology 105: 516-532

Komor SC, Valley JW, Brown PE, Collini B (1988) Fluid inclusions in granite from the Siljan Ring structure and surrounding regions. In: Boden A, Eriksson KS (eds) Deep drilling in crystalline bedrock, Volume 1. Springer Verlag, Berlin-Heidelberg, pp 180-208

Kring DA, Hildebrand AR, Boynton WV (1991) The petrology of an andesitic melt rocks and a polymict breccia from the interior of the Chicxulub structure, Yucatan, Mexico [abs]. Lunar and Planetary Science 22: 755-756

Kristmannsdottir H (1979). Alteration of basaltic rocks by hydrothermal activity at 100-300°C. In: Mortland MM, Farmer VC (eds) Proceedings of the 6th International Clay Conference. Elsevier, Amsterdam, pp 359-367

Lambert P (1977) Effects of natural and artificial shock waves, and the Rochechouart impact crater (in French). Dissertation, Universite de Sud Paris

Laurén L, Lehtovaara J, Boström R (1978) On the geology of the circular depression at Söderfjärden, Western Finland. Geological Society of Finland Bulletin 297: 5-38

Lehtinen M (1976) Lake Lappajärvi, a meteorite impact site in Western Finland. Geological Survey of Finland Bulletin 282: 1-92

Mannkopff H, Friede G (1975) Principles and methods of spectrochemical emission analysis (in German). Verlag Chemie, Weinheim, 218 pp

Masaitis VL (ed) (1983) Textures and structures of impact breccias and impactites (in Russian). Nedra Press, Leningrad, 159 pp

Masaitis VL (1999) Impact structures of northeastern Eurasia: The territories of Russia and adjacent countries. Meteoritics & Planetary Sciences 34: 691-711

Masaitis VL, Mashchak MS (1996) Recrystallization and blasthesis in shocked rocks of impact structures (in Russian). Zapiski Vsesoyuznogo Mineralogicheskogo Obshchestva 125/4: 1-18

Masaitis VL, Naumov MV (1993) Principal model of hydrothermal circulation in impact craters (in Russian). Doklady (Transactions) of Russian Academy of Sciences 333: 70-72

Masaitis VL, Pevzner LA (eds) (1999). Deep drilling in the Puchezh-Katunki impact structure (in Russian). VSEGEI Press, St.Petersburg, 399 pp

Masaitis VL, Danilin AN, Mashchak MS, Raikhlin AI, Selivanovskaya TV, Shadenkov EM (1980) The geology of astroblemes (in Russian). Nedra Press, Leningrad, 231 pp

Masaitis VL, Mashchak MS, Raikhlin AI, Selivanovskaya TV, Shafranovsky GI (1998) Diamond-bearing impactites of the Popigai astrobleme (in Russian). VSEGEI Press, St.Petersburg, 182 pp

Masaitis VL, Naumov MV, Mashchak MS (1999) The anatomy of the Popigai impact crater, Russia. In: Dressler B, Sharpton VL (eds) Large Meteorite Impacts and Planetary Evolution II. Boulder, CO, Geological Society of America Special Paper 339: 1-17

Mashchak MS (1990a) Geological conditions of time of formation of impact craters on Pai-Khoi (in Russian). In: Masaitis VL (ed) Impact craters on MZ-KZ boundary. Nauka Press, Leningrad, pp 24-37

Mashchak MS (1990b) Morphology and structure of Kara and Ust-Kara astroblemes (in Russian). In: Masaitis VL (ed) Impact craters on MZ-KZ boundary. Nauka Press, Leningrad, pp 37-55

Mashchak MS, Selivanovskaya TV (1988) Breccia and impactites on the southwestern edge of the Popigai astrobleme (in Russian). Meteoritika 47: 178-188

McCarville P, Crossey LJ (1996) Post-impact hydrothermal alteration of the Manson impact structure. In: Koeberl C, Anderson RR (eds) The Manson Impact Structure, Iowa: anatomy of an impact crater. Boulder, CO, Geological Society of America Special Paper 302: 347-376

Metzler A, Ostertag R, Redeker H-J, Stöffler D (1988) Composition of the crystalline basement and shock metamorphism of crystalline and sedimentary rocks at the Haughton impact crater, Devon Island, Canada. Meteoritics 23: 197-207

Milton DJ, Barlow BC, Brett R, Brown AR, Glikson AY, Manwaring EA, Moss FJ, Sedmik ECE, Van Son J, Young G A (1972) Gosses Bluff impact structure, Australia. Science 175: 1199-1207

Morad S, Fillipidis A, Aldahan AA (1989) Stellerite and Sr-containing stilbite in granitic rocks from the Siljan Ring structure, central Sweden. Bulletin of the Geological Institutions of the University of Uppsala 12: 143-149

Motuza GB, Gailius RP (1980) The Mizarai meteorite crater (in Russian). Transactions of Lithuanian University, Geologija 1: 78-88

Movshovich AE, Milyavsky AE (1990) Morphology and inner structure of the Kamensk and Gusev astroblemes (in Russian). In: Masaitis VL (ed) Impact craters on MZ-KZ boundary. Nauka Press, Leningrad, pp 110-146

Naumov MV (1993) Zeolite mineralization in impact craters (in Russian). Zapiski Vsesoyuznogo Mineralogicheskogo Obshchestva 122/4: 1-12

Naumov MV (1996) Basic regularities of post-impact hydrothermal process (in Russian). Astronomicheskii Vestnik 30: 25-32

Naumov MV (1999) The hydrothermal-metasomatic mineral formation (in Russian). In: Masaitis VL, Pevzner LA (eds) Deep drilling in the Puchezh-Katunki impact structure. VSEGEI Press, St.Petersburg, pp 276-286

Newsom HE (1980) Hydrothermal alteration of impact melt sheets with implications for Mars. Icarus 44: 207-216

Newsom HE, Graup G, Sewards T, Keil K (1986) Fluidization and hydrothermal alteration of the suevite deposit at the Ries crater, West Germany, and implications for Mars. Journal of Geophysical Research 91: E239-E251

Nikolsky AP (1991) Geological structure of iron ore deposit Pervomayskoe and its transformation caused by meteorite impact (in Russian). Nedra Press, Moscow, 71 pp

Offield TW, Pohn HA (1977) Deformation at the Decaturville impact structure, Missouri. In: Roddy DJ, Pepin RO, Merrill RB (eds) Impact and explosion cratering. Pergamon Press, New York, pp 321-342

Osinski GR, Spray JG, Lee P (2001) Impact-induced hydrothermal activity within Haughton impact structure, arctic Canada: Generation of a transient, warm, wet oasis. Meteoritics & Planetary Science 36: 731-745

Pagel M, Poty B (1975) Fluid inclusion studies in rocks of the Charlevoix structure (Quebec, Canada). Fortsch Miner 52: 479-489

Pagel M, Wheatley K, Ey F (1985) The origin of the Carswell circular structure. In: Laine R, Alonso D, Swab M (eds) The Carswell ore uranium deposits, Saskatchewan. Canadian Geological Association Special Paper, Toronto, pp 213-222

Palme H, Gobel E, Grieve RAF (1979) The distribution of volatile and siderophile elements in the impact melt of East Clearwater (Quebec). Lunar and Planetary Science 10: 2465-2492

Papunen H (1973) Chemical composition and origin of the shock metamorphic rocks of the Sääksjärvi area, Finland. Geological Survey of Finland Bulletin 45: 29-34

Reimold WU, Oskierski W, Huth J (1987) The pseudotachylite from Champagnac in the Rochechouart meteorite crater, France. Journal of Geophysical Research 92: E737-E748

Reimold WU, Barr JM, Grieve RAF, Durrheim RJ (1990) Geochemistry of the melt and country rocks of the Lake St. Martin impact structure, Manitoba, Canada. Geochimica et Cosmochimica Acta 54: 2093-2111

Reimold WU, Koeberl C, Bishop J (1994) Roter Kamm impact crater, Namibia: geochemistry of basement rocks and breccias. Geochimica et Cosmochimica Acta 58: 2689-2710

Robinson D, Santana de Zamora A (1999) The smectite to chlorite transition in the Chipilapa geothermal system, El Salvador. American Mineralogist 84: 607-619

Rondot J (1983) The Charlevoix astrobleme: a trace after the impact of a large meteorite (in French). Annales Sciences de l'Université de Clermont Ferrand 75: 1-169

Roussel DH (1984) Mineralization in the Whitewater Group. In: Pye EG, Naldrett AL, Giblin PE (eds) The geology and ore deposits of the Sudbury structure. Ontario Geological Survey Special Volume 1, pp 219-233

Rudnik VA, Smyslov AA (eds) (1986) Physical and physical-chemical analytical methods during geochemical explorations (in Russian). VSEGEI Press, Leningrad, 263 pp

Ruhlmann E (1985) Mineralogy and metallogeny of uraniferous occurrences in the Carswell structure. In: Laine R, Alonso D, Swab M (eds) The Carswell ore uranium deposits, Saskatchewan. Canadian Geological Association Special Paper, Toronto, pp 105-120

Ryabenko VA (ed) (1982) The geology and petrology of meteoritic explosion craters (in Russian). Naukova Dumka, Kiev, 226 pp

Rychagov SN, Zhatnuev IS, Korobov AD, Kiryukhin VA, Glavatskikh SF (1993). The structure of a hydrothermal system (in Russian). Nauka Press, Moscow, 298 pp

Rychagov SN, Glavatskikh SF, Goncharenko OP (1994) Thermal regime of secondary mineral formation and structure of temperature field in interior of Baransky volcano (Iturup Island) (in Russian). Volcanolgija i seismologija 6: 96-112

Schuraytz BC, Sharpton VL, Marin LE (1994) Petrology of impact-melt rocks at the Chicxulub multiring basin, Yucatan, Mexico. Geology 22: 868-872

Selivanovskaya TV, Mashchak MS, Ezersky VA, Reshetnyak NB (1982) Impact glasses in suevites of the Kara and Ust-Kara astroblemes (in Russian). Meteoritika 40: 122-130

Sheridan MF, Ragan DM (1976) Compaction of ash-flow tuffs. In: Chilingarian GV, Wolf KH (eds) Compaction of coarse-grained sediments, II (Developments in sedimentology 18B). Elsevier, Amsterdam-Oxford-New York, pp 677-718

Short NM (1970) Anatomy of meteorite impact crater, West-Hawk, Manitoba, Canada. Geological Society of America Bulletin 81: 609-648

Slovtsov IB, Moskaleva GN (1989) Phyllosilicates as probable indicators of PT conditions of hydrothermal process (in Russian). Volcanologija i seismologija 5: 104-110

Stähle V, Otteman J (1977) Deep drilling in the Ries 1973: Zeolitization of glass in suevites and petrography of fall-back suevite and lithic breccias (in German). Geologica Bavarica 75: 191-218

Stanfors R (1969) Lake Mien: an astrobleme or a volcanotectonic structure. Geologiska Föreningen i Stockholm Forhandlingar 91: 73-86

Stearns RC, Wilson GW, Tiedemann HA, Wilcox JT, Marsch PS (1968) The Walls Creek structure, Tennessee. In: French BM, Short NM (eds) Shock metamorphism of natural materials. Mono Book Corporation, Baltimore, pp 323-338

Stöffler D, Ewald U, Ostertag R, Reimold WU (1977) Research drilling Nördlingen 1973 (Ries) - composition and texture of polymict impact breccias. Geologica Bavarica 75: 163-189

Sturkell EFF, Broman G, Forsberg P, Torrsander P (1998) Impact-related hydrothermal activity in the Lockne impact structure, Jämtland, Sweden. European Journal of Mineralogy 10: 589-606

Tatarintsev VI, Krochuck VM, Bondarenko IN, Legkova GV (1981) Sphalerite from an explosion structure on the Ukrainian Shield (in Russian). In: Mineralogy and Petrology of Ore Areas of the Ukrainian Shield. Naukova Dumka, Kiev, Ukraine, pp 41-50

Valter AA, Bobonich FM (1979) Rock-forming clinoptilolite in a meteorite crater deposits (in Russian). Doklady (Transactions) of Russian Academy of Sciences 248: 710-714

Valter AA, Ryabenko VA (1977) Explosion craters on the Ukrainian Shield (in Russian). Naukova Dumka, Kiev, 154 pp

Valter AA, Bobonich FM, Polshin EA (1988) The nature of hisingerite from the Terny astrobleme (in Russian). Mineralogichesky Zhournal 10/5: 50-63

Valter AA, Fedulov KV, Bobonich FM, Solomakha VN (1995) Zeolites and associated secondary minerals in impactites (in Russian). Mineralogichesky Zhournal 17/1: 71-81

Vellon C (1976) Optical atomic emission spectroscopic methods. In: Winefordner JD (ed) Trace analysis: spectroscopic methods for elements. A Wiley-Interscience Publication, New York-Ljndon-Sydney-Toronto, pp 136-199

Yushkin NP, Maslov MA, Miklyaev AS (1976) Zeolites (laumontite, analcime) and associated minerals from the Kara circular depression on the Pai-Khoi ridge (in Russian). In: Minerals and mineral parageneses of rocks. Nauka Press, Leningrad pp 71-75

Zhukov F, Petersell V, Fomin Yu (1987) The evidence of Paleozoic volcanism in Estonia (in Russian). Doklady (Transactions) of the Estonian Academy of Sciences 36: 6-13

Geology and Petrography of the Zapadnaya Impact Crater in the Ukrainian Shield

E.P. Gurov, E.P. Gurova, and T.M. Sokur

Institute of Geological Sciences, National Academy of Sciences of Ukraine,
55 b O. Gonchar Street, 01054 Kiev, Ukraine.

Abstract. The Zapadnaya impact structure formed within the Precambrian crystalline basement of the Ukrainian Shield 165 ± 6 (K/Ar) Ma ago. The structure is covered by Cenozoic sediments of up to 40 m thickness, but is delineated by over 100 holes drilled since the 1970s. It is an elongated (3.2 × 2.3 km) complex structure. An uplifted horst-like block of cataclased basement rocks separates the crater in two unequal depressions and complicates its structure. The south-eastern depression is larger and includes the central uplift, whereas the north-western depression represents a marginal part of the structure. Allochthonous breccias and suevites fill the crater and overlie the central uplift, but are absent on top of the horst. Impact-melt rocks form dykes or dyke-like bodies in the autochthonous breccias and brecciated crystalline rocks of the crater basement. Prominent shock metamorphic effects in quartz, feldspars, and some other minerals characterize impactites, including autochthonous breccias. High-pressure phases in impact rocks consist of coesite and impact diamond; the latter occur in glass-rich suevites and impact-melt rocks.

1 Introduction

The Zapadnaya is a complex impact structure that is located in the western part of the Ukrainian Shield (centered at 49°44'N; 29°04'E), about 65 km northeast of the city of Vinnitsa (Fig. 1). The impact origin of the structure was first recognized in 1975 by A.A. Valter and E.P. Gurov, who studied core samples from a circular negative gravity anomaly at that time of unknown origin. Their discovery of impact diamonds lead to extensive drillings, carried out by E.V. Melnitchuk and V.A. Golubev. Since then, only a few papers in the Ukrainian and Russian literature have been published on this structure (Gurov et al. 1985; Valter and Ryabenko 1977; etc.). Several articles have been published in the Ukrainian literature in the 90's (Gurov et al. 1999; Sokur 1999; Valter et al. 1999; etc.). An

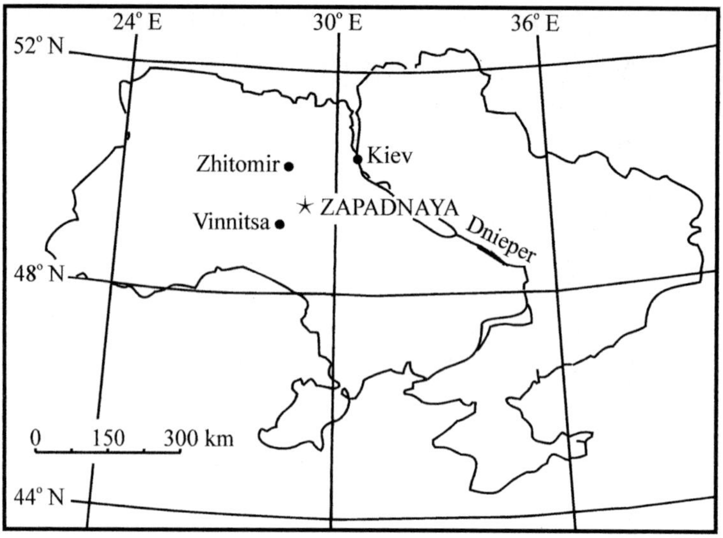

Fig. 1. Map of the Ukraine, with the Zapadnaya impact crater indicated by an asterisk.

attempt to substantiate an endogenic origin of the Zapadnaya impact structure (as well as for the Popigai, Khara, Boltysh and Ilyinets craters) was undertaken by Vaganov et al. (1985).

The goal of this paper is to summarize the geological and petrographical studies that have been performed over the last two decades and to present some new data on the geology and mineralogy of the crater.

2 Geology of the Zapadnaya Impact Structure

In a plan view, the Zapadnaya impact structure has an elongated shape, measuring 3.2 km from northwest to southeast and 2.3 km from northeast to southwest. A gravity survey of the structure indicates a Bouguer anomaly of -4.2 mGal 4.7 km long and 3.9 km across that is elongated in the same direction (B.S. Germanov, unpublished data).

The Zapadnaya impact structure formed in Precambrian crystalline rocks, which consist mainly of garnet-biotite granitoids, rare gneiss and gabbro. While granites predominate in the crater basement, gneisses occur in the south-western part of the structure and partly in its north-western marginal area. Gabbro and some other mafic rocks are occasionally present in the eastern region of the crater. Argillite and siltstone clasts in breccias and suevites indicate that a thin sedimentary cover existed over the crystalline basement at the time of impact, but

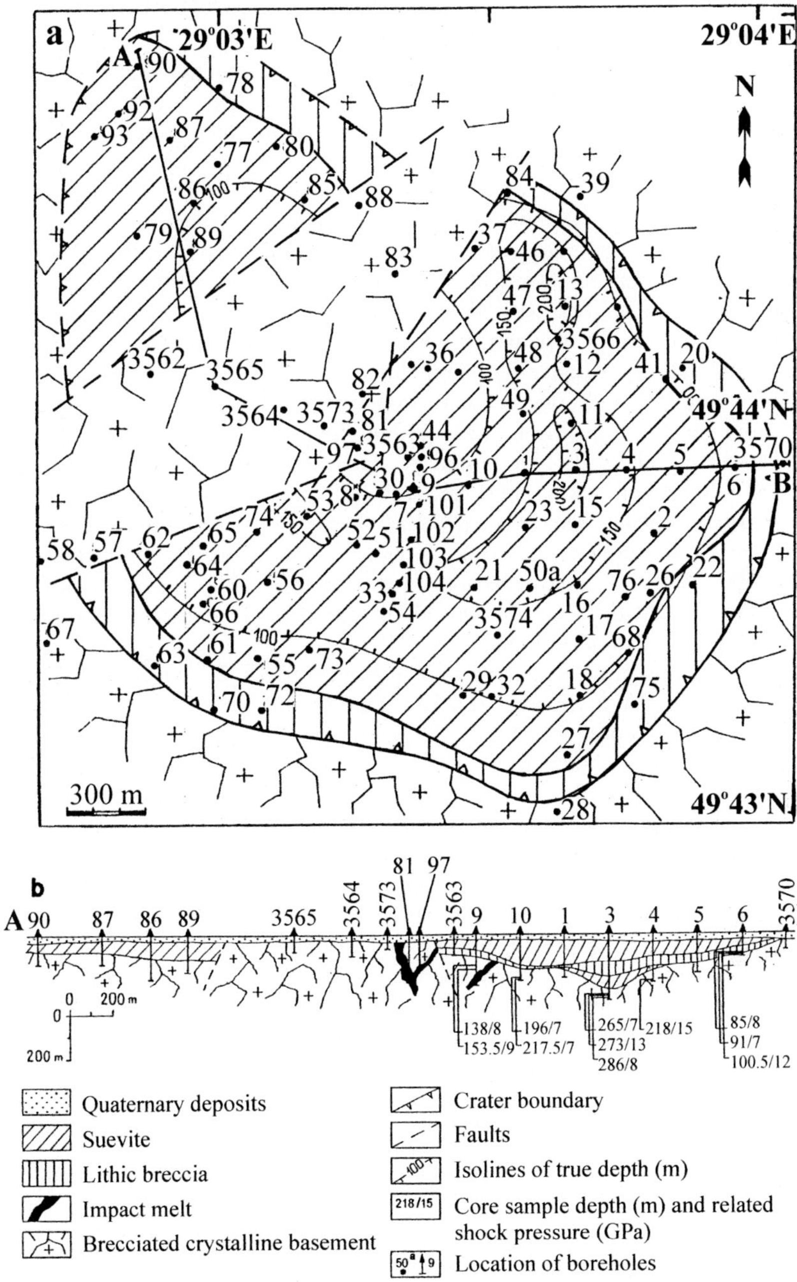

Fig. 2. (a) Schematic map and (b) cross-section of the Zapadnaya impact structure. Cenozoic deposits are omitted from the map. Shock pressures in basement rocks along the latitudinal profile of the eastern part of the crater are shown.

its age and initial thickness are unknown. The K/Ar age of clast-free impact melt of the crater is 165 ± 6 Ma (Valter et al. 1999).

An uplifted horst or horst-like block of cataclased basement rocks separates the crater depression into two unequal parts (Fig. 2a). The larger south-eastern depression contains the main structural elements of a typical complex structure: the central uplift and surrounding trough, while the north-western part is a simple semi-circular depression. A slightly different view on the crater structure was given by Valter et al. (1999), who found a continuous occurrence of suevites along the whole north-eastern marginal part of the structure. However, these rocks are absent from the boreholes 83 and 88 (Fig. 2a).

The horst is an uplifted SW-NE elongated block with a flat surface, cutting across the whole impact structure. It consists of cataclased and weakly brecciated granites and is buried under 35 - 45 m thick cover of the Neogene-Quaternary sediments. Suevites and lithic breccias do not occur on top of the horst. The true crater floor reaches a depth of ~100 - 130 m along the SE boundary of the horst and 100 - 110 m along its NW boundary. The minimum amplitude of the horst uplift is ~70 - 80 m, taking into account the thickness of the Neogene and Quaternary sediments. The horst is 0.6 - 0.8 km wide in its south-western and central parts (according to data from cores 57, 3564, 3565, 3573), decreasing to about 0.4 km in its north-eastern part (cores 83 and 88) (Fig. 2a). We assume that the horst was formed during the late modification stage of the crater formation before its complete stabilization.

The dimensions of the larger SE part of the crater are 2.2 km from NE to SW and 1.6 km from NW to SE (Fig. 2a). The central uplift, about 0.4 × 0.8 km in size and to 200 m high (cores 8, 9, 10, 34, 36 and others), is located in the western part of depression near its boundary with the horst (Fig. 2b). The central uplift has a flat surface with a local rise in its SW part (cores 7 and 8). It is covered with lithic breccia and suevite, and composed of weakly shocked autochthonous coarse breccia and mega-block breccia of crystalline rocks.

An annular trough surrounds the central uplift in the NE, E, SE and S. The depth of the trough is 130 - 150 m in the southern part of the structure (cores 21, 54, 104) and reaches up to 250 - 260 m east of the central uplift (cores 3, 4, 11). The local elevations of the true floor up to 100 m high occur in the trough (cores 1, 3574). The crater rim reaches the pre-Neogene surface at the horizontal distance of about 1.0 km from the base of the central uplift. The trough is filled with lithic breccias and suevites up to 200 - 220 m in thickness.

The NW part of the Zapadnaya structure comprises a semi-circular depression, about 0.9 × 1.3 km in size and up to 100 - 120 m deep (Fig. 2a). The boundary of this part of the crater is poorly known. The depression is filled with suevites up to 100 m thick in its deepest SE part near the contact with the horst (cores 79, 86, 90).

The post-impact sediments are not preserved in the Zapadnaya structure, as it was eroded below the top of the lithic breccia and suevite. The thickness of eroded allochthonous rocks remains unknown, but probably on the order of some tens of meters. According to the size of the gravity anomaly and the relationships proposed by Gurov et al. (1995), the initial crater depth has been estimated to be 350 - 400 m and its original diameter 4 - 5 km.

Autochthonous coarse breccias form the floor of the crater. Their transition to the cataclased crystalline rocks of the crater basement and the lower contact of the breccias with basement rocks is gradual. The thickness of the autochthonous breccias is estimated to be 50 m up to 100 m in the peripheral part of the crater (cores 20, 22, 72, etc.) and up to 100 - 150 m and more in the central uplift (cores 8, 9, 81). The autochthonous breccia is a coarse-fragmented material consisting of crystalline rock clasts and blocks, a few centimeters up to a few meters in diameter, consolidated by finer-grained breccia zones, which are a few centimeters in thickness. Shatter cones occur in clasts and blocks of the breccia.

The allochthonous lithic breccias form the lowermost level of the crater fill (Figs. 2b and 3). Their gradual contact with the underlying autochthonous breccias can be traced as a transition, 20 - 30 m thick, from polymictic breccia (above) to coarse monomictic breccia (below). The thickness of the lithic breccia increases from 30 - 40 m in the peripheral crater zone (cores 5, 6, 27, 92, etc.) to 80 - 90 m and more within the annular trough around the central uplift (cores 3, 4, 11, 18, etc.). Crystalline rocks and mineral clasts in fine-grained matrix compose the lithic breccias. The main rock fragments in the breccias are garnet-biotite granites with rare gneisses and mafic rocks. The size of clasts varies from millimeters to centimeters, but separate blocks larger than 1 m wide may also occur. Shock metamorphic effects are abundant in mineral and rock clasts of the lithic breccias.

Several separate suevite lenses up to 10 m in thickness occur inside of the lithic breccia layer (boreholes 3, 9a, etc.). A suevite lens, up to 60 m thick, underlies the breccia layer within the deepest part of the crater (boreholes 1, 3, 4). Suevites also form the uppermost layer of the crater fill (Figs. 2b and 3). They occur around the central uplift and partly above it within the SE part of the structure, and form the predominant member of the crater fill in the NE part. The thickness of the suevites is about 100 m in the eastern part of the structure (cores 4, 11, 21, 48) and it completely wedges out at a distance of a few hundreds meters from the rim (cores 23, 84). Contacts of suevites and lithic breccias are gradual and can be traced based on the presence of glass in the suevites. Suevites consist of glass, rock and mineral clasts in a fine-grained matrix. Glass content is from a few per cent near the contacts with breccias and increases to some tens of per cents within the uppermost horizons of the suevite layer. Rock fragments in suevites consist of granites and gneisses, and rarely of basic rocks. Argillite and siltstone clasts sometimes occur in the uppermost horizon of suevites. The uppermost suevite horizon is extensively weathered and transformed into a structural weathered crust from 3 to nearly 10 m thick.

Impact-melt rocks in the Zapadnaya structure form dykes and dyke-like bodies in the brecciated rocks of the crater basement (Figs. 2b and 3). Numerous bodies of impact-melted rocks occur in the central uplift (cores 8, 9, 96, 102), they are also common in the SW part of the crater (cores 66, 74) and in some other locations (cores 52, 77). Parameters and shapes of impact-melt rock bodies are complex and remain poorly known. Their visible thickness in boreholes is 1 to 37 m. Inclinations of the dyke contacts vary mainly from 40 to 70°(cores 9, 102).

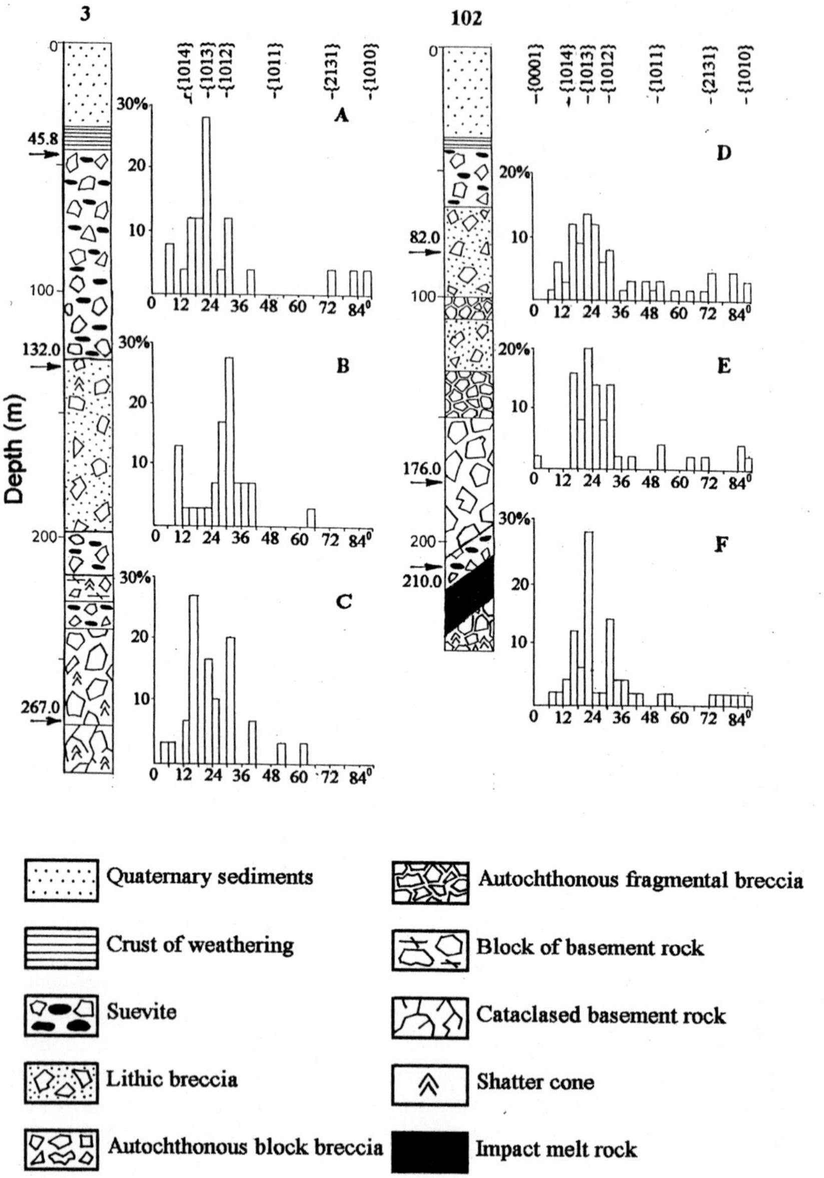

Fig. 3. Stratigraphic columns of drillholes 3 and 102 with insets showing frequencies (in rel%) of crystallographic orientations of poles of PDFs (vertical scale) to the c-axis of quartz crystals (in degrees; horizontal scale) from various stratigraphic intervals (*hole 3*: *A*, 25 sets in 16 grains, sample from 45.8 m; *B*, 30 sets in 20 grains, sample from 132 m; *C*, 30 sets in 24 grains, sample from 276 m; and *hole 102*: *D*, 65 sets in 43 grains, sample from 82 m; *E*, 50 sets in 25 grains, sample from 176 m; *F*, 25 sets in 20 grains, sample from 210 m). Positions of samples measured are shown with arrows).

The central parts of the dykes are composed of clast-poor impact-melt rock, while their margins consist of clast-rich impact-melt rock and impact-melt breccia. The clasts are fragments of shock metamorphosed rocks and minerals. The size of the clasts ranges from millimeters to a few centimeters. Vesicles filled with zeolites and montmorillonite are widespread in impact-melt rocks.

The deutritic alteration products are widespread in impactites of the Zapadnaya crater. Mineralization of zeolites, carbonates, montmorillonite and sulphides occur in suevites, impact-melt rocks and rarely in lithic breccias. Zeolites form veinlets and concentrations up to 3 - 5 mm in size in the hollows and fissures within suevites and lithic breccias. The mineralization of heulandite and erianite is widespread in the impact melts. Crystals and flabellate druses of orange heulandite up to 1 cm in size occur within some cavities of impact melt (cores 9, 96, etc.). Erianite crystals up to 0.5 - 1 mm in size fill the cavities in impact melt and form the vugs on the surface of some vesicles (cores 9, 102, etc.). The rounded vesicles in impact melt are often filled with montmorillonite and chlorite. Calcite veinlets occur in lithic breccias and autochthonous rocks.

3 Samples and Experimental Methods

Many drill cores, especially 1, 3, 4, 6, 8, 9a, 89, 96, and 102 (Fig. 2a), have been investigated visually and microscopically. The stratigraphic columns of the cores 3 and 102 are presented in Fig. 3. The column of core 102 represents a section through the southern part of the central uplift, while the column of core 3 demonstrates a section of the impact structure in its deepest eastern part.

1025 thin sections of core material have been prepared and studied using a polarizing microscope. Orientation of planar deformation features (PDFs) in quartz from cores 3 and 102 (Fig. 3), 1, 9, 10 and some others was measured using a five-axial universal stage. Refractive indices of shocked quartz were determined for estimation of shock pressures using diagrams in Stöffler and Langenhorst (1994) and Grieve et al. (1996).

Impact diamonds were extracted from suevites and impact-melt rocks by high-temperature alkaline dissolution of samples using the method described by Kashkarov and Polkanov (1970). The phase composition of the impact diamonds was studied by X-rays with a Debye-Scherrer camera. The structure and morphology of grain surfaces of impact diamonds was investigated by transmission electron microscopy (TEM), using the platinum-coal replica method.

Values of shock pressures in autochthonous breccias of the crater basement were estimated along a latitudinal profile across the eastern part of the structure, from the top of the central uplift to the eastern crater rim (Fig. 2a). Shock pressures were determined by the method of Hörz and Quaide (1973) (see also Grieve et al. 1996; Stöffler and Langenhorst 1994), using only quartz grains about 0.25 mm in size. In each sample studied, three quartz grains were measured to determine the mean shock pressure.

4
Shock Metamorphism of the Zapadnaya Crater Rocks

All rock types of the Zapadnaya impact structure are characterized by an abundance of shock-metamorphic effects, such as PDFs in quartz and feldspar, kink bands in biotite, fragmentation of garnet and some others. Feldspar diaplectic glasses occur in lithic breccias and suevites, but diaplectic quartz glass and lechatelierite is not preserved in impact rocks. The shock-induced high-pressure phases found in the Zapadnaya crater are coesite and impact diamond. The degree of shock metamorphism increases from the autochthonous breccias of the crater basement to the lithic breccias and further to suevites and clasts within the impact-melt rocks. Alteration has partly destroyed the shock metamorphic effects in the uppermost horizon of suevites.

4.1
Autochthonous Breccias

The shock pressures of about 7 ± 2 GPa predominate in the basement rocks along the profile, but higher pressures, up to 12 - 15 GPa, are present in some fragments and blocks of the autochthonous breccia (Fig 2b). Planar deformation features (PDFs), wavy extinction and fracturing are common in quartz grains from the autochthonous breccia. Measurements of PDFs yield mainly $\{10\bar{1}3\}$ and $\{10\bar{1}4\}$ orientations, and a subordinate occurrence of Brazil twins (0001) and other orientations (Fig. 3). The rhombohedral $\{10\bar{1}4\}$ orientation was first determined in shocked quartz from the craters of Ukrainian Shield (Gurov and Gurova 1991; Gurov et al. 1979). The pole of this system forms an angle of ~18° with the *c*-axis (the theoretical value would be 17°35') and is different from the pole of the $\{10\bar{1}3\}$ and the $\{10\bar{1}2\}$ systems by 5° and 14°, respectively. The $\{10\bar{1}4\}$ orientation was also found later by Goltrant et al. (1992) in TEM investigations of naturally shocked quartz. PDFs in feldspars are rare in the autochthonous breccia. Kink-bands in biotite and thin fragmentation of garnet are abundant in the uppermost section of the cataclasite and autochthonous breccia of the crater basement.

4.2
Lithic Breccias and Suevites

Shock metamorphic effects are widespread in minerals from lithic breccias and suevites. Multiple sets of PDFs, up to 10 sets per grain, occur in quartz. The $\{10\bar{1}2\}$, $\{10\bar{1}3\}$, and $\{10\bar{1}4\}$ orientations are widespread, while $\{10\bar{1}1\}$, $\{10\bar{1}0\}$, (0001) and some other orientations occur rarely. Variation of PDFs orientation in quartz of the lithic breccia and suevite are shown for two drill core sections in Fig. 3. The refractive indices of quartz vary from the values typical of unshocked mineral down to strongly reduced ones. The lowest refractive indices (n_e=1.517, and n_o=1.511), and a birefringence of 0.006, were determined in two quartz grains from the lithic breccia (core 4, depths of 59 and 70 m) (Sokur 1999).

These optical properties correspond to shock pressures of about 28 ± 2 GPa (cf. Stöffler and Langenhorst 1994). The lack of diaplectic quartz glass and lechatelierite in suevites and breccias, which form at higher pressures, is most likely due to the processes of deuteric alteration. Traces of these processes are most evident in the upper horizon of the suevite layer. The alteration products of the shocked quartz and diaplectic quartz glass are fine-grained mosaic aggregates of recrystallized quartz, which in some cases contain relics of PDFs and ballen structures. Lens-like mosaic quartz aggregates with traces of flow structures are probable products of lechatelierite recrystallization. Low contents of coesite have been measured in moderately shocked quartz in suevite (core 10, 130 m) and in quartz from a gneiss clast in impact-melt rock (core 9, 160 m) (Gurov and Gurova 1991). Refractive indices of coesite-bearing quartz from the suevite are: n_e=1.550, n_o=1.541, with a birefringence of 0.009, corresponding to a shock pressure of about 23 ± 3 GPa (Grieve et al. 1996; Stöffler and Langenhorst 1994).

Feldspars in allochthonous breccias, carrying sets of PDFs, have reduced refractive indices and birefringence. Highly shocked microcline is converted to isotropic diaplectic glass. Vesicular structures in some feldspar grains are the result of partial melting. Partial transition of plagioclase into diaplectic glass is defined by the isotropization of one system of polysynthetic twins, while the other preserves its crystalline state. Biotites have kink-bands, and rarely PDFs. Highly shocked biotites are converted into dark-brown masses that partly preserve the initial grain shapes. Shock metamorphism of garnet is represented by thin grain fragmentation. Impact diamonds occur in suevites, mainly in their glass-rich varieties.

Impact glasses are the main components of suevites. The glass content is a few per cent in the lower horizon of the suevite layer near the contacts with lithic breccia and increases up to 50 - 60 % in its uppermost horizon in the eastern part of the structure (cores 2, 3, 48, etc.) and some other locations. Impact glasses are light-gray or greenish-gray particles and/or fragments of irregular form. Sizes of glass bodies vary from parts of a millimeter up to a few centimeters. The glass has a fluidal structure and often contains numerous vesicles and mineral inclusions. Fresh glass is not preserved, but devitrified and replaced by secondary minerals, predominantly zeolite, chlorite, and montmorillonite.

4.3 Impact Melts

The dykes of impact-melt rocks in the crater basement are dark-gray to black dense cryptocrystalline rocks with abundant inclusions of shocked rocks and minerals. Impact-melt rocks from the core of dykes are composed of prismatic crystallites of feldspars in a fine-grained matrix substituted by chlorite and montmorillonite. They rarely include quartz with ballen structure. Impact-melt rocks from the margins of dykes contain abundant inclusions of shocked rocks and minerals, whereas granite clasts are composed of quartz with ballen structure and devitrified vesicular feldspar glass.

5
Impact Diamonds

Impact diamonds, a high-pressure paramorph of initial graphite from the crystalline country rocks, occur in impact-melt rocks and suevites of the Zapadnaya structure (Grieve and Masaitis 1994; Gurov et al. 1996; etc.). Formation of impact diamonds in impact structures is caused by shock metamorphism of graphite-bearing country rocks (Masaitis et al. 1972, 1995, 1998; etc.).

Experiments on shock loading of graphite carried out by McQuine and Marsh (1967) have shown that the transition of graphite to the dense high-pressure phase begins at 40 GPa, and is completed at a pressure of ≥60 GPa. Occurrence of impact diamond in glass-rich suevites and impact-melt rocks of the Zapadnaya structure is in agreement with the fact that melting of granitic rocks happened at shock pressures of 55 - 60 GPa (French 1998; etc.), required to form impact diamonds. The tabular shape and size of diamond grains (Fig. 4a) are similar to those of the original graphite crystals. However, the transition from graphite ($\rho \approx 2200\ kgm^{-3}$) to diamond ($\rho \approx 3500\ kgm^{-3}$) is accompanied by a decrease in volume by a factor of 1.6. Diamond grains ranging from 0.06 to 0.5 mm in diameter and from 0.02 to 0.06 mm thick were extracted from suevites (core 1, 55 - 68 m; core 2, 48 - 60 m; core 3558, 60 - 80 m; etc.) and impact-melt rocks (core 9, 155 -160 m; core 8, 105 - 108 m; core 102, 206 - 209 m; etc.). The diameter to thickness ratio of the diamond grains is predominantly about 10 : 1. The diamonds are colourless, white to light yellow, brown or black. All the diamond grains are anisotropic, their interference colours vary from grey to yellow, blue, etc., and their birefringence ranges from 0.002 to 0.020. Sometimes the diamonds are polysynthetically twinned (Fig. 4b). It was suggested (Masaitis et al. 1998) that twins of impact diamond are inherited features. Up to three systems of opaque planar structures intersecting at an angle of 60° occur in some grains (Fig. 4c). The planar structures have been described in impact diamonds from some other impact structures, such as Popigai (Valter et al. 1992; Vishnevsky et al. 1997).

Optical properties of diamond grains, their extinction, birefringence and development of planar deformation structures are similar to the properties of single crystals. Only X-ray and electron microscopy show the complex composition and cryptocrystalline structure of diamonds. The phase composition was studied using the X-ray powder pattern method for individual grains. The character of the individual diamond grain patterns is similar to the powder patterns that reveal the cryptoaggregate structure of the "monocrystalline" grains of diamond. All grains investigated were binary mixtures of lonsdaleite and cubic diamond, while the graphite phase was never observed. The relative content of the

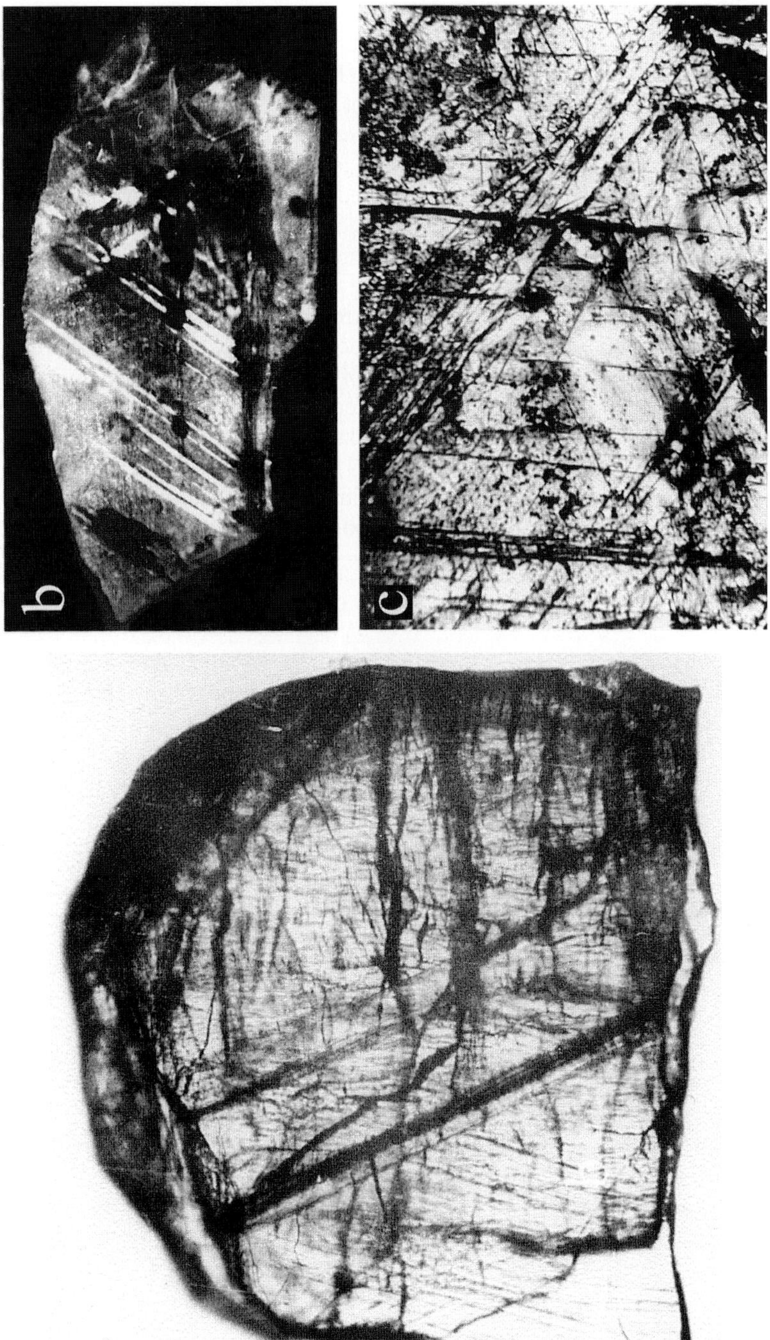

Fig. 4. Photomicrographs of impact diamonds from the Zapadnaya crater. (a) Tabular transparent diamond grain 0.5 mm in diameter (core 9; sample from 155 m; parallel polars). (b) Polysynthetic twins in diamond grain 0.35 mm long (core 9; sample from 155 m; crossed polars). (c) Three systems of opaque planar deformation structures in impact diamond grain (core 1; sample from 60 m, width of the photo 0.6 mm; parallel polars).

different phases was determined by comparison of the intensity of reflection 0.218 nm, corresponding to (100) of lonsdaleite with the reflection intensity of 0.206 nm, corresponding to the total intensity of (111) of cubic diamond and (002) of lonsdaleite, using the relationship of Valter et al. (1992). Investigations of 42 grains of impact diamonds have shown that the relative content of the lonsdaleite phase ranges between 10 and 78 % (Gurov et al. 1996), and diamonds with a lonsdaleite content of about 40 - 50 % are predominant (Fig. 5). A direct but indistinct relationship of diamond birefringence concerning the phase composition was established: increase of the lonsdaleite content by ~10 % causes an increase in the birefringence to ~0.002.

TEM studies of impact diamond surfaces corresponding to the (0001) surface of original graphite crystals reveal its blocky structure (Fig. 6a). The blocks are in the order of 0.5 - 1.0 μm in size. At the scale of TEM images, impact diamond displays lamellar structure (Fig. 6b). Parallel oriented are 0.03 - 0.1 μm thick and 0.3 - 0.9 μm in diameter. The angles of lamellae in individual blocks are not constant. Accordingly to Kurdyumov et al. (1984), the lamellae are crystallites of lonsdaleite that form prior to the cubic phase in synthetic diamond. Thus, it is suggested that the lamellae in the impact diamond of the Zapadnaya crater are crystallites of lonsdaleite.

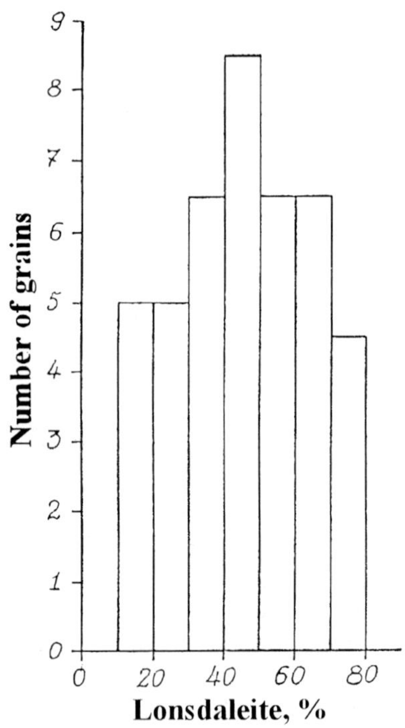

Fig. 5. Variation of lonsdaleite phase content in impact diamonds of the Zapadnaya crater based on X-ray analyses of 42 grains.

6
Summary and Conclusions

The Zapadnaya impact crater is an extensively eroded structure, 3.2 × 2.3 km in diameter and up to 260 m deep. It is the smallest known complex impact structure in crystalline target according to the lists of terrestrial impact craters published by Grieve (1991) and Grieve et al. (1995). The list of complex craters in crystalline target is followed, by increased diameter, by Gow Lake and Ilyinets (4 km), Kursk, Mien, and Sääksjärvi (5 km).

The structure of the Zapadnaya crater is complicated by the horst of crystalline basement rocks that separates it in two unequal parts: the larger one in the SE with the central uplift, and a smaller one in the NW part. A similar complex block tectonics of the crater floor was established for the Logoisk impact structure in Byelorussia (Gurov and Gurova 1991).

The Zapadnaya crater formed in the crystalline basement rocks of the Ukrainian shield. The distribution of the sedimentary rock clasts within suevite layer indicates the occurrence of sedimentary deposits on the surface of crystalline rocks at the time of impact. These sediments are not preserved in the whole western part of the shield, and their age is unknown. It is interesting to note that the same situation is known for the Ilyinets impact structure, located about 65 km to the south of the Zapadnaya crater (Gurov et al. 1998). A search for pollen in the argillite clasts from suevites of the Ilyinets crater did not yield any results.

The high-pressure phases found in impact rocks of the crater are coesite and impact diamond. The source of impact diamonds in the Zapadnaya structure was graphite from the granites and gneisses of the country rocks. Impact diamonds occur in impact-melt rocks and glass-rich suevites (Gurov et al. 1996, 1999; Valter et al. 1999), but are newer observed in the autochthonous breccias of the crater basement and lithic breccias. The occurrence of impact diamonds in impact-melt rocks is due to their formation at the same shock pressures that shock melting of granitic rocks.

Impact diamonds of the Zapadnaya crater are composed of complex cryptocrystalline aggregates of hexagonal and cubic phases, whereas the graphite phase was not observed. Contents of the hexagonal phase in impact diamonds from some craters vary from 0 in diamonds from the Lappajärvi (Vishnevsky and Raitala 2000) and Ries structures (Hough et al. 1995; Vishnevsky et al. 1997), up to 30 % in diamonds from the Popigai structure (Vishnevsky et al. 1997). The relatively low content of the lonsdaleite phase in diamonds from the Popigai structure in comparison with diamonds from the Zapadnaya crater is explained by the low stability of the hexagonal phase (Kurdyumov et al. 1984), which anneals easily into the cubic phase due to the long cooling of impact-melt rocks and suevites within this large crater (Valter et al. 1992; Vishnevsky et al. 1997). While the graphite phase was not observed in the diamonds of the Zapadnaya structure, the presence of graphite in the Popigai diamonds was explained by two different processes: incomplete transition of initial graphite into diamond and partial annealing of diamond in the high-temperature impactites (Masaitis et al. 1998).

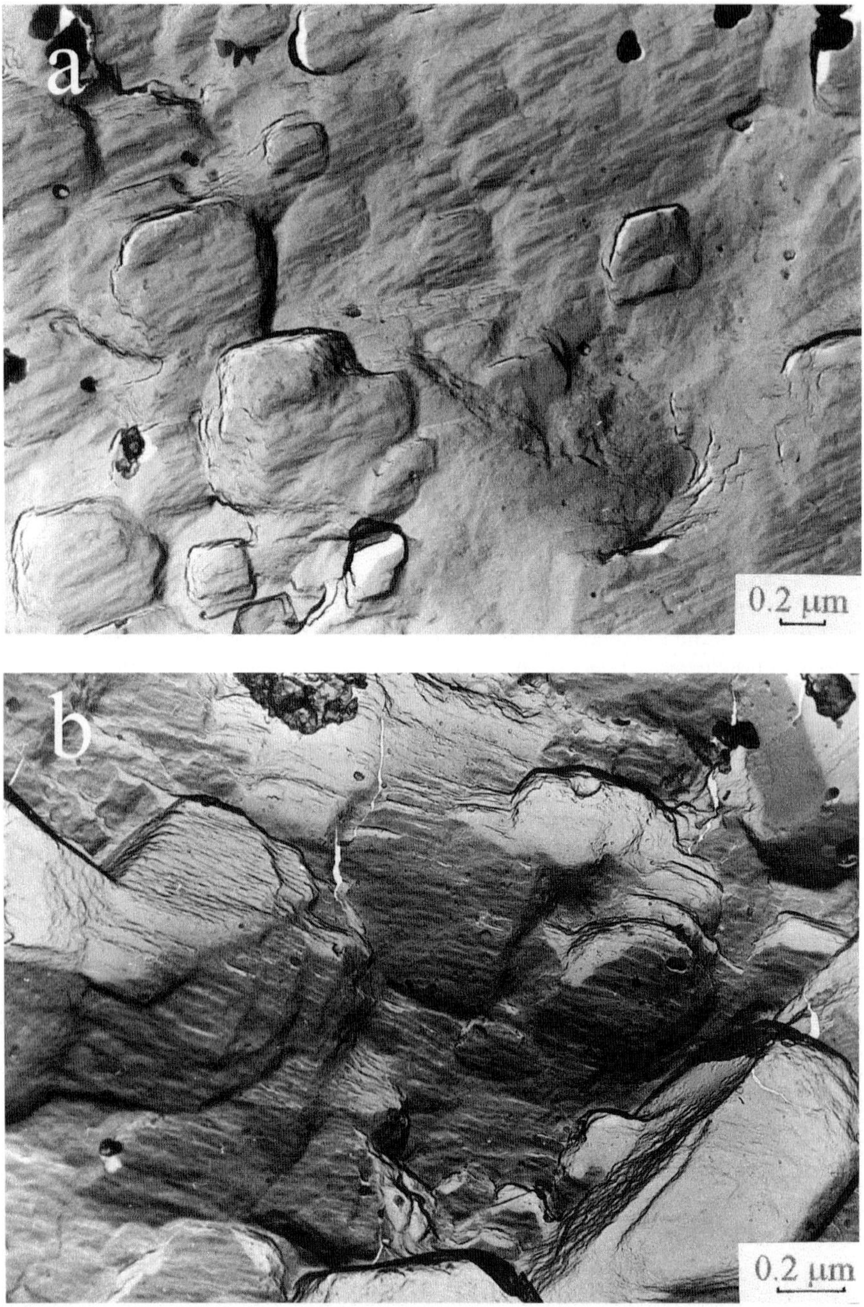

Fig. 6. TEM images of impact diamond grain surface that corresponds to (0001) plane of initial graphite crystal. (a) Blocky structure of grain surface (platinum-coal replica). (b) Lamellar structure of impact diamond (platinum-coal replica).

Acknowledgements

The authors are grateful to E.V. Melnitchuk from the Pravoberezhnaya Expedition at SEVUKRGEOLOGIA Ent. for providing us with the cores of the Zapadnaya crater. Detailed and helpful reviews by A. Therriault, J. Plado, and an anonymous reviewer led to a considerable improvement of the manuscript. We thank R.B. Rakitskaya for X-ray analyses of shock metamorphosed minerals and D.P. Demenko for TEM analyses of impact diamonds at the Institute of Geological Sciences National Academy of Sciences of Ukraine, Kiev.

References

French BM (1998) Traces of Catastrophe: A handbook of shock-metamorphic effects in terrestrial meteorite impact structures. LPI Contribution 954, Lunar and Planetary Institute, Houston, 120 pp

Goltrant O, Leroux H, Doukhan J-C, Cordier P (1992) Formation mechanisms of planar deformation features in naturally shocked quartz. Physics of the Earth and Planetary Interiors 74: 219-240

Grieve RAF (1991) Terrestrial impact: The record in the rocks. Meteoritics 26: 175-194

Grieve RAF, Masaitis VL (1994) The economic potential of terrestrial impact craters. International Geology Review 36: 105-151

Grieve RAF, Rupert J, Smith J, Therriault AM (1995) The record of terrestrial impact cratering. GSA Today 5: 189-196

Grieve RAF, Langenhorst F, Stöffler D (1996) Shock metamorphism of quartz in nature and experiment: II. Significance in geoscience. Meteoritics & Planetary Science 31: 6-35

Gurov EP, Gurova EP (1991) Geological structure and rock composition of impact structures (in Russian). Naukova Dumka Press, Kiev, 160 pp

Gurov EP, Gurova EP, Rakitskaya RB (1979) Orientation of planar elements in quartz from the rocks of meteorite impact craters (in Russian). Zapiski Vsesoyuznogo Mineralogicheskogo Obstchestva 108: 578-584

Gurov EP, Melnitchuk EV, Metalidy SV, Ryabenko VA, Gurova EP (1985) Some peculiarities of the geological structure of eroded astrobleme in the western part of the Ukrainian Shield (in Russian). Dokladi Academii Nauk Ukraini 1: 9-12

Gurov EP, Gurova EP, Yamnichenko AYu (1995) The structure of complex impact craters and estimation of their preservation stage [abs]. Lunar and Planetary Science 26: 535-536

Gurov EP, Gurova EP, Rakitskaya RB (1996) Impact diamonds in the Zapadnaya crater: phase composition and some properties [abs]. Meteoritics & Planetary Science 31: A56

Gurov EP, Koeberl C, Reimold WU (1998) Petrography and geochemistry of target rocks and impactites from the Ilyinets Crater, Ukraine. Meteoritics & Planetary Science 33: 1317-1333

Gurov EP, Gurova EP, Sokur TM (1999) Impact diamonds in the rocks of the Zapadnaya astrobleme (in Russian). Mineralni resursi Ukraini 3: 30-32

Hörz F, Quaide WL (1973) Debye-Scherrer investigation of experimentally shocked silicates. The Moon 6: 45-82

Hough RM, Gilmour I, Pillinger CT, Arden JW, Gilkes KWR, Yuan J, Milledge HJ (1995) Diamond and silicon carbide in impact melt rocks from the Ries crater. Nature 378: 41-44

Kashkarov IF, Polkanov YuA (1970) Discovery of some rare minerals by chemical dressing of heavy concentrates (in Russian). Razvedka y ochrana nedr 12: 49-50

Kurdyumov AV, Ostrovskaya NF, Golubev AS (1984) Mode of origin, stability and real structure of lonsdaleite (in Russian). Sverkhtverdye materialy 4: 17-25

Masaitis VL, Futergendler SI, Gnevushev MA (1972) Diamonds in the impactites of the Popigai meteorite crater (in Russian). Zapiski Vsesoyuznogo Mineralogicheskogo Obshchestva 101: 108-113

Masaitis VL, Shafranovsky GI, Fedorova IG (1995) Apographitic impact diamonds from the Ries and Popigai astroblemes (in Russian). Zapiski Vsesoyusnogo Mineralogicheskogo Obshchestva 124: 12-19

Masaitis VL, Mashchak MS, Raikhlin AI, Selivanovskaya TV, Shafranovsky GI (1998) Diamondiferrous impactites of the Popigai crater (in Russian). VSEGEI Press, St. Petersburg, 179 pp

McQueen RG, Marsh SP (1967) Hugoniots of graphites of various initial densities and equation of state of carbon. In: Behavior of dense media under high dynamic pressures: Comportement des milieux denses sous hautes pressions dynamiques. IUTAM Press, Paris: 208-216

Sokur TM (1999) Shock metamorphism of quartz in rocks from the Zapadnaya impact structure (in Russian). Geologicheski Zhurnal 1: 87-93

Stöffler D, Langenhorst F (1994) Shock metamorphism of quartz in nature and experiment: I. Basic observations and theory. Meteoritics 29: 155-181

Vaganov VI, Ivankin PF, Kropotkin PN, Truchalev AI, Semenenko NP, Tsimbal SN, Tatarintsev VI, Gluchovsky MZ, Bulgakov EA (1985) The Explosive Ring Structures of Shields and Platforms (in Russian). Nedra Press, Moscow, 200 pp

Valter AA, Ryabenko VA (1977) The impact craters on the Ukrainian shield (in Russian). Naukova Dumka Press, Kiev, 154 pp

Valter AA, Yeremenko GK, Kvasnitsa VN, Polkanov YuA (1992) Shock metamorphosed carbon minerals (in Russian). Naukova Dumka Press, Kiev, 171 pp

Valter AA, Gurs'ky DS, Yeremenko GK, Bochko OV (1999) Impact diamonds - the new kind of raw material in Ukraine (in Russian). Mineralni resursi Ukraini 3: 16-22

Vishnevsky SA, Raitala J (2000) Impact diamonds as indicators of shock metamorphism in strongly-reworked Pre-Cambrian impactites. In: Gilmour I, Koeberl C (eds) Impacts and the Early Earth. Springer Verlag, Berlin-Heidelberg, pp 230-247

Vishnevsky SA, Afanasiev VP, Argunov KP, Pal'chik NA (1997) Impact diamonds: their features, origin and significance (in Russian and English). In: Vdovikin GP (ed) Russian Academy of Sciences Siberian Branch Press, Novosibirsk: 55 pp

Bosumtwi Impact Crater, Ghana: A Remote Sensing Investigation

Rüdiger Wagner[1], Wolf Uwe Reimold[2], and Dion Brandt[2]

[1]Anglo Technical Division, 45 Main Street, Marshalltown 2107, Johannesburg, South Africa. (rwagner@angloamerican.co.za)
[2]Impact Cratering Research Group, Department of Geology, University of the Witwatersrand, Private Bag 3, P.O. Wits 2050, Johannesburg, South Africa. (065wur@cosmos.wits.ac.za)

Abstract. The Bosumtwi crater in Ghana is a ~1.05 million year old, very well-preserved complex meteorite impact structure of approximately 10.5 km rim-to-rim diameter. The interior of the structure is largely filled by Lake Bosumtwi, and the crater rim and the environs of the crater are covered by dense tropical rainforest. Thus, direct geological study of this structure is not easy; but remote sensing and geophysical data can be used to better constrain the surface and subsurface geology. Landsat Thematic Mapper (TM) data reveal detailed structure of the crater rim, including arrays of radial, concentric, and tangential or oblique fractures. In all data sets, the presence of an apparent ring feature at ~10 - 11 km from the crater center is visible. Digitised topographic information and, to a lesser extent, ERS-1/2 radar data provide an improved picture of the crater morphology. A digital elevation model (DEM) for the wider Bosumtwi crater area also reveals the presence of a subdued elevation feature at about 10 - 11 km radial distance from the lake center, which, in the western area, coincides with the distribution pattern of vegetation. Perspective views of the DEM and other remote sensing data suggest the existence of a further structural feature at 17 - 19 km distance from the crater center, and which is mainly visible in the southeast sector. An "outer rim" topography is also known from other impact structures, for example, the Elgygytgyn crater in Russia; the origin of this feature is still debated, but the existence of two apparent ring features may suggest that they represent ring faults caused by the impact compression.

1 Introduction

The Bosumtwi Structure in Ghana, centered at 06°30´N and 01°25´W, is a crater structure of a rim-to-rim diameter of ~10.5 km. The topographically spectacular feature (the crater rim rises >300 m above the lake level) is characterised by a steep and complexly faulted rim. The interior of the structure is mostly filled with the ~8.5 km wide Lake Bosumtwi, for which a current maximum water depth of some 80 meters (e.g., Karp et al. 2000a, b) has been determined. Recent publications providing comprehensive overviews of the geology and geophysics of the Bosumtwi structure and detailing the mineralogical and geochemical studies performed on related lithologies include, inter alia, Koeberl and Reimold (1996), Koeberl et al. (1997a, 1998), Reimold et al. (1998), and Plado et al. (2000).

The origin of the structure has long been controversial (see, e.g., Jones 1985a or Koeberl et al. 1998). Fergusson (1902) favored a volcanic origin, whereas Kitson (1916) proposed that the feature represented a subsidence feature. Maclaren (1931) was the first to suggest an impact origin for Bosumtwi, whereas Rohleder (1936) preferred a cryptoexplosion process. Since the 1960s, outcrops of suevitic breccia have been known from Bosumtwi (Chao 1968; Jones et al. 1981), and reports of coesite, Ni-rich iron spherules, and baddeleyite from suevite and from vesicular glass from the crater rim (Littler et al. 1961; El Goresy 1966; El Goresy et al. 1968) provided support for the impact hypothesis. Since then shock metamorphic effects have been described in clasts in suevitic breccia (Koeberl and Reimold 1996; Koeberl et al. 1998). All this evidence conclusively affirms the impact origin of the Bosumtwi structure.

Substantial interest in the Bosumtwi impact crater has been generated by studies of the Ivory Coast tektites, which were first reported by Lacroix (1934) and occur in an about 40 km wide area in Cote d'Ivoire and, as microtektites, over a large area of the Atlantic near West Africa. As reviewed by, for example, Jones (1985b), Koeberl (1994), or Koeberl et al. (1997a, 1998), Bosumtwi crater has long been regarded as the likely source of these tektites, a view which was recently confirmed when very similar ages for Ivory Coast tektites and Bosumtwi Crater samples were established by Koeberl et al. (1997a).

Geological studies of the region around Bosumtwi have been carried out since the 1930s (Junner 1937; Woodfield 1966; Moon and Mason 1967; Jones et al. 1981), but detailed mapping has been hampered by the thick vegetation cover. More recently, some geological work, especially along a then well-exposed road section across the western crater rim and on exposures in the sector around the northern and northeastern parts of the crater, has been reported by Koeberl and Reimold (1996) and Reimold et al. (1998). The region around Bosumtwi is largely covered by dense, tropical rain forest, which has, to date, only permitted detailed studies of rare exposures and road sections.

Figure 1 summarizes the regional geology around Bosumtwi. The crater lake is surrounded by a narrow annulus of lake beds and talus, which is derived from the steep slopes of the crater rim. The crater was formed in lower greenschist facies metasediments (meta-graywacke, meta-arkoses, meta-quartzites, shales, and phyllite) of the 2.0 - 2.2 Ga Birimian Supergroup (Wright et al. 1985; Leube et al. 1990; Davis et al. 1994; Hirdes et al. 1996; Oberthür et al. 1998), which are well exposed in the Obuom mountain range just to the southeast of the crater. Further southeast, metasedimentary rocks of the approximately contemporaneous (~2.1 - 2.13 Ga; Hirdes and Nunoo 1994) Tarkwaian Supergroup have been mapped.

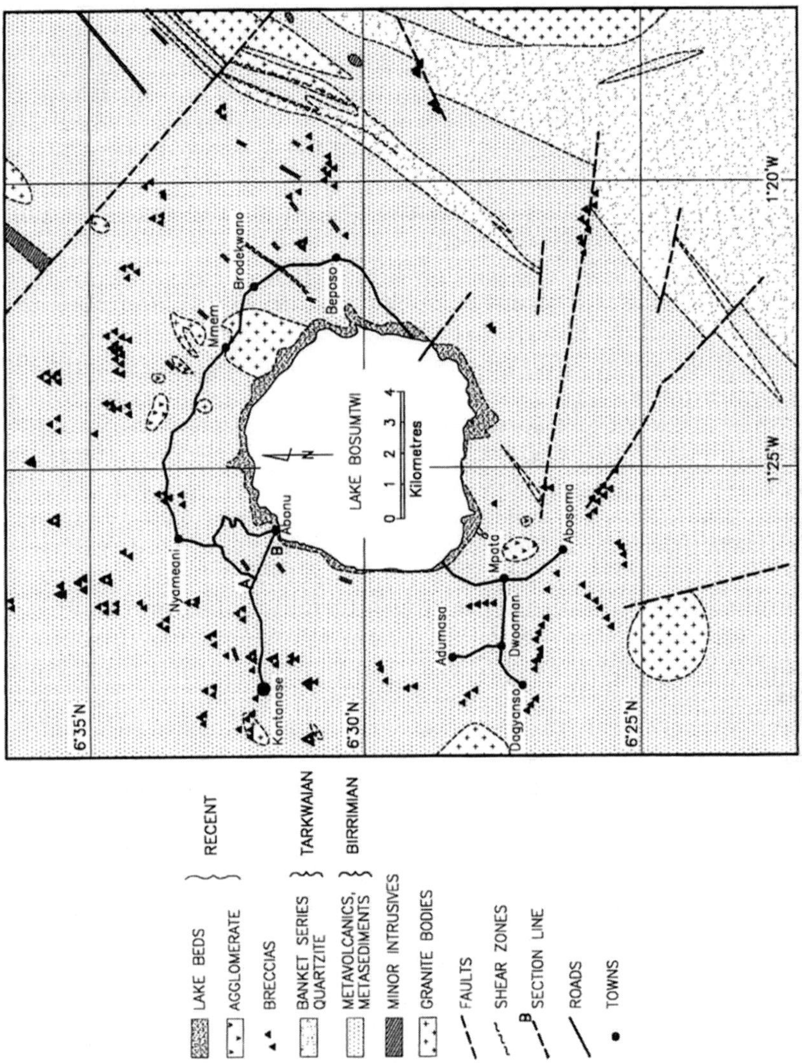

Fig. 1. Schematic geological map of the region around Bosumtwi impact crater (modified after Reimold et al. 1998).

These basement strata are characterised by a strong, northeast-trending fabric with steep dips to either the northwest or southeast. Local variations from this trend have been observed and are regarded as the result of local folding. In the immediate environs of the crater, the lithology is dominated by meta-graywackes and -sandstones, but some shale and mica schist have also been observed, as reported by Reimold et al. (1998), particularly in the northeastern and southern rim sectors. Biotite and amphibole granites have been mapped in this region (see Fig. 1) by Junner (1937) and Moon and Mason (1967). Reimold et al. (1998) reported the observation of numerous up to 1-m-wide dikes of biotite-granitoid, which crosscut metasedimentary strata in many occurrences along the crater rim. However, these authors estimated the overall granitoid component in the region at no more than 2 %.

Numerous breccia exposures have been described in the past from the environs of the crater. It appears likely that many, if not most, of these represent post-impact accumulations of lateritic and other weathering-related breccia. Laterites of >10 m thickness have been observed in many areas around this crater. However, the presence of suevite deposits has been confirmed to the north and to the southwest of the crater. Due to the patchy exposure in the strongly vegetated terrane and the accelerated erosion and weathering processes in this tropical environment it has not been possible to map the likely present remnants of the ejecta blanket from this relatively recent impact event in a comprehensive way.

Because of the limited exposures and the crater fill by Lake Bosumtwi, detailed geophysical studies are required to obtain surface and subsurface information. First magnetic studies across the structure were carried out by Hunting Surveys Limited on behalf of the Geological Survey Department of Ghana (Jones et al. 1981). A central negative anomaly of ~40 nT, flanked on the northern side by a ~20 nT positive anomaly that intersects regional magnetic anomalies, was described by these authors and attributed to the presence of a breccia lens below the lake sediments. A cursory gravity study around the lake by Jones et al. (1981) only reflected regional trends and could not be used to further constrain the magnetic results.

In 1997, a high-resolution, low-altitude airborne survey of the Bosumtwi Structure was carried out by the Geological Survey of Finland in collaboration with several other groups (Koeberl et al. 1997b; Ojamo et al. 1997; Pesonen et al. 1998; Plado et al. 2000) and included total magnetic field, electromagnetic field, and gamma radiation measurements. A first analysis of these new magnetic data has been presented by Plado et al. (2000). A circumferential magnetic halo outside the lakeshore of ~12 km diameter was reported. A central negative magnetic anomaly in the north-central part of the lake was detected, together with smaller positive anomalies to the side as well as to the north and south of it. This pairing of anomalies is regarded as being typical for magnetised bodies occurring at shallow latitudes. In addition, a few weaker anomalies were described from the eastern and western parts of the lake. This group of anomalies was interpreted as encircling the central uplift feature below the lake and thought to be associated with strongly remanently magnetised impact melt or melt-rich suevite bodies.

Koeberl et al. (2000) discussed multi-disciplinary evidence of an outer rim feature at about 10 km distance from the crater center.

Little is known about the deep structure of the crater. At 10.5 km rim-to-rim diameter, Bosumtwi should be a complex impact structure, with a significant central uplift. Since late 1998, two teams of geophysicists from Germany, Ghana, and the United States of America have pursued gravity, refraction and reflection seismic analysis of this impact structure. First results of this team also known as the Lake Bosumtwi Working Group (LBWG) were reported by Karp et al. (2000a-c) and Scholz et al. (2000). According to the LBWG, the gravity study showed a minimum over the lake area, resulting from the water in the lake, the sedimentary lake beds, low density impact formations and brecciated basement rocks. Their first seismic results indicate a complex subsurface structure and the existence of a central uplift, below the lake.

2
This Work

The availability of topographical data sets published by the Ghanaian mapping authorities allows the application of GIS based processing techniques. Furthermore, Landsat Thematic Mapper (TM) and ERS (European Remote Sensing Satellite) radar data were processed to enhance geomorphological and geological information. Digital Elevation Model (DEM) generation and a combination of various data sets allows, for the first time, perspective analysis of the geomorphology of the Bosumtwi Crater integrating satellite imagery.

Through this integrated approach it was basically possible to analyze previous research results that have often been based on single, two-dimensional data sets in three dimensions and, additionally, provide a digital data base for future research. Preliminary results of this study were presented by Wagner et al. (2000).

3
Data and Methodology

3.1
Landsat TM 5

A mosaic over the Bosumtwi Crater area was produced from subsets of the scenes 194/56 (acquired 20/12/1986) and 194/55 (acquired 27/01/1986). To reduce the effects caused by different seasonal acquisition of the data, the two scenes were merged manually after statistical enhancement.

Landsat TM 5 provides data in six bands spread unequally between 0.45 and 2.35 μm with approximately 30 m ground resolution and in one band covering the range between 10.4 and 12.5 μm, with approximately 120 m ground resolution.

3.2 ERS Radar Data

The ERS sensor acquires data in the C-band (5.7 cm) of the electromagnetic spectrum with a VV polarisation and a resulting ground resolution of approximately 28 m. Unlike the passive Landsat system producing an image detecting solar reflectance from the earth surface, the active radar system carries the source of the pulse that it detects after reflection from the sensed surface. Due to operating in the microwave portion of the electromagnetic spectrum, radar data are mostly weather-independent and can be used in highly clouded areas. Furthermore, it can penetrate through the canopy and, therefore, can be used advantageously in highly vegetated areas. The use of radar data in arid areas, using the ability of microwaves to penetrate through dry sand to detect paleochannels, has been well documented (e.g., McCauley et al. 1982).

The scenes 11399/3465 (acquired 25/06/1997), 11399/3483 (acquired 25/06/1997), and 03111/3483 (acquired 24/11/1995) were used for the mosaic. Before combining the individual scenes to a mosaic, the single scenes were processed to reduce speckle effects (salt-and-pepper effect), using an iterative combination of various filters (e.g., the Lee filter) with different kernel sizes. As there is a playoff between noise reduction and loss of resolution, speckle reduction has to be carried out carefully.

Due to acquisition geometry, geo-referencing is not sufficient for radar data to be represented in a map projection. Ortho-rectification does not only allow the data to be registered in map coordinates but also accounts for effects such as foreshortening, layover or shadow, and removes these effects. This procedure results in the formation of „holes“ in the output data, which have to be filled. Unfortunately, the algorithm that interpolates to fill up these gaps in data is not very sophisticated in the ERDAS software and produces „spaghetti-like“ structures, which are, for instance, visible along the crater rim. To reduce this effect, various filter applications were applied with mediocre success.

3.3 Methodology

Lineament analysis was carried out on Landsat TM (Fig. 2) and SPOT satellite images of the Bosumtwi area. Topocadastral maps at the 1 : 50 000 scale, obtained from the Ghanaian mapping authority in Accra, as well as the geological maps by Woodfield (1966) and Moon and Mason (1967) were used as base maps for the compilation of drainage patterns in the immediate and wider vicinity of the structure, respectively. A SPOT satellite image of the Bosumtwi area, identical to the one used by Garvin et al. (1992), was made available by R.P. Viljoen (University of the Witwatersrand, Johannesburg).

The topographic information was digitized and a DEM model was generated using the Arcinfo (version 8.1) software. Image processing and data integration were undertaken using the ERDAS Imagine (version 8.4) software. Several processing steps were applied to the Landsat TM data to enhance lithologies and mineralogy, as expected with very little success due to the dense vegetation.

Figure 2a shows the Landsat TM image RGB 741 highlighting vegetation in green color and Fe-rich rocks or, more likely, lateritic soils in red tones. Clearly visible are the villages and roads.

More promising results could be expected from vegetation indices, assuming that vegetation somehow reflects underlying geology. Through the GIS-based approach, different perspective views could be generated using the remote sensing data as overlays. Various perspective views of the DEM for the Bosumtwi crater are shown below.

4 Results

Figure 2b shows a lineament map of the area around the impact crater, derived from visual analysis of SPOT and Landsat TM imagery. In comparison to large-scale aerial photography, analysis of which is severely hampered by the thick vegetation in this region, these small-scale data provide enhanced structural information on the rim zone and surrounding areas. It is obvious that mega-lineaments occur in an only up to 3 km wide zone around the crater rim, whereas the wider environs are basically devoid of such features. These lineaments, presumably representing fractures and faults, have predominantly radial orientations to the crater center. Oblique, tangential, and concentric orientations are present but are of lesser abundance.

This pattern correlates well with ground observations emphasizing the dramatic structural deformation of the crater rim (Figs. 3a and b). In 1997, some of us (DB and WUR) observed apparent displacements of rim blocks radially inwards, off the rim, and along what are most probably listric faults downwards and inwards. The zone characterized by a high-density of lineaments is not concentric with respect to the center of the crater lake. Instead, it is clearly seen that the zone of perceived macro-deformation is broader on the northwest, north and northeast sides than in the other sectors around the crater. The reason for this asymmetry effect is not yet known.

The regional northeast-southwest trend of the Birimian basement is most obvious in the Obuom Range to the southeast of the Bosumtwi structure. Radar data and Landsat TM data in conjunction with the topographic data (Fig. 4, as well as the digital elevation model - DEM - shown in Fig. 5a) suggest the existence of a system of large-scale lineaments of the same trend to the northeast of the crater. The lineament map reflects the regional trend, in addition to the obvious lack of strong macro-deformation that could be related to the impact event, in this area.

Subjective separation of isopachs presumed to be part of the watershed and others belonging to the crater rim indicated to Jones et al. (1981) a roughly circular crater shape and rim-to-rim diameter of about 11 km. The drainage maps derived from analysis of topocadastral maps in comparison with SPOT and Landsat TM interpretation (Figs. 6a and b) demonstrate the intense radial drainage into the crater lake and, to a much lesser degree, radially outwards off the crater rim. The pattern in the wider environs of the structure is more complex, but crater-tangential and –concentric drainage is still obvious. The regional northeast-southwest fabric trend is discernable, though not very pronounced. To the southeast of the crater structure, the Obuom mountain range represents a strong watershed, off which a significant number of drainage lines runs perpendicular to the extension of this range. A relationship between drainage pattern and crater structure is most obvious in a zone of 0.5 - 0.7 crater diameters outside of the topographic crater rim.

A ring-shaped depression surrounds the crater rim with a maximum width in the western and northern sectors of about 5 km (compare also Fig. 7). It is in this depression that the drainage pattern attains a most pronounced annular trend (as already remarked upon by Junner 1937). At about 8 - 11 km from the center of the

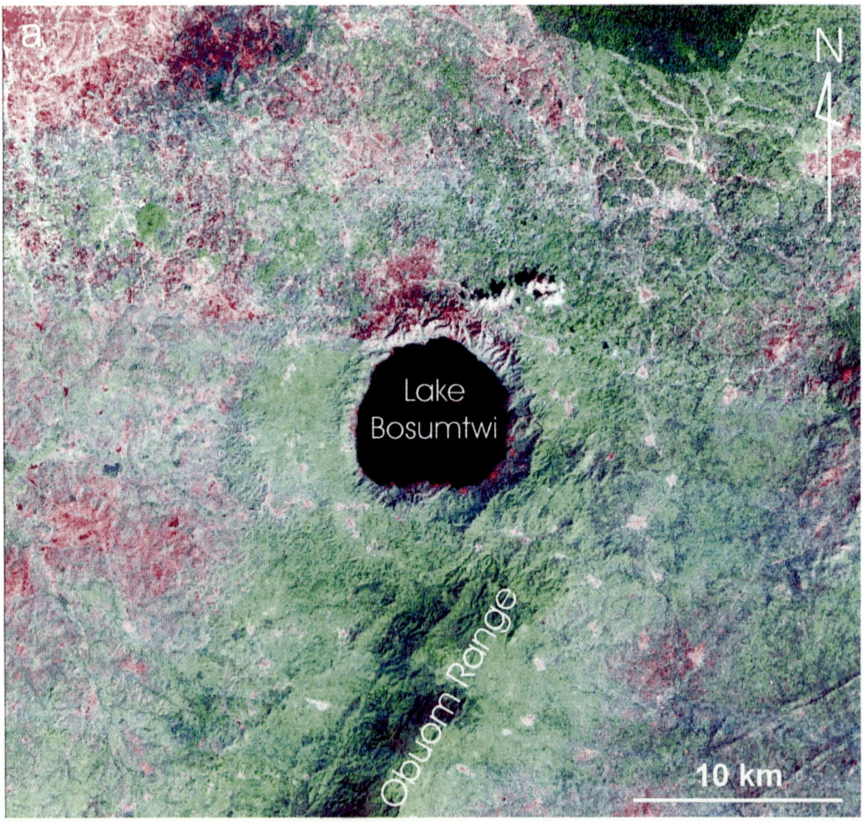

Fig. 2. (above, a) Landsat Thematic Mapper (bands 7, 4 and 1) image of Bosumtwi and surroundings. Some structural detail of the inner crater rim is evident, as well as the strong northeast-southwest trend of the Obuom Range to the southeast of the crater structure. Green colors denote thick vegetation, and purplish color is thought to denote iron-rich (lateritic) soils and/or human habitat. (opposite, b) Lineament map derived from visual analysis of Landsat Thematic Mapper and SPOT satellite imagery.

crater, a somewhat elevated, circular feature (30 - 80 m elevation above the level of the depression) is noted.

Jones et al. (1981) drew attention to the apparent coincidence of the main crater rim with the major regional watershed, the Obuom Range, in the southeast sector of the study area. Following the slopes of the Obuom Range to the southwest, the regional watershed distances itself from the crater rim by several kilometers. This is thought to be mainly a combined effect of the morphologically higher Obuom Range and the almost east-west trending ridge that represents the watershed to the south, which can be connected with the otherwise more dominant expression of the outer rim. In an earlier publication on the use of remote sensing data for impact crater studies, Garvin et al. (1992), using SPOT data, concluded that the extension of the watershed to the south might be a product of major slumping of the inner crater wall. This was also the cause of the not entirely circular shape of the crater structure.

The Landsat TM image in Fig. 2a shows the shape of the crater almost as an upside-down heart with an indentation in the southern rim. In accordance with Garvin et al. (1992), this is most likely due to block faulting and tilting of the crater wall. It is, however, remarkable that none of the data sets used show major slumping or faulting on either side of the main block. Therefore, reconstructive modeling suggests an originally flatter shape of the crater in the south. It is likely that this was due to the massive Obuom Range resisting major impact deformation in this area.

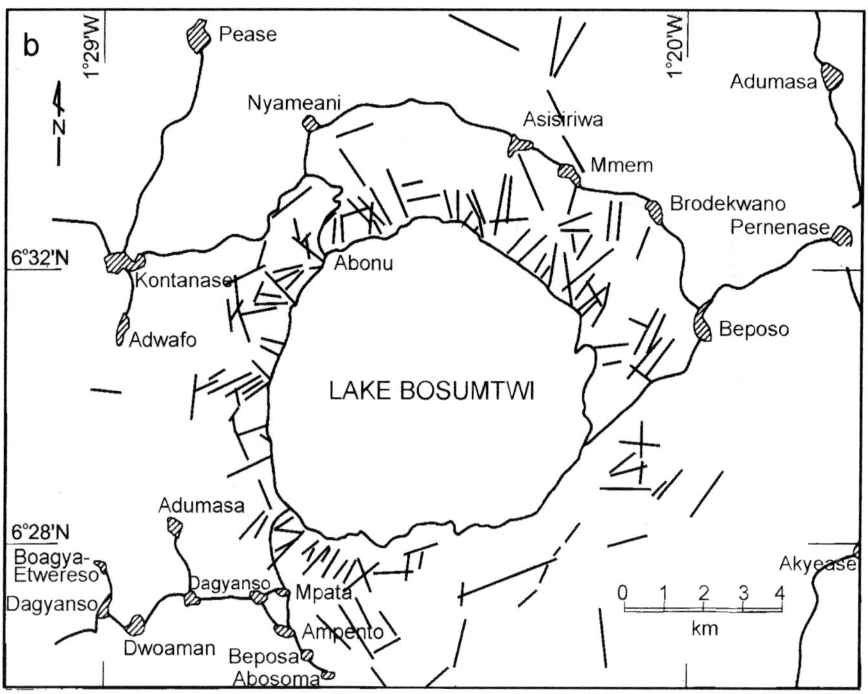

Fig. 3. (a) Photograph of the northwestern crater rim, showing some of the very prominent radial fractures. In the foreground a section of the tar-road winding down to the lake, along which Reimold et al. (1998) found excellent exposures. (b) Part of the western crater rim, with an array of major faults of radial, rim-parallel, and oblique orientations.

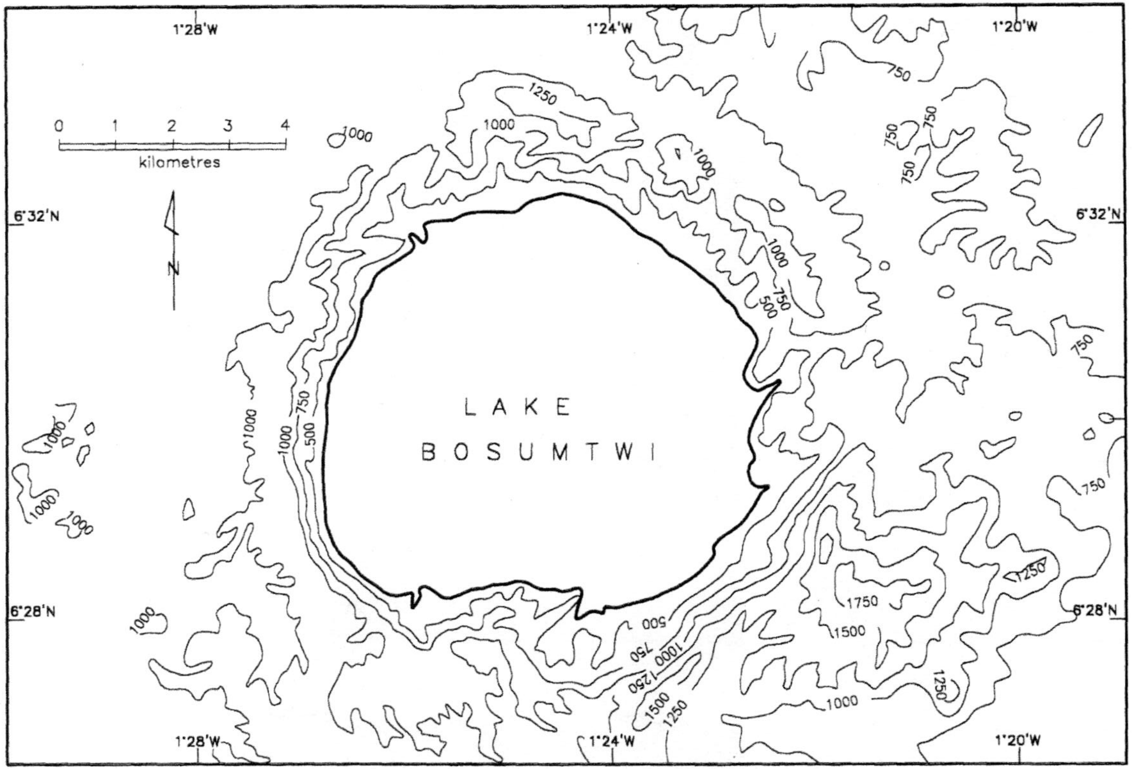

Fig. 4. Topographic map of the Bosumtwi crater area, as derived from contour analysis of 1 : 50 000 topocadastral maps of the Ghanaian mapping authorities.

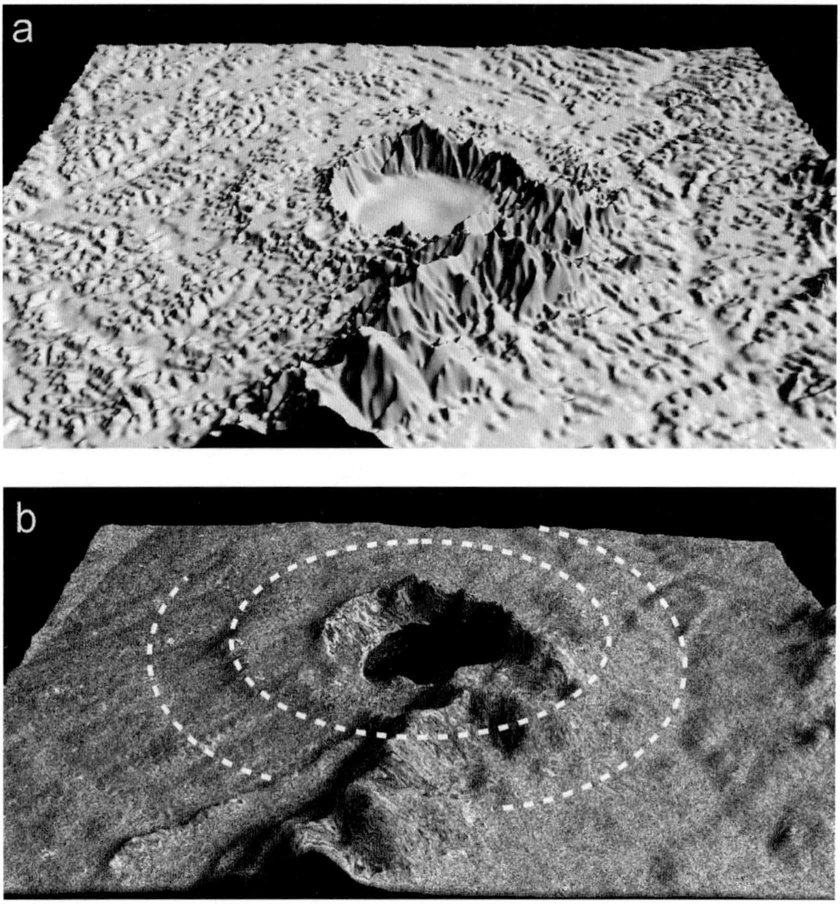

Fig. 5. (a) Digital Elevation Model (DEM) for the Bosumtwi crater area. (b) Combined imaging of the DEM and the ERS radar data. The two outer ring features identified in this work are emphasized by lines overlain onto the image.

Garvin et al. (1992), furthermore, discussed that „subtle variations in either type or density of vegetation or change in local slope" appeared to enhance some features in the image that could relate to the crater geomorphology. An „outer rim" or „annular ridge" feature was discernable in the SPOT image at an approximate distance of 9 - 10 km from the lake center. Garvin et al. (1992) also claimed that it was possible to relate some known, localised suevite occurrences to the north and south of the crater with apparently distinctive areas in the satellite image and, thus, concluded that remote sensing analysis could, even in densely vegetated regions, be of considerable geological value in guiding ground-based exploration. Even though the latter statement is, in principle, correct, neither our Landsat TM nor radar data provide any evidence that could allow localisation of these suevite occurrences in our images. Even though the Landsat TM sensor has a better spectral resolution than the SPOT sensor, it is, due to dense vegetation and, possibly, the worse geometrical resolution, impossible to derive information that could contribute to lithological mapping. Additionally, even though vegetation indices enhance geomorphological features, these images show little correlation with known geology. This is most probably also severely hampered through the development of vast and thick laterite deposits.

Garvin et al. (1992) mentioned that vegetation patterns highlight geomorphological features such as, in the Bosumtwi case, the outer rim zone. Neither our Landsat TM nor radar data support this finding, except for a small strip on the western side of the crater where vegetation obviously follows an annular trend for less than ten kilometres. As for the rest of the image, vegetation is interrupted in numerous areas by human activity and shows differences in density or type (e.g., locally known cocoa plantations appear as patches of slightly lighter coloration) in the northwestern part of the area, which do not follow the geomorphological trend of the outer rim. We suggest that the presence of the optically dominant crater itself deceives the eye to interpret annular structures in the vegetation pattern. The rather unconventional method of covering the „crater proper“ with a filled square and in such hiding the circular feature allows a more unbiased interpretation of the patterns in the vicinity of the crater rim. This leads, in the absence of detailed regional geological information that could be related to vegetation patterns, to the conclusion that, except for the small part to the west of the crater, patterns in the vegetation are following trends set by local human activity.

However, analysis of the drainage system reveals clearly an annular pattern and combined with various perspective views of the DEM, as, for example, shown in Figs. 8a and b, supports the findings by Jones et al. (1981) and allows more detailed interpretation.

The perspective views of the exaggerated DEM suggest the continuation of the outer rim into the Obuom Range. Garvin et al. (1992) reported a faint expression of this rim visible in the southeastern quadrant of the examined SPOT image, in the area of the Obuom Range. Assuming that this outer rim maintains more or less equidistance from the crater center, it is expected to merge frontally with the northernmost extents of the Obuom Range. A weak concentric geometry is evident and seems to support the reasonable assumption that the crater edge is superimposed on the northeast trending mountain range. Figures 8a - a view of the DEM towards the south - and 8b - a view from the south - clearly show a secondary, semi-annular feature in the southeast sector at further distance from the Obuom Range. This feature is marked by a topographic high and a river following its trend. Admittedly, it is rather difficult to trace this second „rim“ all around the crater.

The four topographical profiles across the Bosumtwi impact structure illustrated in Fig. 7 show the dramatic rise of the crater rim, followed on the outside by a plateau-like feature extending for several kilometers, before the ring depression is reached. It is thought that this zone along the outer rim represents, to a large degree, talus debris, which at least locally ought to cover remnants of impact ejecta. The apparent width of the crater rim is asymmetric with values around 1.5 km determined for the western side and of up to 2.5 - 3 km for the northern and eastern sides. Following the main ridge, it becomes evident that the crater shape was most probably not that circular after all. It is, also, noticeable in some of the profiles that a second „rim" structure seems to occur at about 6 - 7 km outside of the „outer rim". The zone between these two „rim" features has also been observed in the aerogeophysical data by the Bosumtwi Working Group (e.g., Pesonen et al. 1998), for example in the form of a broad anomaly in the airborne radiometric data for U and K.

Garvin et al. (1992) discussed that if the first „outer diameter" represented the actual crater rim of this impact structure, the catastrophic effect of this impact

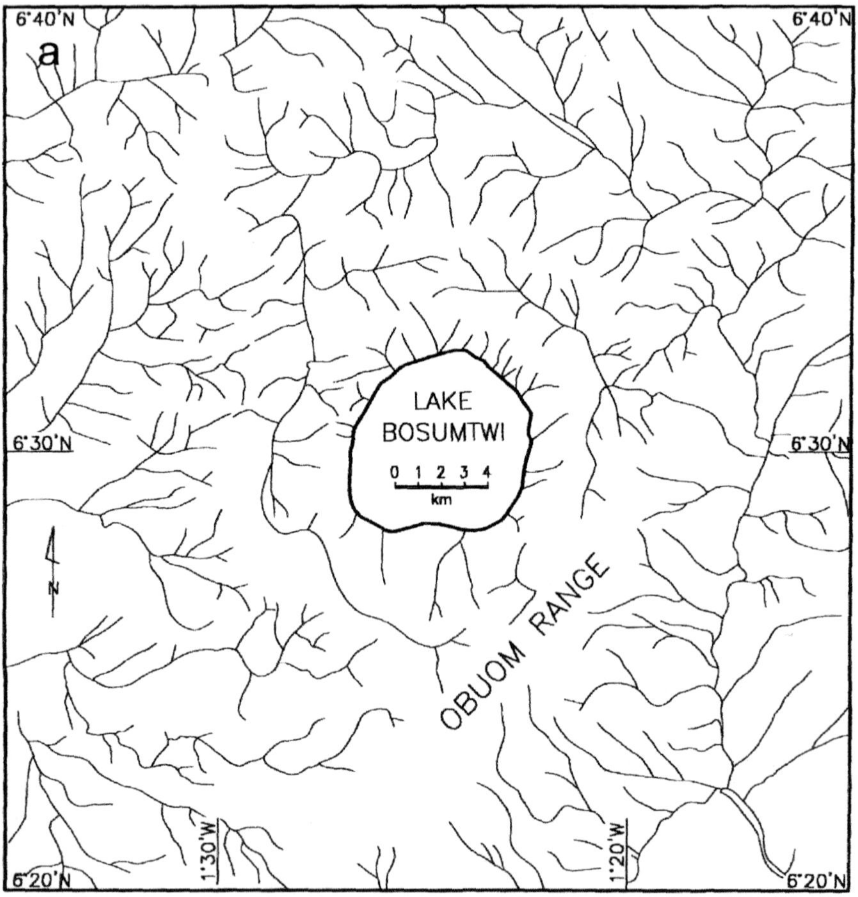

Fig. 6. (above, a; and opposite, b) Drainage maps derived from analysis of topocadastral and geological maps by the Ghanaian mapping authority and Geological Survey Department of Ghana at 1 : 62 500 and 1 : 50 000.

event could have caused violent disruption of photosynthesis in equatorial regions. In this context it may be relevant that Dupont et al. (2001) recently reported paleo-environmental data from the Congo fan, approximately 1500 km southeast of Bosumtwi. These data do not indicate a significant disturbance at around 1.05 Ga (the approximate age of both the Bosumtwi crater and the Ivory Coast tektites - Koeberl et al. 1997) that one would attempt to correlate with the Bosumtwi impact event. Dupont et al. (2001), however, noted a general (i.e., in relation to longer term variation) change in the paleoclimate cycles (glacial-interglacial variation) starting at about that time.

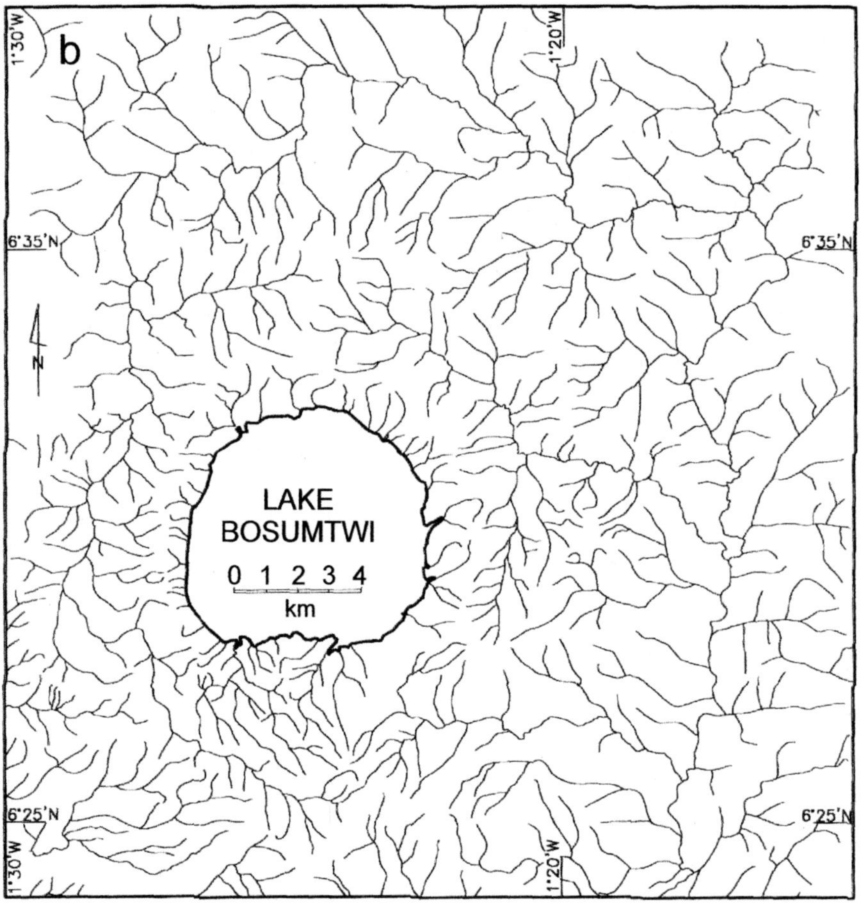

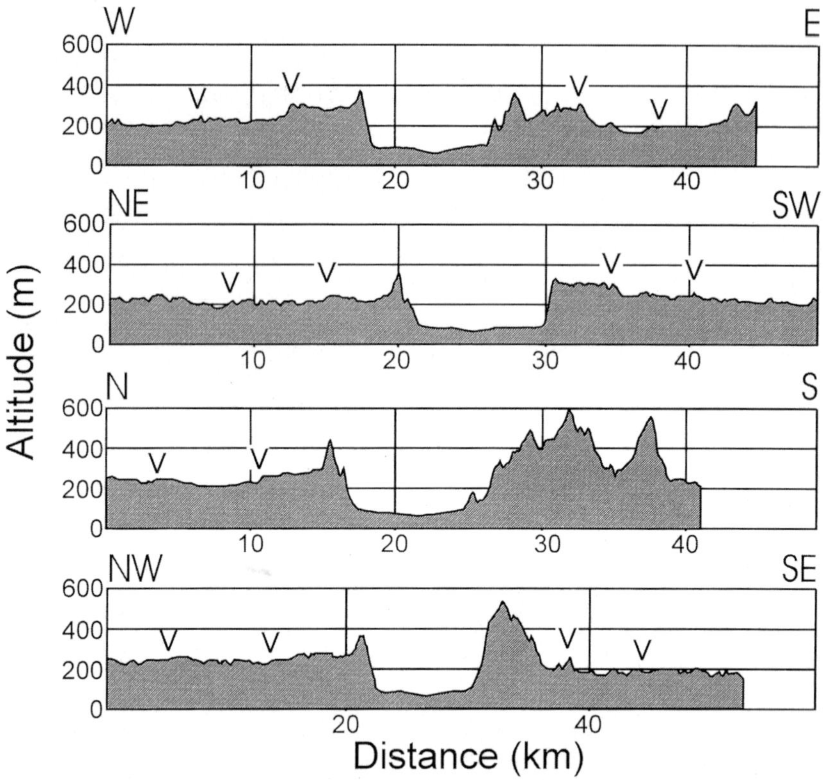

Fig. 7. Several topographic profiles across Bosumtwi crater, emphasizing the plateau-like outer rim zone and the subdued topographic feature at the outer edge of the ring depression. The arrows indicate the positions of the outer topographic ring features, as discussed in the text.

5 Discussion

Remote sensing-based geomorphological analysis proved to be most advantageous on heavily vegetated terranes, such as the tropical region around Lake Bosumtwi. Macro-lineament analysis based on satellite imagery demonstrates that intense macro-deformation, involving faulting and block faulting, is delimited to a relatively narrow zone of approximately 3 km outside of the crater rim and, thus, must be related to the impact event. Including the width of the crater rim zone of ~1.5 - 2.5 km, this indicates a maximum width of the macro-deformation zone of approximately 0.5 crater diameter.

Drainage pattern analysis at and around the impact structure shows that in the rim zone and also in a wider zone, surrounding the rim, of ~0.7 crater diameter

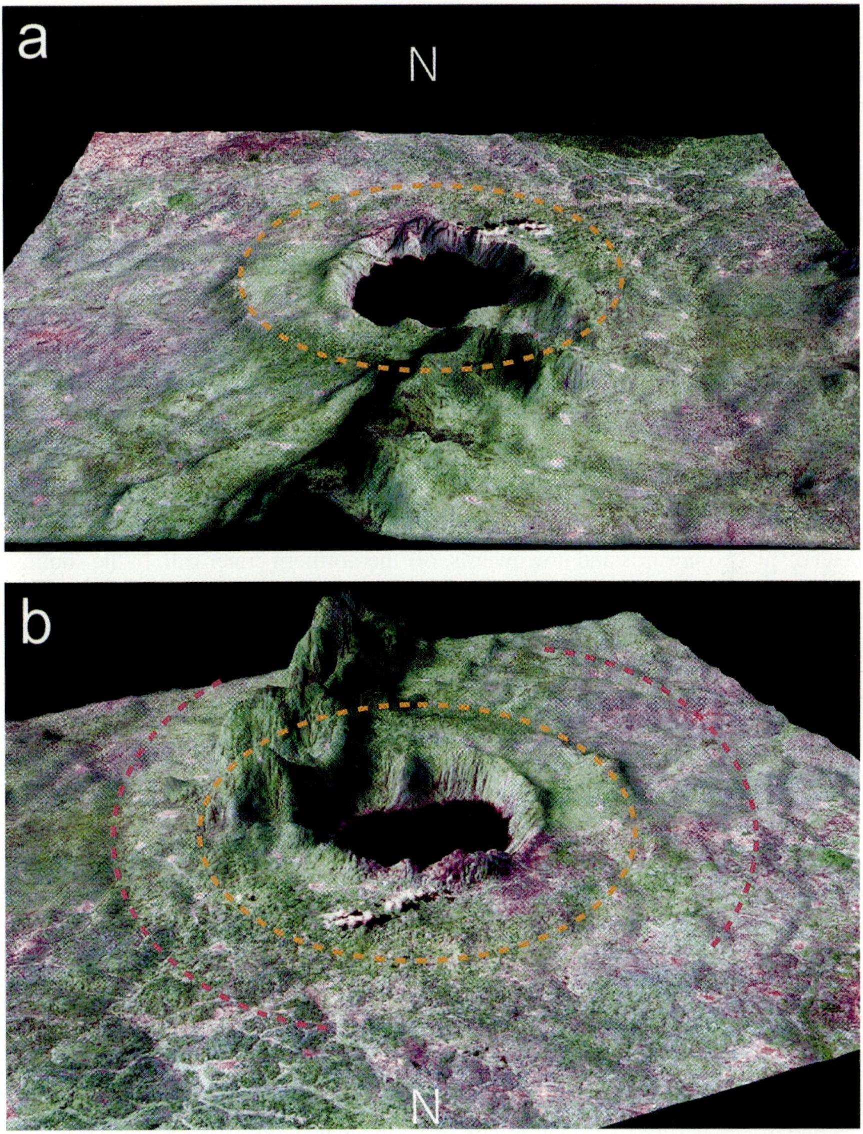

Fig. 8. (a) The DEM combined with Landsat TM bands 7, 4, and 1 data. The inner "outer ring feature" is emphasized by a circular line. (b) View to the southeast for the same data combination utilised in the generation of Fig. 8a. See text for discussion. Both outer ring features are emphasized.

width strong structural deformation related to the impact event can be recognised. It is possible that this zone mirrors the reasonably thick part of the impact ejecta blanket formed about 1 Ma ago, as a strong, radially outward directed, attitude of drainage lines is noted at this distance from the crater rim. With regard to both the lineament and drainage line distribution patterns, it is quite easy to separate the obviously impact-related features from the regional background (southwest-northeast).

The topography at and around Bosumtwi crater (Fig. 7) can be divided into five sections: (1) the actual crater rim, which stands out with elevations of - locally - more than 300 m above the surrounding country and more than 330 m above the lake level; (2) a plateau-like feature with gentle slopes dipping outwards, away from the crest of the crater rim; (3) the so-called ring depression, which is not very pronounced and probably partially obscured by talus off the outer crater rim; (4) an outer feature (rim) of very slight elevation (generally <100 m, mostly <20 m) at the outer margin of the ring depression at a distance of about 9 - 10 km from the center of the structure and (5) a subdued morphological feature at about 15 - 20 km from the crater center.

The highest point of the southern crater rim can be found immediately to the south of the lake edge, whereas in the north a relatively broader talus zone, shallowly dipping towards the lake level, is present. Along the inner crater rim, parallel slump scarps are noted (Fig. 3b), suggesting that a series of slivers has been displaced inwards and downwards, off the northern crater rim. In the south, a large block seems to have slumped inwards off the crater rim. As for the crater shape, the rim position in the south and in the north indicates an almost oval or pear-shaped geometry of the entire crater structure.

Outer ridges have been described from a number of other impact structures in the size range from 10 - 24 km diameter (Deep Bay, 10 km diameter - M. Dence, pers. commun. 2000; Innes et al. 1964; Elgygytgyn, 18 km diameter, e.g., Gurov and Yamnichenko 1995; Ries Crater, 24 km, e.g., Johnson et al. 1964; Jones et al. 1981). They seem to occur at distances between approximately 1.6 and 1.9 crater radii from the center of a structure. Gurov et al. (1995) interpreted this outer ridge of an impact structure as the edge of a so-called "outer crater", whereby they considered the "inner crater" as that part of an impact structure that is characterised by a pronounced crater rim. They noted the presence of a second, subdued, peripheral rim in a number of impact structures and proposed that its radial distance from the center is a function of the degree of erosion of a given impact structure. These authors did not discuss the likely origin for this outer ridge (peripheral rim).

A number of processes/effects could be envisaged to be possible causes of such an outer ridge. The outer (peripheral) morphological feature could represent (1) the actual crater rim; (2) the presence of a ring fault outside of the actual crater, possibly formed due to slumping of the whole block between the edge of the transient crater and the position of the peripheral elevation during the crater modification stage; (3) the outer limit of the original ejecta blanket, which has been, by now, preferentially eroded in certain areas - an effect that could have been controlled by underlying structures. The latter might also be detectable in the drainage pattern, which has a certain annular aspect to it.

The outer rim features at about 9 - 10 and 17 - 19 km from the center of the crater structure appear to coincide with a distinct pattern in some of the airborne geophysical data (e.g., the U, Th, and K radiometric anomaly maps - Pesonen et al. 1998 and L.J. Pesonen, pers. commun., 1997; also Koeberl et al. 2000). It is possible that the "outer rim" at 9 - 10 km from the center marks the extent of the original ejecta blanket. According to Koeberl et al. (1998), several of the presumed target components, namely shale and some granitic rocks, contain significant amounts of K (compare with Boamah and Koeberl this volume), and these, as well as melt fragments analysed from suevite from the outside of the crater rim, contain appreciable U and Th contents. Clearly, some ejecta have been eroded and it cannot be excluded that the original distribution may have reached further away from the crater. As discussed by Melosh (1989), the average width of ejecta blankets around impact craters, in general, and for terrestrial impact craters, in particular, is not all that well constrained. Only limited information has been obtained for terrestrial craters, only few of which are young and retained portions of their ejecta blankets, but information cited by Melosh suggests that for lunar craters, continuous ejecta blankets ought to be roughly two crater radii wide. This would correspond reasonably well to the situation interpreted for Bosumtwi.

The fact that a second apparent ring feature at 17 - 19 km from the crater center was identified in the DEM could be interpreted as indicating the existence of a probable series of ring faults around the crater structure sensu strict, probably caused by structural adjustments of the basement as a reaction to compression caused by the impact event.

6 Conclusions

Due to extensive soil and vegetation cover around Bosumtwi impact crater, the local geology is not very well-known. Even though remote sensing and geophysical analysis are especially important in this case, their true contribution lies within the geomorphological context rather than a lithological one.

The regional structural fabric, as well as the impact-induced macrodeformation, is recognisable in the topography and drainage patterns around Bosumtwi impact crater. The presence of a subdued but distinct peripheral elevation referred to as the "outer rim" has been confirmed and several causes for its presence are discussed. At this stage, it is not possible to provide an unequivocal explanation for this feature, but for geophysical (radiometrics) and drainage reasons, it could be assumed that this elevation could represent the outer limit of the ejecta blanket. Shallow drilling by Boamah and Koeberl (this volume) did, however, fail to detect ejecta at this distance from the crater rim. A second apparent and weak elevation feature was discovered further out in the digital elevation model. Both these topographical ring features could be the traces of ring faults formed during the compression phase of the impact process.

Acknowledgements

W.U. Reimold and D. Brandt gratefully acknowledge financial support of their research from the National Research Foundation of South Africa (to WUR) and the Research Office of the University of the Witwatersrand. R.J. Wagner is grateful to Anglo Technical Division and EAD for the opportunity to carry out this research and the permission to publish the results. Alex Crowie is thanked for the generation of the Digital Elevation Model. Lyn Whitfield and Di Du Toit provided expert drafting support. Critical comments by Lutz Bischoff, James Garvin, and Jüri Plado on an earlier version of this manuscript have been much appreciated. This paper represents University of the Witwatersrand Impact Cratering Research Group Contribution No. 28.

References

Boamah D, Koeberl C (2001) Geochemistry of soils from the Bosumtwi impact structure, Ghana, and relationship to radiometric airborne geophysical data. This volume.

Chao ECT (1968) Pressure and temperature histories of impact metamorphosed rocks - based on petrographic observations. In: French BM, Short NM (eds) Shock Metamorphism in Natural Materials. Mono Book Corp., Baltimore, pp 135-158

Davis DW, Hirdes W, Schaltegger U, Nunoo EA (1994) U-Pb age constraints on deposition and provenance of Birimian and gold-bearing Tarkwaian sediments in Ghana, West Africa. Precambrian Research 67: 89-107

Dupont LM, Donner B, Schneider R, Wefer G (2001) Mid-Pleistocene environmental change in tropical Africa began as early as 1.05 Ma. Geology 29: 195-198

El Goresy A (1966) Metallic spherules in Bosumtwi crater glasses. Earth and Planetary Science Letters 1: 23-24

El Goresy A, Fechtig H, Ottemann T (1968) The opaque minerals in impactite glasses. In: French BM, Short NM (eds) Shock Metamorphism in Natural Materials. Mono Book Corporation, Baltimore, pp 531-554

Fergusson M (1902) Lake Bosumtwi. Geographical Journal 19: 370-372

Gurov EP, Yamnichenko AYu (1995) Morphology of rim of complex terrestrial craters [abs]. Lunar and Planetary Science 26: 533-534

Gurov EP, Gurova EP, Yamnichenko AYu. (1995) Structure of complex impact craters. Geological Journal (Academy of Science of the Ukraine) 1: 22-24

Hirdes W, Nunoo B (1994) The Proterozoic paleoplacers at Tarkwa gold mine, SW Ghana: Sedimentology, mineralogy, and precise age dating of the Main Reef and West Reef, and bearing of the investigations on source area aspects. In: Oberthür T (ed) Metallogenesis of selected gold deposits in Africa. Geologisches Jahrbuch D100: 247-311

Hirdes W, Davis DW, Lüdtke G, Konan G (1996) Two generations of Birimian (Paleoproterozoic) volcanic belts in northeastern Cote d'Ivoire (West Africa): Consequences for the "Birimian Controversy". Precambrian Research 80: 173-191

Innes MJ, Pearson WJ, Geuer JW (1964) The Deep Bay Crater. Publications Dominion Observatory, Ottawa 31: 21-51

Johnson GG, Vand V, Dachille F (1964) Additional rims around the Ries Kessel meteoritic crater. Nature 201: 592-593

Jones WB (1985a) The origin of the Bosumtwi crater, Ghana - an historical review. Proceedings Geological Association London 96: 275-284

Jones WB (1985b) Chemical analyses of Bosumtwi crater target rocks compared with Ivory Coast tektites. Geochimica et Cosmochimica Acta 49: 2569-2576
Jones WB, Bacon M, Hastings DA (1981) The Lake Bosumtwi impact crater, Ghana. Geological Society of America Bulletin 92: 342-349
Junner NR (1937) The geology of the Bosumtwi caldera and surrounding country. Gold Coast Geological Survey Bulletin 8: 1-38
Karp T, Milkereit B, Janle P, Bussat S, Danuor SK, Pohl J, Scheu B, Froehler M, Berckhemer H, Baier B, Zacher G, Scholz CA (2000a) Geophysical investigation of the Lake Bosumtwi impact crater [abs]. In: Plado J, Pesonen LJ (eds) Meteorite Impacts in Precambrian Shields. Geological Survey of Finland and University of Helsinki, Espoo, p 75
Karp T, Milkereit B, Janle P, Bussat S, Danuor S, Pohl J, Scheu B, Froehler M, Berckhemer H, Baier B, Zacher G, Scholz CA (2000b) Geophysical investigation of the Lake Bosumtwi impact crater [abs]. In: Catastrophic Events and Mass Extinctions: Impacts and Beyond. Houston, Lunar and Planetary Institute Contribution No 1053, p 89
Karp T, Milkereit B, Janle P, Danuor SK, Pohl J, Berckhemer H, Scholz CA (2000) Geophysical signature of the Bosumtwi impact crater [abs]. EOS (Transactions American Geophysical Union) 81, suppl. to No. 48: F800
Kitson AK (1916) The Gold Coast: Some consideration of its structure, people and natural history. Geographical Journal 48: 369-392
Koeberl C (1994) Tektite origin by hypervelocity asteroidal or cometary impact: Target rocks, source craters, and mechanisms. In: Dressler BO, Grieve RAF, Sharpton VL (eds) Large Meteorite Impacts and Planetary Evolution, Geological Society of America Special Paper 293: 133-151
Koeberl C, Reimold WU (1996) Lake Bosumtwi, Ghana, impact structure: Petrography and geochemistry of target rocks [abs]. Meteoritics & Planetary Science 31: A72
Koeberl C, Bottomley RJ, Glass BP, Storzer D (1997a) Geochemistry and age of the Ivory Coast tektites and microtektites. Geochimica et Cosmochimica Acta 61: 1745-1772
Koeberl C, Reimold WU, Pesonen LJ, Brandt D (1997b) New studies of the Bosumtwi impact structure, Ghana: The 1997 field season [abs]. Meteoritics & Planetary Science 32: A72-A73
Koeberl C, Reimold WU, Blum JD, Chamberlain CP (1998) Petrology and geochemistry of target rocks from the Bosumtwi impact structure, Ghana, and comparison with Ivory Coast tektites. Geochimica et Cosmochimica Acta 62: 2179-2196
Koeberl C, Boamah D, Reimold WU (2000) Geological, geophysical, and geochemical studies of an outer rim at the Bosumtwi impact structure, Ghana [abs]. Geological Society of America Abstracts with Programs 32 (7): A-161
Lacroix A (1934) Sur la découverte de tectites à la Côte d'Ivoire. Comptes Rendus de l'Academie des Sciences Paris 199: 1539-1542
Leube A, Hirdes W, Mauer R, Kesse GO (1990) The Early Proterozoic Birimian Supergroup of Ghana and some aspects of its associated gold mineralization. Precambrian Research 46: 136-165
Littler J, Fahey JJ, Dietz RS, Chao ECT (1961) Coesite from the Lake Bosumtwi crater, Ashanti, Ghana [abs]. Geological Society of America Special Paper 68: p 218
Maclaren M (1931) Lake Bosumtwi, Ashanti. Geographical Journal 78: 270-276
McCauley JF, Schaber GG, Breed CS, Grolier MJ, Haynes CU, Issawi B, Elachi C, Bloom R (1982) Subsurface valleys and geoarchaeology of the eastern Sahara revealed by Shuttle radar. Science 218: 1004-1019
Melosh HJ (1989) Impact Cratering - A Geologic Process. Oxford University Press, Oxford, 245 pp
Moon PA, Mason D (1967) The geology of 1/4° field sheet s nos. 129 and 131, Bompata S.W. and N.W. Ghana Geological Survey Bulletin 31: 1-51
Oberthür T, Vetter U, Davis DW, Amanor JA (1998) Age constraints on gold mineralization and palaeoproterozoic crustal evolution in the Ashanti belt of southern Ghana. Precambrian Research 89: 129-143

Ojamo H, Pesonen LJ, Elo S, Hautaniemi H, Koeberl C, Reimold WU, Plado J (1997) The Bosumtwi impact structure, Ghana: International geophysical co-operation at the best [abs]. In: Kaikkonen P, Komminaho K, Salmirinne H (eds) Sovelletun geofysiikan XI neuvottelupäivät. Oulun yliopisto, Oulu, p 10-11

Pesonen LJ, Koeberl C, Ojamo H, Hautaniemi H, Elo S, Plado J (1998) Aerogeophysical studies of the Bosumtwi impact structure, Ghana [abs]. Geological Society of America Abstracts with Programs 30 (7): A-190

Plado J, Pesonen LJ, Koeberl C, Elo S (2000) The Bosumtwi meteorite impact structure, Ghana: A magnetic model. Meteoritics & Planetary Science 35: 723-732

Reimold WU, Brandt D, Koeberl C (1998) Detailed structural analysis of the rim of a large, complex impact crater: Bosumtwi Crater, Ghana. Geology 26: 543-546

Rohleder HPT (1934) Über den Fund von Vergriesungs-Erscheinungen und Druckstrukturen am Kesselrand des kryptovulkanischen Bosumtwi-Sees, Ashanti. Zentralbl Miner Geol Palaeontol, Jahrgang 1934: 316-318

Scholz CA, Brooks KM, Arko JA, Amoakao PYO (2000) Multichannel seismic reflection imaging of the Lake Bosumtwi impact crater, Ghana [abs]. EOS (Transactions American Geophysical Union) 81, suppl. to No. 48: F800

Wagner RJ, Reimold WU, Brandt D (2000) A remote sensing investigation of the Bosumtwi impact structure, Ghana [abs]. In: Plado J and Pesonen LJ (eds) Meteorite Impacts in Precambrian Shields, Geological Survey of Finland and University of Helsinki, Espoo, p 24

Woodfield PD (1966) The geology of the ¼ field sheet 91, Fumso N.W., Ghana Geological Survey Bulletin 30: 66 pp

Wright JB, Hastings DA, Jones WB, Williams HR (1985) Geology and Mineral Resources of West Africa. George Allen and Unwin, London, pp 38-45.

Geochemistry of Soils from the Bosumtwi Impact Structure, Ghana, and Relationship to Radiometric Airborne Geophysical Data

Daniel Boamah and Christian Koeberl

Institute of Geochemistry, University of Vienna, Althanstrasse 14, A-1090 Vienna, Austria.
(christian.koeberl@univie.ac.at)

Abstract. In a high-resolution aerogeophysical survey across the Bosumtwi impact structure, a pronounced high radiometric signal (high equivalent concentration of potassium) was found around the crater rim and to the north of the crater rim. Here we report data on major and trace element chemical compositions of soils from the structure, which have been investigated to determine the provenance, influence of source area weathering, and possible cause of the high equivalent concentration of potassium in the airborne gamma radiation data. Bosumtwi, an impact crater of 1.07 Ma age and a diameter of ~10.5 km, is located in early Proterozoic Birimian-Tarkwaian rocks of the semi-equatorial climatic region of Ghana, West Africa. The area forms part of the tropical rainforest environment, where warm climate, high rainfall, and high organic activity prevail, and chemical weathering is intense, leading to the formation of lateritic soils, which can be up to tens of meters thick.

Fifty-four soil samples, including forty-four from 30 cm depth and ten from 100 cm depth, were used in the study. The concentrations of both major and trace elements in the soil samples show considerable variation, as is expected from a chemical weathering environment. Compared to the underlying parent rocks, the soils are generally enriched in SiO_2, Fe_2O_3, Zr, Nb, Hf, Ta, W and Sb, and depleted in Na, K, and the rare earth elements (REE). Compared to average upper continental crust composition, the Bosumtwi soils are more siliceous and ferruginous. At this advanced stage of weathering, with an average chemical index of alteration (CIA) of 83, the soils have a rather simple mineralogical composition that is dominated by quartz, Fe- and Al-oxides and kaolinite, in addition to a few refractory minerals (e.g., zircon).

Pronounced positive Ce anomalies, associated with high Fe_2O_3, are noted in the soils. A positive correlation of REE and Al_2O_3 contents in the soils suggests that REE are contained principally in phyllosilicates and clays. Discrimination based on abundances of Co, Ni, and Cr, elemental ratios (Th/Sc, Cr/Zr, Zr/Hf, Co/Th, La/Sc), and La-Th-Sc and Th-Hf-Co plots, shows the importance of mafic rock components in the soils.

Soils collected from areas that show a high radiometric potassium signal in the aerogeophysical maps are characterized by relatively higher K_2O contents. This may reflect K mobilization due to the impact event, or source rocks for these sediments with variable concentrations of K.

1 Introduction

The study area is located just north of the Bosumtwi impact crater, outside of the crater rim. The ~1.07 Ma Bosumtwi crater in Ghana, West Africa (centered at 06°30'N and 01°25'W) is a well preserved, complex impact structure, which displays a pronounced rim and is largely filled by the 8-km-diameter Lake Bosumtwi. The crater has a steep rim of about 10.5 km diameter, which is surrounded by a slight and irregular circular depression and an outer ring of minor topographic highs with a diameter about 20 km (Jones et al. 1981; Wagner et al. 2001).

The Bosumtwi impact event has been demonstrated to be associated with the Ivory Coast tektite strewn field (Fig. 1a), based on similarities in chemical and isotopic composition, and identical ages of Ivory Coast tektites and impact glasses from the crater area (Koeberl et al. 1997a).

The study area falls into the semi-equatorial climatic region of Ghana. This region is covered by tropical rainforest characterized by warm temperatures, high rainfall, and high organic activity; chemical weathering is intense. The intense chemical weathering of the rocks in the area has produced lateritic soils. Bowell (1993) reported that the weathering profile in some parts of the tropical rainforest of Ghana includes a saprolite zone ~70 m thick with an overlying soil thickness of 1 to more than 10 m, depending on lithology, topography, and drainage.

Recently, research activities have been intensified around the Bosumtwi crater. These include studies of the petrology and geochemistry of the target rocks (Koeberl et al. 1998), structural analysis of the crater rim (Reimold et al. 1998), a high-resolution aerogeophysical survey across the structure (Koeberl et al. 1997b; Ojamo et al. 1997; Pesonen et al. 1998, 1999; Plado et al. 2000) and shallow drilling outside of the crater rim (Boamah and Koeberl 1999). In addition, U.S. and German research teams are currently engaged in detailed land- and lake-based geophysical studies. The gamma radiation results of the airborne geophysical survey exhibit anomalous equivalent concentrations of K around the crater rim and in an outer zone mainly evident to the north of the crater. The apparent resistivity and Th and U distribution maps have features similar to those of the K map (Pesonen et al. 1998), but not as pronounced as the K radiometry data.

This study was initiated to investigate the geochemistry of soils from the Bosumtwi impact structure in terms of provenance (source rock), source area weathering, and the possible cause of the radiometric anomalies.

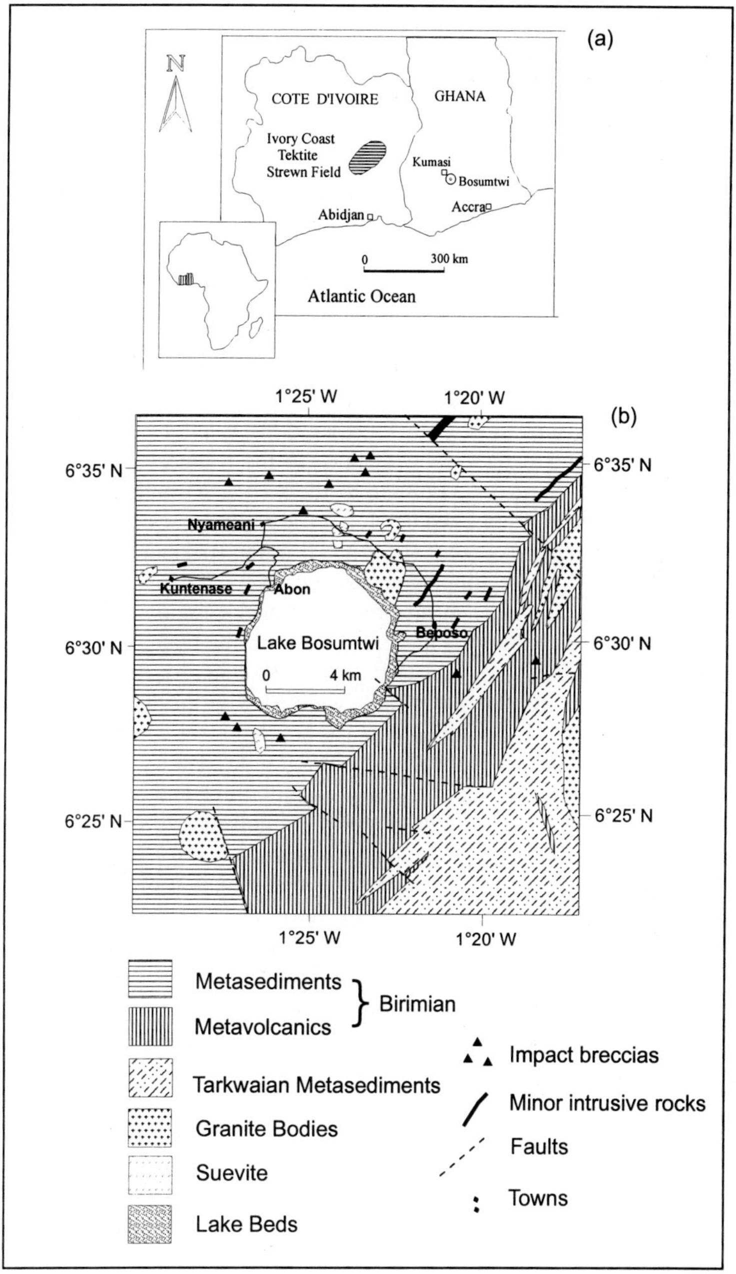

Fig. 1. (a) Geographic location of the Bosumtwi impact crater, Ghana, in relation to the Ivory Coast tektite strewn field, after Koeberl et al. (1998). (b) Geological map of the area around Lake Bosumtwi, showing the provenance of different target rocks (after Jones et al. 1981; Reimold et al. 1998).

2 Geological Setting

The Bosumtwi crater was excavated in lower greenschist facies metasediments of the 2.1 - 2.2 Ga Birimian Supergroup (cf. Wright et al. 1985; Leube et al. 1990). The early Proterozoic Birimian Supergroup has historically been subdivided into the Lower Birimian, dominated by metasediments, and the Upper Birimian, dominated by greenstone-type metavolcanics (e.g., Junner 1937). New geochronological data indicate that there is no evidence for a separation of Lower and Upper Birimian in time, with rocks from both groups having a Sm-Nd isochron age of 2.17 ± 0.07 Ga (cf. Leube et al. 1990; Taylor et al. 1992; Davis et al. 1994; Hirdes et al. 1996). These authors argued that both lavas and sedimentary rocks were deposited contemporaneously. The metasediments comprise interbedded phyllites and meta-tuffs together with meta-graywackes, quartzitic graywackes, shales and slates. Metavolcanics to the southeast of the crater also contain altered basic intrusives. Further to the east and southeast of the crater Tarkwaian sediments are found, which are clastic sedimentary rocks formed from the erosion of Birimian rocks (Fig. 1b). The Birimian rocks are generally folded, with a general NE-SW strike, dipping at high angles.

Several Proterozoic granitic intrusions occur in the region around the crater, and some strongly weathered granitic dikes occur in the crater rim (Junner 1937; Jones 1985; Reimold et al. 1998). The Pepiakese Granite, located to the northeast of the crater, which is considered to be associated with the Kumasi Batholith to the north of the crater (Moon and Mason 1967), comprises a range of rock types including amphibole diorite, biotite-muscovite granite, and an almost pure albite rock. There are a few dikes of dolerite, amphibolite, and intermediate rocks around the crater. Some acid and basic dikes are considered to represent the last phase of the Birimian orogeny. Recent deposits include the Bosumtwi lake beds, as well as soils and breccias associated with the formation of the crater (Junner 1937; Kolbe et al. 1967; Woodfield 1966; Moon and Mason 1967; Jones et al. 1981; Jones 1985; Koeberl et al. 1997a). Suevite deposits have been observed just outside of the northern and the southwestern crater rim sections (Fig. 1b).

In recent times, geological activity has been restricted to the reduction of the topography by chemical decomposition of the rocks and mechanical removal of material. Under present climatic conditions, the former process is more rapid than the latter, with the result that deep weathering profiles have been produced by lateritization, and outcrops, except in steep hill country, are few and usually restricted to stream beds.

3 Bosumtwi Weathering Profile

The weathering profile of the Bosumtwi study area (Fig. 2), which is locally up to 50 m thick, is similar to the lateritic profiles in Gabon described by Lecomte

(1988). The profile is dominated by a structured saprolite zone (H3), in which the original rock features are preserved. Towards its top, the saprolite becomes increasingly leached and the bedrock structures are no longer visible. Upwards, it becomes increasingly ferruginous, until this horizon passes into a stone line (H2). The stone line is characterized by the presence of lateritic nodules, rock fragments, and angular fragments of quartz and duricrust, in a ferruginous clay matrix. The thickness of this layer is variable (0 to 3 m). Above this horizon is a friable homogeneous clay horizon (H1). The top few centimeters of dark-brownish coloration are due to the presence of organic matter. This surface unit (lateritic soil) is composed essentially of reddish-brown to yellowish-brown sandy clay with some quartz grains and a rare ferruginous nodules.

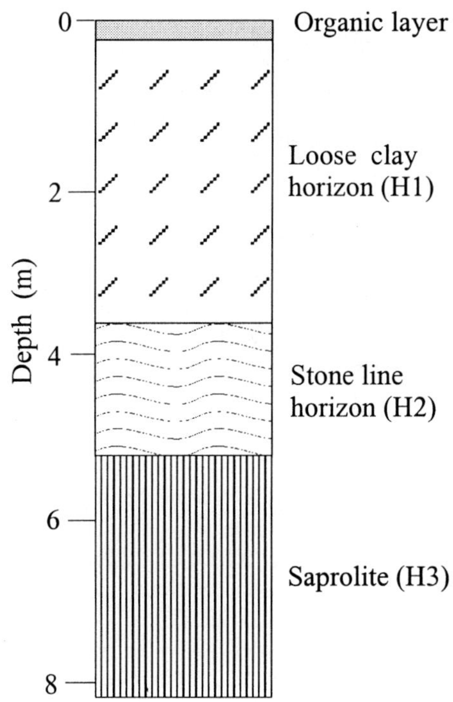

Fig. 2. Schematic weathering profile for the Bosumtwi crater area showing the various horizons from the top organic-rich layer to the saprolite.

4
Sampling and Analytical Methods

4.1
Sampling

The sampling was planned in order to collect soils from areas of both high and low K concentrations (from airborne radiometry). As expected from (few) previous calibrations of aeroradiometry data (e.g., Gregory and Horwood 1961; Multala 1981; Grasty 1987; Chepik et al. 1998), the radiometry signal most likely originates, from the top meter of ground or below surface, depending on some parameters such as water content. Thus, sampling was restricted to the top meter of laterite or soil, after the top 20 to 30 cm have been removed. The soil sampling was carried out in two phases. The first phase took place during the 1999 shallow drilling program around the Bosumtwi impact crater (Boamah and Koeberl 1999); the second one was carried out in February 2000. The sample locations are shown in Fig. 3, which shows part of the radiometric map for equivalent concentration of K.

During the drilling program, 10 pits, 1 m × 1 m, with a depth of about 1 m, were excavated to collect soil samples for geochemical studies. Twenty soil samples, each weighing about 500 g, were taken from the 10 pits. Two samples were taken from each pit, one from the top (~30 cm) and the other from the bottom of the pit (~100 cm).

In February 2000, further soil sampling was carried out along three traverses to the north of the crater (Fig. 3) to cover a wider area (locations in the area 1°20' - 1°28'W and 6°32' - 6°37'N). Two traverses were made along roads and a third along footpaths. Thirty-four soil samples were collected from depths of about 30 - 40 cm, i.e., below the top organic-rich layer. At each site, about 500 g of soil material was collected. Descriptions of the soils are given in the Appendix.

4.2
Analytical Methods

The samples were air-dried and ground with a mechanical agate mill and analyzed by X-ray fluorescence (XRF) and instrumental neutron activation analysis (INAA). Qualitative X-ray diffraction (XRD) analysis of fifteen of the samples from areas of both high and low K concentrations was carried out to check the influence of mineralogy on the radiometric anomalies. The measurements were made using a Philips PW 1710 powder X-ray diffractometer. Major element oxides and some trace element (Rb, Sr, Y, Zr, Nb, Co, Ni, Cu, Zn, V, Cr and Ba) concentrations were determined on powdered samples by X-ray fluorescence (XRF) spectrometry at the Department of Geology, University of Witwatersrand, Johannesburg, South Africa. Details on procedures, precision and accuracy are given by Reimold et al. (1994). Other trace elements (Sc, Cr, Co, Ni, Zn, As, Se, Br, Rb, Sr, Zr, Sb, Cs, Ba, Hf, Ta, W, Ir, Au, Th and U) and the REEs were determined by instrumental neutron activation analysis (INAA). These analyses

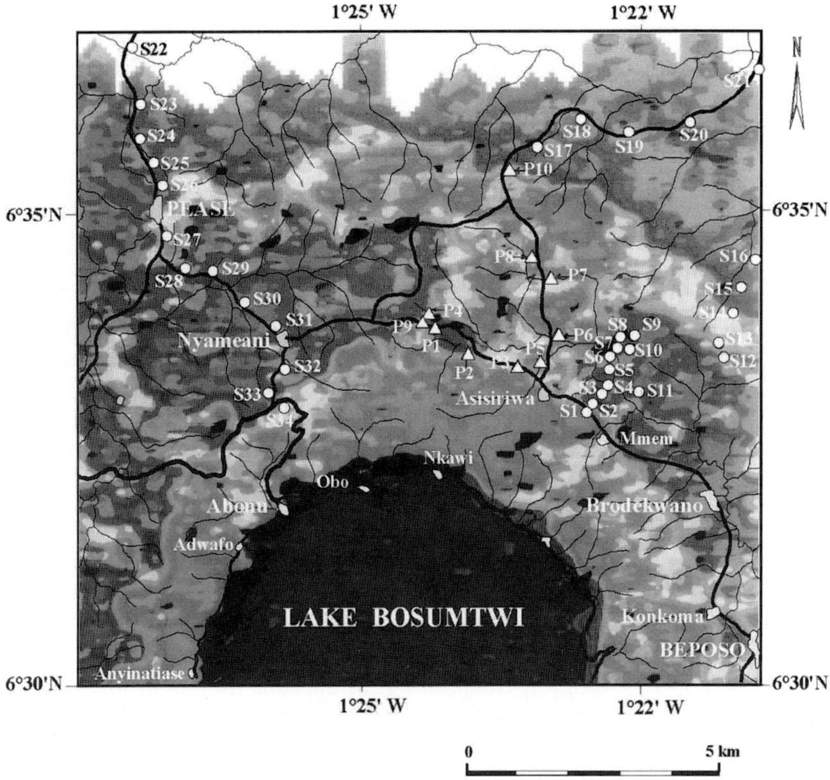

Fig. 3. Soil sample location map, showing part of the radiometric map. Circles represent soils collected in February 2000 and triangles represent locations of the pits. The radiometric map (equivalent concentrations of potassium) was produced from a high-resolution airborne geophysical survey carried out in the area (cf. Ojamo et al. 1997; Pesonen et al., in preparation).

were carried out at the Institute of Geochemistry, University of Vienna, Austria. About 100 to 150 mg of powdered samples were weighed into clean polyethylene vials that were then heat-sealed. Together with geological standard reference materials (AC-E Granite, G-2 Granite and Allende) (Govindaraju 1989), which were used to monitor the quality of the analyses, they were irradiated for 8 hours at a flux of 2×10^{12} n cm^{-2}s^{-1} at the TRIGA Mark II type reactor of the Atominstitut der Österreichischen Universitäten in Vienna. Details on the analytical procedures (including standard data) are given in Koeberl (1993).

5
Results

Minerals identified in the soils from both high and low K radiometry areas on the radiometric map are given in Table 1. Major and trace element concentrations in the soils are given in Table 2. For comparison, mean values of chemical compositions of parent rocks (Koeberl et al. 1998) and upper continental crust (UCC) (Taylor and McLennan 1985) are given in Table 3. The results for the pit samples (Table 4) were treated separately to compare the variations of the elements in the top (30 to 40 cm depth) and bottom (100 cm depth) zones. Table 5 compares the relationship between the various elements in terms of correlation coefficients. For the discussion of the results, areas on the radiometric map (equivalent concentrations of potassium) with K values <0.8 % were treated as low K radiometry areas (O), and areas of values >0.8 % as high K radiometry areas (I). The concentrations of both major and trace elements in the soil samples vary somewhat, as is expected from a chemical weathering environment, but are similar to the compositions obtained for the source or parent rocks (compare, e.g., Koeberl et al. 1998).

5.1
Mineralogy

A comparison of mineralogical compositions of the soils from both low and high K radiometry areas, and with increasing distance from the crater rim are given in Table 1. Both the soils from low and high K radiometry areas, and irrespective of distance from the crater rim have similar mineralogical composition, dominated by quartz and muscovite. Other minerals identified are clinochlore and albite.

5.2
Major Elements

Major element concentrations of the soils show considerable variations in abundances (Table 2). These reflect the mineralogical composition, which is dominated by quartz, muscovite, clinochlore and albite (Table 1). Apart from sample S2, which has a very high value of 2.22 wt% P_2O_5, all other soil samples have P_2O_5 values <0.1 wt%. Also, apart from samples S1 and S4, all others have SiO_2 contents greater than 60 wt% - as high as 93.6 wt% SiO_2 in sample S19. Compared to upper continental crust (UCC) composition, these soils are more siliceous and ferruginous, with considerably lower levels of other major cations. The soils have high LOI, with an average of 6 wt%.

Harker variation diagrams, showing the scatter in the major element compositions, are displayed in Fig. 4. Figure 4a shows a negative correlation between Al_2O_3 and SiO_2. Similar trends are shown on the Fe_2O_3, MgO, and K_2O versus SiO_2 plots. The samples collected from low K radiometry areas on the radiometric map seem to have high SiO_2. Figure 4f shows that samples from the high K radiometry areas have high K_2O. The other oxides, Na_2O and TiO_2, show no correlation with SiO_2.

5.3 Trace Elements

The trace element data show large variations similar to the results for the major elements (Tables 2). Apart from Zr, Hf, Nb and Ta, which show positive correlations with SiO_2, all other trace elements show negative correlations or no correlation at all. There are relatively higher concentrations of trace elements in soils with higher Al_2O_3, compared to those with higher SiO_2.

5.3.1 Large Ion Lithophile Elements (LILE): K, Rb, Cs, Ba, Sr, Th, U

The contents of these elements in the soils are variable (Table 2), perhaps due to their mobility during weathering or other secondary processes. Compared to upper continental crust (UCC), the soils are slightly depleted in these elements. Except

Table 1. Mineralogical composition of selected soil samples from low and high K radiometry areas

Sample No.	%K [a]	K_2O (wt%)	Description	Mineralogy
S1	1.2-2.0	2.88	Reddish-brown, clay with weathered rock fragments	Quartz, muscovite, clinochlore
S8	1.2-2.0	1.62	Brownish, silty clay	Quartz, muscovite, albite
S13	0.8-1.2	1.83	Brownish clay with weathered fragments	Quartz, muscovite, clinochlore
S16	0.0-0.4	0.50	Yellowish-brown, silty clay	Quartz, muscovite
S17	0.0-0.4	0.95	Chocolate brown, silty sand	Quartz, muscovite, clinochlore
S20	0.0-0.4	0.02	Brown, silty clay with some ferruginous nodules	Quartz, albite, muscovite, goethite(?)
S22	0.0-0.4	0.23	Brownish-gray sand	Quartz, clinochlore
S25	0.4-0.8	0.71	Light brown silt	Quartz, muscovite, clinochlore
S27	0.8-1.2	0.8	Reddish-brown gravel	Quartz, muscovite, clinochlore
S30	0.4-0.8	0.31	Light brown, silty clay	Quartz, albite
S32	0.4-0.8	0.99	Light brownish, silty sand	Quartz, albite, muscovite, clinochlore
S34	0.8-1.2	2.12	Light brown, silty sand	Quartz, muscovite, clinochlore
T7	1.2-2.0	1.01	Brownish, gravel with some clay	Quartz, albite, muscovite, clinochlore
B7	1.2-2.0	1.51	Reddish-brown sand with some clay	Quartz, albite, muscovite, clinochlore
T10	0.8-1.2	1.45	Light brown, sandy clay with some gravel and weathered graywacke	Quartz, albite, muscovite, clinochlore

[a] %K from the aero-radiometric data (Ojamo et al. 1997; Pesonen et al., in preparation)

for Th and U, which are weakly correlated with Al_2O_3, the other elements have significant positive correlations with Al_2O_3 ($r > 0.5$) (Table 5). The abundances of Rb, Ba, and Cs are closely linked with that of K, as seen from the very good correlation between these elements ($r = 0.95$, 0.93 and 0.83, respectively). Thorium and U show similar geochemical behaviour, as is attested by the relatively high positive correlation coefficient ($r = 0.71$). The Th/U ratios are variable ($1.25 < Th/U < 5.51$), but the average Th/U ratio of 2.63 is lower than the average for Post-Archean sediments ($Th/U = 5.0 \pm 0.7$; Nance and Taylor 1977) or the average of 3.8 for upper continental crust (Taylor and McLennan 1985). There is no apparent difference in the Th/U ratios between the silica-rich and silica-poor samples.

5.3.2
High Field Strength Elements (HFSE): Zr, Hf, Ta, Nb, Y, Ti

Compared to UCC, the soils are generally depleted in these elements. However, those soils with very high SiO_2 contents are relatively enriched in Zr and Hf. There is a distinct correlation between TiO_2 and Nb ($r = 0.79$) and between TiO_2 and Ta ($r = 0.7$). The correlations between Ti - Zr and Ti - Hf are less pronounced, although high values of Zr and Hf generally correspond to high TiO_2 values. Zr and Hf behave similarly, as attested by their high correlation coefficient ($r = 0.91$). The Zr/Hf ratios in the soils range from 28 to 44 (average: 37), which is similar to the normal ratio in zircon from various geological terrains (Murali et al. 1983; Condie 1991; Götze and Lewis 1994). The absolute concentrations of Zr in most samples are, however, lower than that of upper continental crust (190 ppm). There are significant positive correlations between SiO_2 and Zr and Hf ($r = 0.78$, 0.77, respectively) (Table 5). The contents of Ta, TiO_2 and Nb also show slight positive correlation with the SiO_2 abundance.

5.3.3
Transition Trace Elements (TTE): Cr, Co, Ni, Zn, Sc, V, Cu, Fe, Mn

All transition elements show positive correlations with Al_2O_3 (Table 5). The abundances of the transition trace elements are highly variable, as seen from Table 2. Cobalt, Ni and Cr are enriched relative to UCC, but are within the range of the parent rocks for these soils (cf. Koeberl et al. 1998). Sample S20 clearly differs from the rest of the soil samples, especially in the contents of Cr (3417 ppm), Ni (315 ppm) and Co (82 ppm). This sample also has high contents of Fe_2O_3 (14.8 wt%) and MgO (2.46 wt%). There are significant correlations between Fe_2O_3 and Cr, V, Co, Sc, Zn and Ni. Those soil samples with high MgO contents contain more of the ferromagnesian trace elements (Co, Ni, Cr) (e.g., S4, S5, and S9).

Table 2. Chemical composition of soils from the Bosumtwi impact crater area, Ghana. **(a)** Samples collected at 30 cm depth from high K radiometry areas on the radiometric map (high equivalent concentration of potassium)

Sample	S1	S7	S8	S12	S13	S14	S27	S34
SiO_2	59.52	67.57	76.23	73.57	66.63	67.72	72.58	66.48
TiO_2	0.76	0.76	0.47	0.77	0.72	0.61	0.75	0.79
Al_2O_3	18.87	14.32	11.05	10.94	14.34	15.89	9.86	16.30
Fe_2O_3	8.95	6.46	2.52	6.34	8.09	5.45	10.27	6.02
MnO	0.06	0.06	0.04	0.06	0.04	0.04	0.05	0.07
MgO	1.85	1.43	1.55	0.70	1.03	1.61	0.46	1.48
CaO	0.02	0.33	1.46	0.01	0.05	0.05	<0.01	0.01
Na_2O	0.16	0.99	2.32	0.36	0.15	1.08	0.08	0.69
K_2O	2.88	1.76	1.62	1.25	1.85	1.51	0.80	2.12
P_2O_5	0.08	0.06	0.05	0.05	0.07	0.07	0.06	0.08
LOI	7.20	6.63	3.08	6.31	7.52	6.38	5.49	6.16
Total	100.35	100.38	100.38	100.38	100.5	100.42	100.4	100.18
Sc	18.9	13.3	8.48	11.3	16.5	13.7	6.42	10.3
V	201	132	73	152	170	123	162	146
Cr	186	163	130	183	227	155	263	140
Co	18.9	20.5	8.95	21.0	17.2	17.2	7.18	14.5
Ni	68	54	36	37	50	60	24	52
Cu	35	16	<2	4	16	9	12	15
Zn	77	60	30	38	60	64	26	72
As	20.2	26.5	6.77	13.3	16.1	8.01	12.3	3.72
Se	0.16	0.14	0.07	0.21	0.23	0.09	0.14	0.07
Br	0.42	0.4	0.51	0.62	0.88	0.54	0.57	0.4
Rb	109	91.5	66	74.5	105	70.2	36.9	72.2
Sr	104	137	255	93	82	165	51	167
Y	10	25	10	11	17	15	10	13
Zr	133	149	175	162	175	139	158	125
Nb	9	10	8	10	11	8	10	9
Sb	1.22	1.36	0.48	0.96	1.19	0.32	0.37	0.09
Cs	3.81	4.36	2.07	3.02	4.53	3.19	1.84	2.71
Ba	728	515	526	414	506	533	390	618
La	7.5	27.9	13.9	10.5	10.3	18.7	7.71	7.85
Ce	24.1	60.2	26.3	38	47.9	37.1	27.0	27.1
Nd	5.12	39.0	13.2	8.72	10.5	16.6	7.85	7.89
Sm	1.23	5.63	2.43	1.56	2.15	3.01	1.24	1.29
Eu	0.4	1.56	0.66	0.44	0.67	0.84	0.38	0.42
Gd	1.6	5.94	2.46	1.64	2.32	3.19	1.60	0.77
Tb	0.23	0.82	0.31	0.39	0.49	0.43	0.30	0.25
Tm	0.14	0.37	0.18	0.23	0.37	0.28	0.31	0.26
Yb	1.41	2.47	1.19	1.60	2.44	1.80	1.33	1.30
Lu	0.22	0.36	0.18	0.24	0.37	0.25	0.20	0.20
Hf	3.39	4.17	5.24	4.80	5.01	3.59	3.95	3.18
Ta	0.44	0.55	0.32	0.54	0.66	0.42	0.51	0.45
W							1.03	1.32
Ir (ppb)	<2.5	0.4	<1.2	<1.7	<2.4	<2.1	1.6	0.3
Au (ppb)	7.2	11.9	<7.4	4.7	12.8	15.3	4.8	1.4
Th	3.2	4.75	2.69	3.42	5.02	3.79	3.97	2.58
U	1.29	2.37	1.24	1.36	2.19	1.11	1.12	0.77
CIA	85	78	57	85	86	82	91	83
K/U	18525	6162	10841	7627	7010	11288	5927	22846
Cr/Zr	1.40	1.09	0.74	1.13	1.30	1.12	1.66	1.12
Zr/Hf	39.2	35.7	33.4	33.8	34.9	38.7	40.0	39.3
La/Th	2.34	5.87	5.17	3.07	2.05	4.93	1.94	3.04
Hf/Ta	7.70	7.58	16.4	8.89	7.59	8.55	7.75	7.07
Th/U	2.48	2.00	2.17	2.51	2.29	3.41	3.54	3.35
La_N/Yb_N	3.59	7.63	7.89	4.43	2.85	7.02	3.92	4.08
Gd_N/Yb_N	0.92	1.95	1.68	0.83	0.77	1.44	0.97	
Eu/Eu*	0.87	0.82	0.82	0.84	0.92	0.83	0.82	
Ce/Ce*	1.74	0.92	0.92	1.84	2.21	0.99	1.66	1.65

Major elements in wt%, trace elements in ppm, except as noted. Total Fe as Fe_2O_3. Blank spaces: not determined. Chemical Index of Alteration (CIA) = $[Al_2O_3/(Al_2O_3 + CaO + Na_2O + K_2O)] \times 100$ in molecular proportions. $Eu/Eu^* = Eu_N/(Sm_N \times Gd_N)^{0.5}$; $Ce/Ce^* = Ce_N/[(La_N)^{0.6667} \times (Nd_N)^{0.3333}]$

Table 2. (cont.)
(a) Samples collected at 30 cm depth from high K radiometry areas

Sample	T3	T5	T6	T7	T8	T9	T10
SiO_2	71.88	77.05	70.28	71.21	62.57	70.68	63.57
TiO_2	0.93	0.89	0.7	0.84	0.87	0.67	0.71
Al_2O_3	12.39	11.04	14.14	10.23	19.97	10.09	18.88
Fe_2O_3	5.33	3.80	5.29	10.34	8.63	4.68	7.46
MnO	0.07	0.05	0.05	0.07	0.05	0.05	0.05
MgO	0.66	0.75	1.22	0.57	0.46	0.56	1.51
CaO	0.07	0.09	0.17	0.05	0.01	0.31	0.06
Na_2O	0.40	0.55	0.90	0.38	1.02	0.48	0.34
K_2O	1.39	1.42	1.83	1.01	2.44	1.04	1.45
P_2O_5	0.04	0.04	0.04	0.04	0.03	0.05	0.02
LOI	5.7	4.66	5.83	5.69	4.04	11.65	6.40
Total	98.86	100.35	100.45	100.44	100.09	100.27	100.44
Sc	12.6	11.0	11.1	9.25	8.28	16.0	12.8
V	119	100	111	127	78	119	121
Cr	143	124	131	200	92	164	141
Co	20	15	10	18	10	17	15
Ni	42	27	36	31	27	57	43
Cu	7	<2	3	12	3	25	7
Zn	48	43	55	42	34	101	60
As	5.54	4.69	7.15	8.29	6.72	1.25	7.16
Se	0.17	0.10	0.10	0.28	0.14	0.08	0.11
Br	0.56	0.60	0.57	0.67	0.55	0.32	0.51
Rb	67	60	69	48	39	75	73
Sr	96	84	123	57	137	90	91
Y	11	14	12	10	7	34	12
Zr	171	187	158	145	129	145	162
Nb	13	13	11	11	10	10	12
Sb	0.29	0.23	0.14	0.65	0.33	0.21	0.25
Cs	2.95	3.10	2.93	2.13	1.99	3.69	3.36
Ba	516	484	587	415	409	518	575
La	8.21	11.8	11.8	7.33	7.56	51.7	7.01
Ce	39.8	23.0	21.2	41.6	25.2	77.6	22.0
Nd	8.87	11.1	8.81	7.08	6.99	46.9	5.09
Sm	1.32	1.8	1.67	1.47	1.25	8.31	1.12
Eu	0.39	0.6	0.5	0.41	0.38	2.52	0.37
Gd	1.61	3.02	2.3	1.9	1.66	9.49	1.44
Tb	0.29	0.32	0.22	0.25	0.19	0.87	0.27
Tm	0.21	0.23	0.18	0.15	0.12	0.30	0.18
Yb	1.30	1.58	1.18	1.15	0.85	2.36	1.23
Lu	0.20	0.24	0.18	0.20	0.13	0.36	0.19
Hf	4.30	4.81	3.56	3.95	3.39	2.97	4.28
Ta	0.75	0.56	0.41	0.63	0.35	0.45	0.41
W	0.76	2.68	0.36	0.73	0.89	<1	1.33
Ir (ppb)	0.8	2.5	1.5	<1.7	<1.5	1.3	0.53
Au (ppb)	3.4	2.5	3.0	4.1	5.0	9.8	5.3
Th	3.31	2.65	2.57	3.19	2.58	3.42	2.83
U	1.53	1.36	0.96	1.62	1.04	0.84	1.25
CIA	87	84	83	88	85	85	91
K/U	7539	8664	15818	5173	19468	10274	9626
Cr/Zr	0.84	0.66	0.83	1.38	0.71	1.13	0.87
Zr/Hf	39.8	38.9	44.4	36.7	38.1	48.8	37.9
La/Th	2.48	4.45	4.59	2.30	2.93	15.12	2.48
Hf/Ta	5.73	8.59	8.68	6.27	9.69	6.60	10.44
Th/U	2.16	1.95	2.68	1.97	2.48	4.07	2.26
La_N/Yb_N	4.27	5.05	6.76	4.31	6.01	14.8	3.85
Gd_N/Yb_N	1.00	1.55	1.58	1.34	1.58	3.26	0.95
Eu/Eu*	0.82	0.79	0.78	0.75	0.81	0.87	0.89
Ce/Ce*	2.26	0.95	0.95	2.74	1.64	0.74	1.67

Major elements in wt%, trace elements in ppm, except as noted. Total Fe as Fe_2O_3. Blank spaces: not determined. Chemical Index of Alteration (CIA) = $[Al_2O_3/(Al_2O_3 + CaO + Na_2O + K_2O)] \times 100$ in molecular proportions. Eu/Eu* = $Eu_N/(Sm_N \times Gd_N)^{0.5}$; Ce/Ce* = $Ce_N/[(La_N)^{0.6667} \times (Nd_N)^{0.3333}]$

Table 2. (cont.)
(b) Samples collected from 30 cm depth from low K radiometry areas

Sample	S2	S3	S4	S5	S6	S9	S10	S11
SiO_2	83.61	63.55	58.07	72.47	75.72	64.35	70.41	64.66
TiO_2	0.58	0.68	0.82	0.73	0.58	0.67	0.44	0.64
Al_2O_3	6.17	15.88	17.24	12.60	10.18	15.23	13.61	16.75
Fe_2O_3	2.45	5.51	10.52	3.87	5.74	5.46	3.81	6.61
MnO	0.04	0.05	0.04	0.05	0.03	0.04	0.06	0.09
MgO	0.59	1.73	2.07	1.34	0.56	3.65	1.28	0.93
CaO	0.22	1.48	0.00	0.40	0.15	0.77	2.03	0.12
Na_2O	0.80	0.77	0.25	1.51	0.46	2.45	1.97	0.36
K_2O	0.46	0.84	2.07	1.13	0.82	1.14	0.99	2.34
P_2O_5	2.22	0.06	0.05	0.07	0.05	0.06	0.06	0.07
LOI	3.21	9.18	9.26	6.26	6.08	6.24	5.70	7.83
Total	100.34	99.76	100.41	100.43	100.39	100.06	100.36	100.41
Sc	5.95	14.0	19.5	10.6	8.55	17.7	11.8	15.5
V	73	117	222	92	134	148	101	158
Cr	118	250	295	133	179	648	96	205
Co	4.84	15.5	17.4	12.8	7.75	12.4	12.0	23.0
Ni	16	81	82	38	36	105	24	291
Cu	<2	4	25	<2	<2	<2	<2	45
Zn	20	54	68	50	30	73	37	54
As	3.18	5.22	20.9	4.69	13.2	10.1	7.50	36.1
Se	0.08	0.05	0.11	0.15	0.14	0.09	0.11	0.16
Br	0.43	0.51	0.60	0.46	0.58	0.40	0.77	0.70
Rb	27.6	54	95.3	56.3	52.3	54.8	53.5	108
Sr	71	236	78	203	80	394	281	139
Y	8	16	20	14	13	14	10	13
Zr	192	160	146	175	197	94	146	147
Nb	8	9	10	9	10	8	7	10
Sb	0.28	0.41	1.32	0.53	0.57	0.18	0.53	0.99
Cs	1.34	3.30	4.00	2.78	2.35	2.40	2.69	4.39
Ba	221	456	576	428	284	566	441	646
La	7.12	21.0	20.0	16.2	12.2	24.4	15.0	13.2
Ce	14.9	38.4	76.1	33.0	46.8	33.9	33.5	37.9
Nd	7.89	21.4	24.1	16.0	14.0	18.3	15.5	11.8
Sm	1.33	4.16	4.05	2.86	2.48	3.42	2.56	2.01
Eu	0.40	1.23	0.97	0.79	0.61	0.94	0.79	0.59
Gd	1.45	3.41	4.00	2.90	2.75	3.54	2.51	2.40
Tb	0.19	0.52	0.54	0.40	0.34	0.39	0.35	0.31
Tm	0.13	0.21	0.28	0.18	0.21	0.23	0.22	0.23
Yb	1.16	1.70	2.06	1.67	1.47	1.36	1.26	1.63
Lu	0.18	0.26	0.31	0.25	0.22	0.18	0.17	0.25
Hf	5.39	3.86	3.85	5.94	5.34	2.57	4.57	4.16
Ta	0.34	0.39	0.49	0.52	0.47	0.30	0.33	0.46
W								
Ir (ppb)	1	<2.2	2.1	<2	0.2	1.3	0.5	1.8
Au (ppb)	2.6	7.6	6.1	4.9	8.5	3.3	11.5	11.6
Th	2.32	3.89	4.29	2.99	3.62	2.37	2.31	3.81
U	0.92	1.62	3.44	1.59	1.90	0.43	0.75	1.88
CIA	74	77	87	74	84	70	63	83
K/U	4149	4303	4993	5897	3581	21999	10953	10328
Cr/Zr	0.61	1.56	2.02	0.76	0.91	6.89	0.66	1.39
Zr/Hf	35.6	41.5	37.9	29.5	36.9	36.6	31.9	35.3
La/Th	3.07	5.4	4.66	5.42	3.37	10.3	6.49	3.46
Hf/Ta	15.9	9.90	7.86	11.4	11.4	8.57	3.80	9.04
Th/U	2.52	2.40	1.25	1.88	1.91	5.51	3.08	2.03
La_N/Yb_N	4.15	8.35	6.56	6.56	5.61	12.1	8.04	5.47
Gd_N/Yb_N	1.01	1.63	1.57	1.41	1.52	2.11	1.61	1.19
Eu/Eu*	0.88	1.00	0.74	0.84	0.71	0.83	0.95	0.82
Ce/Ce*	0.97	0.87	1.71	0.98	1.75	0.73	1.06	1.42

Major elements in wt%, trace elements in ppm, except as noted. Total Fe as Fe_2O_3. Blank spaces: not determined. Chemical Index of Alteration (CIA) = $[Al_2O_3/(Al_2O_3 + CaO + Na_2O + K_2O)] \times 100$ in molecular proportions. $Eu/Eu^* = Eu_N/(Sm_N \times Gd_N)^{0.5}$; $Ce/Ce^* = Ce_N/[(La_N)^{0.6667} \times (Nd_N)^{0.3333}]$

Table 2. (cont.)
(b) Samples collected from 30 cm depth from low K radiometry areas

Sample	S15	S16	S17	S18	S19	S20	S21
SiO_2	76.84	80.01	71.84	89.98	93.60	67.00	84.54
TiO_2	0.67	0.59	0.70	0.54	0.72	0.91	0.83
Al_2O_3	9.27	7.26	8.19	3.77	2.55	6.59	6.28
Fe_2O_3	6.38	5.49	12.46	1.72	1.05	14.78	2.29
MnO	0.07	0.06	0.06	0.05	0.04	0.20	0.04
MgO	0.62	0.54	0.60	0.27	0.17	2.46	0.25
CaO	0.01	0.29	<0.01	0.05	<0.01	0.20	0.01
Na_2O	0.27	0.75	0.09	0.23	<0.01	0.03	0.04
K_2O	1.08	0.50	0.95	0.19	0.04	0.02	0.10
P_2O_5	0.05	0.07	0.05	0.05	0.05	0.08	0.06
LOI	5.22	4.79	5.27	3.51	2.21	8.03	5.93
Total	100.48	100.35	100.22	100.36	100.44	100.29	100.37
Sc	10.0	6.62	8.56	2.66	2.13	11.5	4.60
V	131	113	200	53	42	267	64
Cr	140	100	253	64	48	3417	82
Co	14.0	11.8	12.4	4.28	2.81	82.1	4.28
Ni	24	24	27	11	12	315	11
Cu	3	<2	7	<2	<2	47	<2
Zn	33	37	26	19	13	51	44
As	12.2	20.5	15.0	4.91	3.0	14.4	5.79
Se	0.20	0.12	0.27	0.04	0.13	0.20	0.15
Br	0.72	0.77	0.62	0.54	0.44	0.55	0.49
Rb	57.1	37.7	60.7	17.6	13.8	15.6	22.7
Sr	54	102	45	37	19	41	27
Y	11	11	8	8	7	12	14
Zr	175	155	195	233	261	188	274
Nb	10	9	10	9	10	11	14
Sb	0.84	0.74	1.33	0.18	0.13	0.81	0.18
Cs	2.17	1.86	2.65	0.96	0.93	1.24	1.67
Ba	387	251	307	198	86	173	96
La	11.7	11.8	4.59	6.07	4.56	13.60	10.50
Ce	42.7	28.8	14.5	14.1	8.79	51.2	25.4
Nd	9.74	13.2	3.57	4.21	4.7	13.1	7.92
Sm	2.17	2.48	1.08	0.97	0.72	2.67	1.71
Eu	0.54	0.66	0.32	0.25	0.19	0.60	0.45
Gd	2.27	2.66	1.06	1.28	0.71	2.26	1.86
Tb	0.34	0.38	0.20	0.18	0.14	0.34	0.34
Tm	0.26	0.18	0.16	0.21	0.19	0.43	0.30
Yb	1.79	1.62	1.46	0.89	0.88	1.67	1.88
Lu	0.25	0.25	0.23	0.15	0.14	0.27	0.29
Hf	5.75	5.63	5.72	5.33	5.99	6.13	7.35
Ta	0.62	0.55	0.57	0.41	0.59	0.65	0.80
W				0.78	0.9	1.16	1.12
Ir (ppb)	0.8	0.7	<2	<1	0.5	<2.7	1.0
Au (ppb)	6.8	6.4	7.9	2.6	3.9	3.6	5.8
Th	3.57	2.96	4.13	1.91	1.93	3.66	4.05
U	2.05	2.07	1.85	0.74	0.99	1.58	1.72
CIA	85	76	87	85	98	94	97
K/U	4372	2004	4261	2131	335	105	482
Cr/Zr	0.80	0.65	1.30	0.27	0.18	18.2	0.30
Zr/Hf	30.4	27.5	34.1	43.7	43.6	30.7	37.3
La/Th	3.28	3.99	1.11	3.18	2.36	3.72	2.59
Hf/Ta	9.27	10.2	10.0	13.0	10.2	9.43	9.19
Th/U	1.74	1.43	2.23	2.58	1.95	2.32	2.35
La_N/Yb_N	4.42	4.92	2.12	4.61	3.50	5.50	3.77
Gd_N/Yb_N	1.03	1.33	0.59	1.17	0.65	1.10	0.80
Eu/Eu*	0.74	0.79	0.91	0.69	0.81	0.75	0.77
Ce/Ce*	1.85	1.12	1.64	1.25	0.91	1.82	1.27

Major elements in wt%, trace elements in ppm, except as noted. Total Fe as Fe_2O_3. Blank spaces: not determined. Chemical Index of Alteration (CIA) = $[Al_2O_3/(Al_2O_3 + CaO + Na_2O + K_2O)] \times 100$ in molecular proportions. $Eu/Eu^* = Eu_N/(Sm_N \times Gd_N)^{0.5}$; $Ce/Ce^* = Ce_N/[(La_N)^{0.6667} \times (Nd_N)^{0.3333}]$

Table 2. (cont.)
(b) 30 cm depth samples from low K radiometry areas

Sample	S22	S23	S24	S25	S26	S28	S29
SiO_2	85.27	76.29	92.87	83.85	63.38	69.57	82.99
TiO_2	1.22	0.88	1.29	1.22	0.92	0.82	0.87
Al_2O_3	6.51	11.29	2.81	7.05	10.96	10.95	7.82
Fe_2O_3	2.16	3.31	1.23	2.45	16.36	11.67	2.12
MnO	0.03	0.08	0.03	0.07	0.05	0.08	0.04
MgO	0.25	0.69	0.22	0.41	0.59	0.42	0.46
CaO	<0.01	0.16	<0.01	0.04	<0.01	0.00	0.07
Na_2O	0.09	0.98	0.16	0.18	0.23	0.18	1.44
K_2O	0.23	1.09	0.13	0.71	1.18	1.10	0.58
P_2O_5	0.06	0.06	0.04	0.06	0.07	0.07	0.07
LOI	4.58	5.34	1.71	4.31	6.70	5.59	3.95
Total	100.40	100.18	100.50	100.36	100.44	100.47	100.40
Sc	5.24	7.7	2.7	6.08	8.19	7.76	4.34
V	64	101	40	81	202	163	72
Cr	65	115	42	101	387	307	67
Co	4.12	11.6	2.80	10.2	11.6	13.9	3.94
Ni	10	25	4	17	32	29	15
Cu	<2	<2	<2	<2	28	16	<2
Zn	30	40	11	32	42	29	24
As	2.06	5.63	2.5	3.9	16.9	13.1	6.86
Se	0.13	0.12	0.21	0.20	0.21	0.19	0.11
Br	0.34	0.48	0.36	0.58	0.54	0.49	0.37
Rb	20.3	49.6	17.7	47.9	58.1	54.6	31.9
Sr	29	129	33	61	83	81	44
Y	13	15	15	16	16	15	15
Zr	244	224	247	245	166	144	181
Nb	13	12	15	15	11	11	12
Sb	0.13	0.20	0.10	0.12	0.49	0.37	0.09
Cs	2.76	2.34	0.88	2.09	2.7	1.97	1.51
Ba	165	415	85	260	354	397	244
La	9.67	13.2	6.03	9.61	5.73	10.9	8.52
Ce	18.7	29.8	12.5	25.1	26.2	37.6	16.3
Nd	8.4	12.3	3.85	7.48	8.32	10.1	7.77
Sm	1.46	2.20	0.95	1.43	1.21	1.63	1.36
Eu	0.43	0.60	0.34	0.47	0.40	0.48	0.41
Gd	1.12	1.98	1.1	1.98	1.45	1.81	<0.8
Tb	0.27	0.31	0.3	0.45	0.26	0.30	0.25
Tm	0.26	0.26	0.26	0.36	0.36	0.29	0.25
Yb	1.56	1.67	2.11	2.82	1.68	1.45	1.44
Lu	0.24	0.26	0.29	0.43	0.26	0.23	0.22
Hf	6.28	5.55	6.22	6.12	4.39	3.60	4.21
Ta	0.84	0.64	0.95	0.93	0.49	0.44	0.54
W	0.89	0.36	1.39	1.39	1.28	1.20	0.57
Ir (ppb)	0.6	<1.8	<1.5	0.3	2.0	0.4	1.0
Au (ppb)	4.1	4.5	2.9	4.4	1.2	2.9	3.1
Th	3.23	2.90	2.41	3.40	4.44	4.21	2.07
U	1.35	1.22	1.01	1.55	1.89	1.12	0.67
CIA	94	79	87	86	87	88	71
K/U	1414	7414	1068	3801	5181	8150	7183
Cr/Zr	0.27	0.51	0.17	0.41	2.33	2.13	0.37
Zr/Hf	38.9	40.4	39.7	40	37.8	40	43
La/Th	2.99	4.55	2.50	2.83	1.29	2.59	4.12
Hf/Ta	7.48	8.67	6.55	6.58	8.96	8.18	7.80
Th/U	2.39	2.38	2.39	2.19	2.35	3.76	3.09
La_N/Yb_N	4.19	5.34	1.93	2.30	2.30	5.08	4.00
Gd_N/Yb_N	0.58	0.96	0.42	0.57	0.70	1.01	
Eu/Eu*	1.03	0.88	1.02	0.85	0.92	0.85	
Ce/Ce*	0.97	1.10	1.15	1.36	1.93	1.69	0.94

Major elements in wt%, trace elements in ppm, except as noted. Total Fe as Fe_2O_3. Blank spaces: not determined. Chemical Index of Alteration (CIA) = $[Al_2O_3/(Al_2O_3 + CaO + Na_2O + K_2O)] \times 100$ in molecular proportions. $Eu/Eu^* = Eu_N/(Sm_N \times Gd_N)^{0.5}$; $Ce/Ce^* = Ce_N/[(La_N)^{0.6667} \times (Nd_N)^{0.3333}]$

Table 2. (cont.)
(b) 30 cm depth samples from low K radiometry areas

Sample	S30	S31	S32	S33	T1	T2	T4
SiO_2	85.10	71.45	78.94	62.02	73.56	63.08	72.06
TiO_2	0.97	0.88	0.94	0.67	0.95	0.89	0.71
Al_2O_3	6.26	11.68	9.65	18.86	10.49	18.50	8.33
Fe_2O_3	2.51	6.79	3.87	7.04	6.46	11.40	11.64
MnO	0.06	0.05	0.06	0.06	0.04	0.05	0.04
MgO	0.28	0.39	0.57	1.32	0.30	0.67	0.37
CaO	0.06	0.25	0.02	0.00	0.41	0.33	0.09
Na_2O	1.48	0.13	0.87	0.49	0.13	1.50	0.06
K_2O	0.31	0.86	0.99	1.92	0.57	0.50	0.18
P_2O_5	0.06	0.16	0.05	0.09	0.03	0.02	0.06
LOI	3.20	7.43	4.46	7.89	6.39	3.45	6.92
Total	100.31	100.06	100.44	100.38	99.33	100.41	100.46
Sc	3.25	7.29	6.18	10.8	8.92	6.53	6.40
V	69	131	99	162	107	65	162
Cr	61	150	137	174	153	90	259
Co	5.63	4.44	9.67	14	18	14	13
Ni	11	24	24	56	32	19	30
Cu	<2	2	<2	23	8	<2	11
Zn	16	69	31	68	38	27	31
As	5.08	12.1	4.36	4.43	5.27	2.64	9.96
Se	0.16	0.15	0.11	0.1	0.21	0.13	0.21
Br	0.48	0.55	0.39	0.37	0.71	0.49	0.72
Rb	22.2	44.3	39.8	67	35	31	13
Sr	138	83	128	166	52	175	35
Y	12	14	12	18	13	11	9
Zr	194	193	169	122	241	171	169
Nb	11	12	11	8	14	11	11
Sb	0.11	0.35	0.11	0.13	0.28	0.1	0.37
Cs	0.92	2.31	1.67	2.84	2.14	1.58	0.64
Ba	219	296	387	682	304	345	158
La	7.95	9.69	8.02	20	9.41	8.34	6.07
Ce	16.9	21.3	23.7	35.6	24	21.4	14.7
Nd	7.33	8.32	7.05	18.1	8.6	9.01	4.34
Sm	1.36	1.21	1.32	3.06	1.42	1.42	0.92
Eu	0.47	0.34	0.41	0.83	0.38	0.42	0.25
Gd	1.84	1.51	1.51	3.12	1.70	1.73	0.94
Tb	0.24	0.25	0.26	0.37	0.30	0.23	0.19
Tm	0.26	0.28	0.24	0.32	0.22	0.17	0.17
Yb	1.41	1.33	1.50	1.36	1.51	1.17	1.18
Lu	0.21	0.21	0.23	0.21	0.24	0.18	0.18
Hf	5.86	4.93	4.38	2.98	5.97	4.40	4.20
Ta	0.62	0.63	0.49	0.33	0.73	0.49	0.46
W	0.93	1.13	1.40	1.01	1.38	0.48	1.26
Ir (ppb)	<1.5	0.7	1.5	<4	0.3	0.6	0.3
Au (ppb)	3.3	5.3	3.0	8.1	6.6	5.8	5.4
Th	1.83	4.20	2.22	3.19	3.39	2.37	2.94
U	0.83	1.34	0.90	0.75	1.51	1.25	1.09
CIA	68	88	79	87	90	89	96
K/U	3099	5326	9128	21243	3132	3319	1370
Cr/Zr	0.31	0.78	0.81	1.43	0.63	0.53	1.53
Zr/Hf	33.1	39.1	38.6	40.9	40.4	38.9	40.2
La/Th	4.34	2.31	3.61	6.27	2.78	3.52	2.06
Hf/Ta	9.45	7.83	8.94	9.03	8.18	8.98	9.13
Th/U	2.20	3.13	2.47	4.25	2.25	1.90	2.70
La_N/Yb_N	3.81	4.92	3.61	9.94	4.21	4.82	3.48
Gd_N/Yb_N	1.06	0.92	0.82	1.86	0.91	1.20	0.65
Eu/Eu*	0.91	0.77	0.89	0.82	0.75	0.82	0.82
Ce/Ce*	1.04	1.11	1.47	0.88	1.26	1.20	1.29

Major elements in wt%, trace elements in ppm, except as noted. Total Fe as Fe_2O_3. Chemical Index of Alteration (CIA) = $[Al_2O_3/(Al_2O_3 + CaO + Na_2O + K_2O)] \times 100$ in molecular proportions. $Eu/Eu^* = Eu_N/(Sm_N \times Gd_N)^{0.5}$; $Ce/Ce^* = Ce_N/[(La_N)^{0.6667} \times (Nd_N)^{0.3333}]$

Table 2. (cont.)
(c) 100 cm depth samples from high K radiometry areas

Sample	B3	B5	B6	B7	B8	B9	B10
SiO_2	63.85	70.83	61.16	62.15	67.66	61.36	62.86
TiO_2	0.50	0.84	0.53	0.64	0.74	0.72	0.60
Al_2O_3	17.52	13.54	19.27	18.20	14.57	17.96	17.64
Fe_2O_3	6.58	5.21	7.07	7.87	6.11	7.32	6.66
MnO	0.05	0.04	0.04	0.12	0.06	0.04	0.06
MgO	1.24	0.98	1.55	1.17	1.24	1.27	1.66
CaO	0.07	0.05	0.22	0.06	0.04	0.04	0.11
Na_2O	0.32	0.36	1.04	0.37	0.21	0.26	0.76
K_2O	1.63	1.63	1.65	1.52	1.77	1.18	1.60
P_2O_5	0.04	0.03	0.01	0.03	0.04	0.01	0.02
LOI	6.87	5.40	7.73	8.23	6.79	10.17	8.39
Total	98.68	98.90	100.29	100.36	99.22	100.33	100.37
Sc	13.3	12.3	14.7	15.6	17.2	17.2	12.0
V	123	122	117	140	156	111	134
Cr	141	147	139	177	169	173	139
Co	18	11	14	22	4	17	15
Ni	61	34	46	67	39	55	56
Cu	16	4	11	24	12	23	18
Zn	67	52	68	89	31	88	71
As	4.07	4.26	8.3	4.58	6.74	1.30	4.11
Se	0.07	0.16	0.11	0.10	0.04	0.06	0.06
Br	0.41	0.48	0.37	0.35	0.39	0.30	0.48
Rb	67	63	66	66	100	63	59
Sr	91	67	181	90	84	85	95
Y	9	14	28	18	9	35	17
Zr	119	167	109	129	169	140	147
Nb	9	11	10	11	13	10	9
Sb	0.20	0.23	0.17	0.23	0.45	0.19	0.25
Cs	3.55	3.32	4.06	3.43	3.70	5.80	2.68
Ba	630	532	669	640	654	584	704
La	19.9	13.6	40.2	34.2	7.84	45.3	19.2
Ce	45	27.1	60.6	75.3	20.6	66.7	29.0
Nd	14.6	11.0	37.3	27.5	9.0	47.9	13.7
Sm	2.2	2.1	7.13	3.34	1.5	8.66	2.05
Eu	0.53	0.63	1.84	0.86	0.38	2.42	0.57
Gd	2.51	2.42	8.48	4.53	1.63	8.58	2.48
Tb	0.19	0.35	0.79	0.37	0.26	0.96	0.28
Tm	0.18	0.22	0.28	0.17	0.21	0.39	0.20
Yb	1.18	1.58	2.00	1.35	1.41	2.77	1.24
Lu	0.18	0.25	0.28	0.18	0.22	0.40	0.17
Hf	3.34	4.1	2.65	3.1	4.07	3.37	3.19
Ta	0.46	0.66	0.28	0.37	0.61	0.41	0.35
W	0.6	1.44	<1.6	0.2	0.7	0.95	1.1
Ir (ppb)	<1.5	<1.4	<1.3	0.6	0.7	<1.4	0.4
Au (ppb)	5.6	8.1	14.2	<7	<4.5	7.43	5.28
Th	3.51	2.97	3.91	3.79	4.72	3.37	3.17
U	0.78	1.31	1.15	1.56	0.79	1.21	0.56
CIA	90	87	87	90	88	92	88
K/U	17340	10325	11906	8085	18591	8092	23708
Cr/Zr	1.18	0.88	1.28	1.37	1.00	1.24	0.95
Zr/Hf	35.6	40.7	41.1	41.6	41.5	41.5	46.1
La/Th	11.4	4.58	10.3	9.02	1.66	13.4	6.06
Hf/Ta	7.26	6.21	9.46	8.38	6.67	8.22	9.11
Th/U	4.5	2.27	3.4	2.43	5.97	2.79	5.66
La_N/Yb_N	5.67	5.82	13.6	17.1	3.76	11.1	10.5
Gd_N/Yb_N	1.72	1.24	3.44	2.72	0.94	2.51	1.62
Eu/Eu*	0.69	0.85	0.72	0.68	0.74	0.86	0.77
Ce/Ce*	1.2	1.02	0.74	1.13	1.2	0.69	0.81

Major elements in wt%, trace elements in ppm, except as noted. Total Fe as Fe_2O_3. $Eu/Eu^* = Eu_N/(Sm_N \times Gd_N)^{0.5}$; $Ce/Ce^* = Ce_N/[(La_N)^{0.6667} \times (Nd_N)^{0.3333}]$. Chemical Index of Alteration (CIA) = $[Al_2O_3/(Al_2O_3 + CaO + Na_2O + K_2O)] \times 100$ in molecular proportions

Table 2. (cont.)
(d) 100 cm depth samples from low K radiometry areas

Sample	B1	B2	B4
SiO_2	71.16	73.74	63.32
TiO_2	0.82	0.64	0.70
Al_2O_3	9.27	12.49	14.52
Fe_2O_3	5.52	4.00	8.94
MnO	0.06	0.04	0.04
MgO	0.4	0.60	0.52
CaO	0.11	0.31	0.08
Na_2O	0.02	1.05	0.08
K_2O	0.53	0.82	0.12
P_2O_5	0.03	0.01	0.04
LOI	12.42	4.94	12.04
Total	100.34	98.67	100.31
Sc	14.9	8.11	12.2
V	156	87	137
Cr	208	108	227
Co	20	10	15
Ni	62	32	54
Cu	29	<2	13
Zn	58	38	52
As	4.48	2.26	4.45
Se	0.16	0.10	0.14
Br	0.58	0.54	1.02
Rb	38	40	12
Sr	74	150	34
Y	15	12	9
Zr	183	141	158
Nb	13	11	11
Sb	0.31	0.14	0.35
Cs	2.86	1.92	0.79
Ba	324	444	138
La	17.7	13.6	9.57
Ce	40.6	28.9	27.6
Nd	12.9	12.8	5.55
Sm	2.24	2.55	1.15
Eu	0.52	0.70	0.31
Gd	3.40	2.54	1.82
Tb	0.37	0.32	0.24
Tm	0.23	0.19	0.18
Yb	1.53	1.37	1.25
Lu	0.23	0.21	0.2
Hf	4.61	3.42	3.83
Ta	0.55	0.48	0.38
W	1.61	0.53	<0.9
Ir (ppb)	<1.8	<1.4	<1.5
Au (ppb)	7.7	1.6	7.4
Th	4.11	2.78	3.65
U	1.77	1.13	1.21
CIA	93	85	98
K/U	2485	6021	823
Cr/Zr	1.14	0.77	1.44
Zr/Hf	40.0	41.2	41.3
La/Th	4.31	4.89	2.62
Hf/Ta	8.38	7.13	10.1
Th/U	2.32	2.46	3.02
La_N/Yb_N	7.82	6.71	5.17
Gd_N/Yb_N	1.80	1.50	1.18
Eu/Eu*	0.58	0.84	0.65
Ce/Ce*	1.22	1.04	1.65

Major elements in wt%, trace elements in ppm, except as noted. Total Fe as Fe_2O_3. $Eu/Eu^* = Eu_N/(Sm_N \times Gd_N)^{0.5}$; $Ce/Ce^* = Ce_N/[(La_N)^{0.6667} \times (Nd_N)^{0.3333}]$. Chemical Index of Alteration (CIA) = $[Al_2O_3/(Al_2O_3 + CaO + Na_2O + K_2O)] \times 100$ in molecular proportions

Table 3. Chemical compositions of Bosumtwi crater rocks, average soil and upper continental crust (UCC), for comparison with soil samples

	Soil Average	UCC [a]	Shale Average [b]	Graywacke-Phyllite Average [b]	Granite Dike Average [b]	Pepiakese Granite Average [b]
SiO_2	72.63	66.0	55.56	66.75	68.74	57.81
TiO_2	0.77	0.5	0.84	0.66	0.50	0.46
Al_2O_3	11.60	15.2	19.56	15.27	15.91	16.45
Fe_2O_3	6.18	4.5	8.54	6.37	3.97	6.09
MnO	0.06		0.05	0.03	0.01	0.07
MgO	0.95	2.2	2.9	2.12	1.44	6.63
CaO	0.22	4.2	0.09	0.19	0.31	4.36
Na_2O	0.62	3.9	1	2.26	4.14	6.04
K_2O	1.08	3.4	2.89	1.8	1.92	0.67
P_2O_5	0.11		0.08	0.06	0.06	0.10
LOI	6.01		7.91	4.25	2.98	1.48
Total	100.25		99.41	99.76	99.98	100.15
Sc	9.96	11	23.4	15.5	9.76	17.5
V	124	60	184	134	91	110
Cr	242	35	194	165	127	517
Co	13.7	10	22.6	12.1	9.66	30.4
Ni	49	20	79	48	49	172
Cu	16	25	52	16	11	24
Zn	45	71	143	104	82	90
As	9.37	1.5	8.61	7	14.9	12.7
Se	0.14	50	0.19	0.16	0.21	0.09
Br	0.53		0.23	0.18	0.15	0.16
Rb	53.4	112	104	65.2	69.9	22.4
Sr	111	350	118	152	342	377
Y	14	22	24	19	11	11
Zr	174	190	153	143	129	82
Nb	10.5	25	6.1	5.7	3.7	1.8
Sb	0.45	0.20	0.38	0.2	0.18	0.42
Cs	2.51	3.70	5.57	3.27	4.22	0.87
Ba	401	550	765	454	605	226
La	13.2	30	30.9	23.4	18.8	15.6
Ce	32.2	64	38.4	34.8	39.4	32
Nd	12.5	26	28.2	26.5	19.8	17.5
Sm	2.21	4.50	5.48	5.06	3.74	3.58
Eu	0.62	0.88	1.37	1.29	1.03	1.19
Gd	2.25	3.80	6.55	4.8	3.40	3.07
Tb	0.34	0.64	1.06	0.73	0.59	0.47
Tm	0.48	0.33	0.47	0.35	0.22	0.2
Yb	1.57	2.20	2.70	2.14	1.13	1.18
Lu	0.24	0.32	0.38	0.29	0.16	0.14
Hf	4.64	5.80	4.12	4.04	3.66	1.88
Ta	0.53	2.20	0.53	0.42	0.28	0.19
W	1.03	2	0.5	0.64	0.84	0.54
Ir (ppb)	0.9	0.02				
Au (ppb)	5.9	1.8	11.8	10.3	21.5	18.3
Th	3.27	10.7	4.36	3.94	3.10	2.21
U	1.35	2.80	1.69	1.35	1.75	0.74
CIA	83	46	80	72	63	47
K/U	7597	10076	14190	11064	9104	7513
Cr/Zr	1.49	0.18	1.27	1.15	0.98	6.30
Zr/Hf	37.9	32.8	37.1	35.4	35.2	43.6
La/Th	4.14	2.80	7.09	5.94	6.06	7.06
Hf/Ta	9.11	2.64	7.77	9.62	13.07	9.89
Th/U	2.63	3.82	2.58	2.92	1.77	2.99
La_N/Yb_N	5.68	9.21	7.73	7.39	11.2	8.93
Gd_N/Yb_N	1.17	1.4	1.97	1.82	2.44	2.11
Eu/Eu*	0.85	0.65	0.7	0.8	0.88	1.1
Ce/Ce*	1.29	1.07	0.61	0.68	0.98	0.94

Major elements in wt%, trace elements in ppm, except as noted. Total Fe as Fe_2O_3. Blank spaces: not determined. Chemical Index of Alteration (CIA) = $[Al_2O_3/(Al_2O_3 + CaO + Na_2O + K_2O)] \times 100$ in molecular proportions. $Eu/Eu^* = Eu_N/(Sm_N \times Gd_N)^{0.5}$; $Ce/Ce^* = Ce_N/[(La_N)^{0.6667} \times (Nd_N)^{0.3333}]$. [a] Data from Taylor and McLennan (1985). [b] Data from Koeberl et al. (1998)

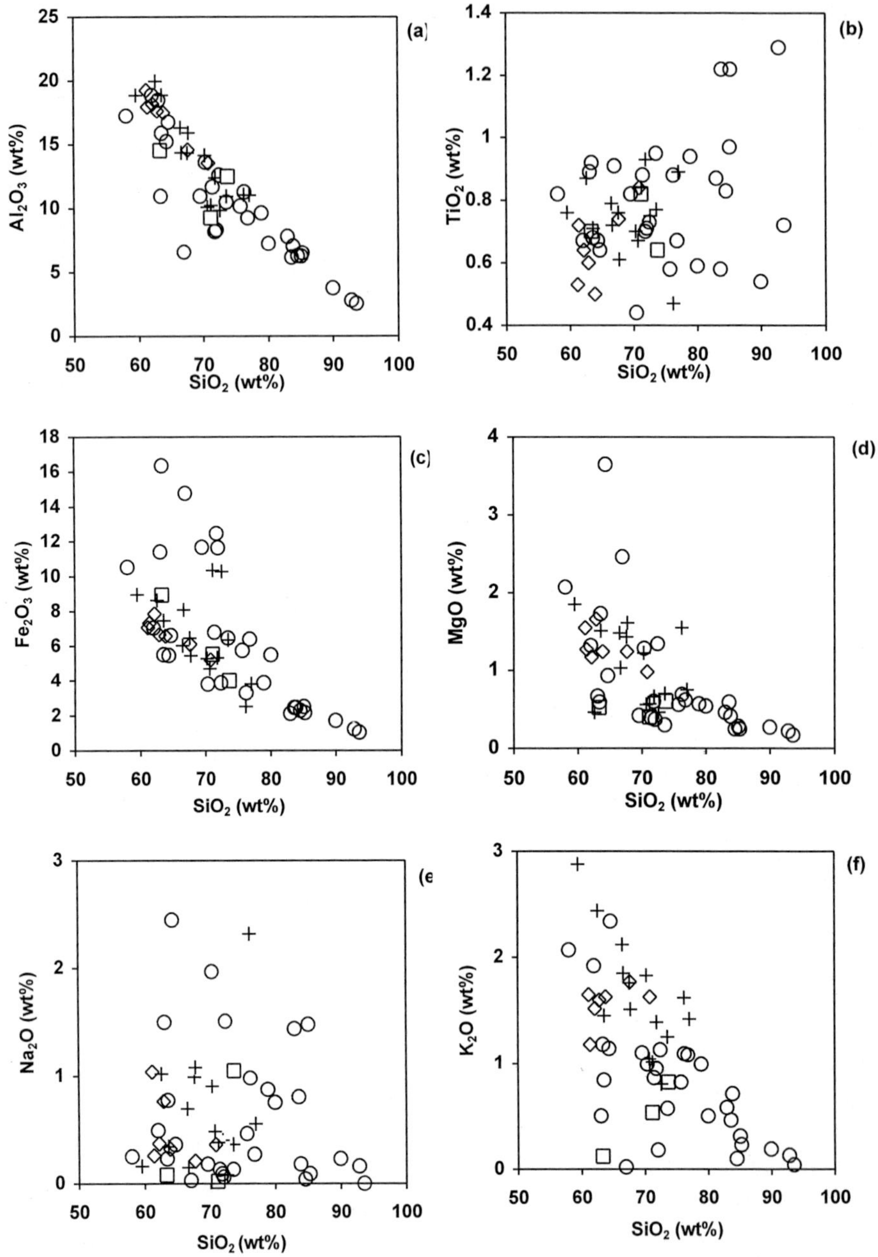

Fig. 4. Harker variation diagrams for soils from the Bosumtwi crater area. Solid circles represent locations of samples collected inside of the anomalous areas of high K abundance at 30 cm depth; open circles are those collected outside of the anomalous areas at 30 cm depth; solid squares represent those from the anomalous areas from 100 cm depth; open squares are those samples from 100 cm depth outside of the anomalous areas.

Table 4. Chemical compositions of pit soils from the Bosumtwi crater area
(a) 30 cm depth samples

Sample	T1	T2	T3	T4	T5	T6	T7	T8	T9	T10
SiO_2	73.56	63.08	71.88	72.06	77.05	70.28	71.21	62.57	70.68	63.57
TiO_2	0.95	0.89	0.93	0.71	0.89	0.70	0.84	0.87	0.67	0.71
Al_2O_3	10.49	18.5	12.39	8.33	11.04	14.14	10.23	19.97	10.09	18.88
Fe_2O_3	6.46	11.4	5.33	11.64	3.80	5.29	10.34	8.63	4.68	7.46
MnO	0.04	0.05	0.07	0.04	0.05	0.05	0.07	0.05	0.05	0.05
MgO	0.30	0.67	0.66	0.37	0.75	1.22	0.57	0.46	0.56	1.51
CaO	0.41	0.33	0.07	0.09	0.09	0.17	0.05	0.01	0.31	0.06
Na_2O	0.13	1.50	0.40	0.06	0.55	0.90	0.38	1.02	0.48	0.34
K_2O	0.57	0.50	1.39	0.18	1.42	1.83	1.01	2.44	1.04	1.45
P_2O_5	0.03	0.02	0.04	0.06	0.04	0.04	0.04	0.03	0.05	0.02
LOI	6.39	3.45	5.70	6.92	4.66	5.83	5.69	4.04	11.65	6.40
Total	99.33	100.4	98.86	100.5	100.4	100.5	100.4	100.1	100.3	100.4
Sc	8.92	6.53	12.6	6.40	11.0	11.1	9.25	8.28	16.0	12.8
V	107	65	119	162	100	111	127	78	119	121
Cr	153	90	143	259	124	131	200	92	164	141
Co	18	14	20	13	15	10	18	10	17	15
Ni	32	19	42	30	27	36	31	27	57	43
Cu	8	<2	7	11	<2	3	12	3	25	7
Zn	38	27	48	31	43	55	42	34	101	60
As	5.27	2.64	5.54	9.96	4.69	7.15	8.29	6.72	1.25	7.16
Se	0.21	0.13	0.17	0.21	0.10	0.10	0.28	0.14	0.08	0.11
Br	0.71	0.49	0.56	0.72	0.60	0.57	0.67	0.55	0.32	0.51
Rb	35	31	67	13	60	69	48	39	75	73
Sr	52	175	96	35	84	123	57	137	90	91
Y	13	11	11	9	14	12	10	7	34	12
Zr	241	171	171	169	187	158	145	129	145	162
Nb	14	11	13	11	13	11	11	10	10	12
Sb	0.28	0.1	0.29	0.37	0.23	0.14	0.65	0.33	0.21	0.25
Cs	2.14	1.58	2.95	0.64	3.10	2.93	2.13	1.99	3.69	3.36
Ba	304	345	516	158	484	587	415	409	518	575
La	9.41	8.34	8.21	6.07	11.8	11.8	7.33	7.56	51.7	7.01
Ce	24.0	21.4	39.8	14.7	23	21.2	41.6	25.2	77.6	22.0
Nd	8.6	9.01	8.87	4.34	11.1	8.81	7.08	6.99	46.9	5.09
Sm	1.42	1.42	1.32	0.92	1.8	1.67	1.47	1.25	8.31	1.12
Eu	0.38	0.42	0.39	0.25	0.6	0.5	0.41	0.38	2.52	0.37
Gd	1.7	1.73	1.61	0.94	3.02	2.3	1.9	1.66	9.49	1.44
Tb	0.3	0.23	0.29	0.19	0.32	0.22	0.25	0.19	0.87	0.27
Tm	0.22	0.17	0.21	0.17	0.23	0.18	0.15	0.12	0.3	0.18
Yb	1.51	1.17	1.30	1.18	1.58	1.18	1.15	0.85	2.36	1.23
Lu	0.24	0.18	0.20	0.18	0.24	0.18	0.20	0.13	0.36	0.19
Hf	5.97	4.40	4.30	4.20	4.81	3.56	3.95	3.39	2.97	4.28
Ta	0.73	0.49	0.75	0.46	0.56	0.41	0.63	0.35	0.45	0.41
W	1.38	0.48	0.76	1.26	2.68	0.36	0.73	0.89	<1	1.33
Ir (ppb)	0.3	0.6	0.8	0.3	2.4	1.5	<1.7	<1.5	1.3	0.5
Au (ppb)	6.6	5.8	3.4	5.4	2.5	3.0	4.1	5.0	9.8	5.3
Th	3.39	2.37	3.31	2.94	2.65	2.57	3.19	2.58	3.42	2.83
U	1.51	1.25	1.53	1.09	1.36	0.96	1.62	1.04	0.84	1.25
CIA	90	89	87	96	84	83	88	85	85	91
K/U	3132	3319	7539	1370	8664	15818	5173	19468	10274	9626
Cr/Zr	0.63	0.53	0.84	1.53	0.66	0.83	1.38	0.71	1.13	0.87
Zr/Hf	40.4	38.9	39.8	40.2	38.9	44.4	36.7	38.1	48.8	37.9
La/Th	2.78	3.52	2.48	2.06	4.45	4.59	2.30	2.93	15.12	2.48
Hf/Ta	8.18	8.98	5.73	9.13	8.59	8.68	6.27	9.69	6.60	10.44
Th/U	2.25	1.90	2.16	2.70	1.95	2.68	1.97	2.48	4.07	2.26
La_N/Yb_N	4.21	4.82	4.27	3.48	5.05	6.76	4.31	6.01	14.8	3.85
Gd_N/Yb_N	0.91	1.20	1.0	0.65	1.55	1.58	1.34	1.58	3.26	0.95
Eu/Eu*	0.75	0.82	0.82	0.82	0.79	0.78	0.75	0.81	0.87	0.89
Ce/Ce*	1.26	1.20	2.26	1.29	0.95	0.95	2.74	1.64	0.74	1.67

Major elements in wt%, trace elements in ppm, except as noted. Total Fe as Fe_2O_3. Blank spaces: not determined. Chemical Index of Alteration (CIA) = $[Al_2O_3/(Al_2O_3 + CaO + Na_2O + K_2O)] \times 100$ in molecular proportions. Eu/Eu* = $Eu_N/(Sm_N \times Gd_N)^{0.5}$; Ce/Ce* = $Ce_N/[(La_N)^{0.6667} \times (Nd_N)^{0.3333}]$

Table 4. Chemical compositions of pit soils from the Bosumtwi crater area
(b) 100 cm depth samples

Samples	B1	B2	B3	B4	B5	B6	B7	B8	B9	B10
SiO_2	71.16	73.74	63.85	63.32	70.83	61.16	62.15	67.66	61.36	62.86
TiO_2	0.82	0.64	0.50	0.70	0.84	0.53	0.64	0.74	0.72	0.60
Al_2O_3	9.27	12.49	17.52	14.52	13.54	19.27	18.2	14.57	17.96	17.64
Fe_2O_3	5.52	4.00	6.58	8.94	5.21	7.07	7.87	6.11	7.32	6.66
MnO	0.06	0.04	0.05	0.04	0.04	0.04	0.12	0.06	0.04	0.06
MgO	0.40	0.60	1.24	0.52	0.98	1.55	1.17	1.24	1.27	1.66
CaO	0.11	0.31	0.07	0.08	0.05	0.22	0.06	0.04	0.04	0.11
Na_2O	0.02	1.05	0.32	0.08	0.36	1.04	0.37	0.21	0.26	0.76
K_2O	0.53	0.82	1.63	0.12	1.63	1.65	1.52	1.77	1.18	1.60
P_2O_5	0.03	0.01	0.04	0.04	0.03	0.01	0.03	0.04	0.01	0.02
LOI	12.42	4.94	6.87	12.04	5.40	7.73	8.23	6.79	10.17	8.39
Total	100.3	98.67	98.68	100.3	98.9	100.3	100.4	99.22	100.3	100.4
Sc	14.9	8.11	13.3	12.2	12.3	14.7	15.6	17.2	17.2	12.0
V	156	87	123	137	122	117	140	156	111	134
Cr	208	108	141	227	147	139	177	169	173	139
Co	20	10	18	15	11	14	22	4	17	15
Ni	62	32	61	54	34	46	67	39	55	56
Cu	29	<2	16	13	4	11	24	12	23	18
Zn	58	38	67	52	52	68	89	31	88	71
As	4.48	2.26	4.07	4.45	4.26	8.3	4.58	6.74	1.30	4.11
Se	0.16	0.10	0.07	0.14	0.16	0.11	0.10	0.04	0.06	0.06
Br	0.58	0.54	0.41	1.02	0.48	0.37	0.35	0.39	0.300	0.48
Rb	38	40	67	12	63	66	66	100	63	59
Sr	74	150	91	34	67	181	90	84	85	95
Y	15	12	9	9	14	28	18	9	35	17
Zr	183	141	119	158	167	109	129	169	140	147
Nb	13	11	9	11	11	10	11	13	10	9
Sb	0.31	0.14	0.2	0.35	0.23	0.17	0.23	0.45	0.19	0.25
Cs	2.86	1.92	3.55	0.79	3.32	4.06	3.43	3.70	5.8	2.68
Ba	324	444	630	138	532	669	640	654	584	704
La	17.7	13.6	19.9	9.57	13.6	40.2	34.2	7.84	45.3	19.2
Ce	40.6	28.9	45.0	27.6	27.1	60.6	75.3	20.6	66.7	29
Nd	12.9	12.8	14.6	5.55	11	37.3	27.5	9.0	47.9	13.7
Sm	2.24	2.55	2.2	1.15	2.1	7.13	3.34	1.5	8.66	2.05
Eu	0.52	0.7	0.53	0.31	0.63	1.84	0.86	0.38	2.42	0.57
Gd	3.40	2.54	2.51	1.82	2.42	8.48	4.53	1.63	8.58	2.48
Tb	0.37	0.32	0.19	0.24	0.35	0.79	0.37	0.26	0.96	0.28
Tm	0.23	0.19	0.18	0.18	0.22	0.28	0.17	0.21	0.39	0.20
Yb	1.53	1.37	1.18	1.25	1.58	2.00	1.35	1.41	2.77	1.24
Lu	0.23	0.21	0.18	0.20	0.25	0.28	0.18	0.22	0.4	0.17
Hf	4.61	3.42	3.34	3.83	4.10	2.65	3.10	4.07	3.37	3.19
Ta	0.55	0.48	0.46	0.38	0.66	0.28	0.37	0.61	0.41	0.35
W	1.61	0.53	0.6	<0.9	1.44	<1.6	0.24	0.76	0.95	1.13
Ir (ppb)	<1.8	<1.4	<1.5	<1.5	<1.4	<1.3	0.6	0.8	<1.4	0.4
Au (ppb)	7.7	1.6	5.7	7.5	8.1	14.2	<7	<4.5	7.4	5.3
Th	4.11	2.78	3.51	3.65	2.97	3.91	3.79	4.72	3.37	3.17
U	1.77	1.13	0.78	1.21	1.31	1.15	1.56	0.79	1.21	0.56
CIA	93	85	90	98	87	87	90	88	92	88
K/U	2485	6021	17340	823	10325	11906	8085	18591	8092	23708
Cr/Zr	1.14	0.77	1.18	1.44	0.88	1.28	1.37	1.00	1.24	0.95
Zr/Hf	40.0	41.2	35.6	41.3	40.7	41.1	41.6	41.5	41.5	46.1
La/Th	4.31	4.89	11.4	2.62	4.58	10.28	9.02	1.66	13.44	6.06
Hf/Ta	8.38	7.13	7.26	10.08	6.21	9.46	8.38	6.67	8.22	9.11
Th/U	2.32	2.46	4.50	3.02	2.27	3.40	2.43	5.97	2.79	5.66
La_N/Yb_N	7.82	6.71	5.67	5.17	5.82	13.58	17.12	3.76	11.05	10.46
Gd_N/Yb_N	1.80	1.50	1.72	1.18	1.24	3.44	2.72	0.94	2.51	1.62
Eu/Eu*	0.58	0.84	0.69	0.65	0.85	0.72	0.68	0.74	0.86	0.77
Ce/Ce*	1.22	1.04	1.20	1.65	1.02	0.74	1.13	1.20	0.69	0.81

Major elements in wt%, trace elements in ppm, except as noted. Total Fe as Fe_2O_3. Blank spaces: not determined. Chemical Index of Alteration (CIA) = $[Al_2O_3/(Al_2O_3 + CaO + Na_2O + K_2O)] \times 100$ in molecular proportions. $Eu/Eu^* = Eu_N/(Sm_N \times Gd_N)^{0.5}$; $Ce/Ce^* = Ce_N/[(La_N)^{0.6667} \times (Nd_N)^{0.3333}]$

Table 5. Correlation coefficients (r) calculated for the soil analyses of Table 2.

r	SiO_2	TiO_2	Al_2O_3	Fe_2O_3	K_2O	Zr	Ba	Th
SiO_2		0.20	-0.88	-0.73	-0.71	0.78	-0.78	0.44
Sc	-0.84	-0.30	0.82	0.69	0.73	-0.70	0.81	0.43
Cr	-0.70	-0.07	0.45	0.83	0.48	-0.46	0.39	0.67
Co	-0.67	-0.15	0.66	0.70	0.65	-0.58	0.61	0.45
Ni	-0.86	-0.26	0.85	0.67	0.75	-0.68	0.74	0.40
Zn	-0.78	-0.12	0.77	0.68	0.62	-0.65	0.71	0.35
Rb	-0.68	-0.25	0.77	0.70	0.95	-0.66	0.87	0.41
Sr	-0.43	-0.30	0.62	-0.02	0.55	-0.60	0.64	0.21
Zr	0.78	0.40	-0.80	-0.49	-0.71		-0.82	0.12
Nb	0.30	0.79	-0.28	-0.01	-0.30	0.51	-0.39	0.18
Cs	-0.68	-0.20	0.76	0.65	0.81	-0.57	0.81	0.46
Ba	-0.78	-0.33	0.91	0.69	0.92	-0.82		0.17
La	-0.45	-0.31	0.52	0.34	0.38	-0.50	0.56	0.19
Ce	-0.54	-0.25	0.48	0.56	0.46	-0.53	0.54	0.51
Yb	-0.05	0.31	-0.03	0.01	0.02	0.11	0.02	0.47
Hf	0.77	0.30	-0.81	-0.46	-0.67	0.91	-0.79	0.11
Ta	0.49	0.70	-0.50	-0.01	-0.38	0.64	-0.52	0.21
Th	-0.44	0.65	0.26	0.10	0.28	-0.12	0.17	
U	-0.10	0.10	0.02	0.36	0.28	0.18	0.01	0.71
La_N/Yb_N	-0.48	-0.47	0.56	0.11	0.35	-0.56	0.58	0.03
Gd_N/Yb_N	-0.43	-0.56	0.51	0.05	0.39	-0.60	0.58	0.08

5.3.4
Au, Ir, Sb, As

There is an enrichment in these elements in the soils relative to upper continental crust. These relatively high concentrations are more pronounced in the samples enriched in ferruginous nodules. From Table 2, it can be seen that the soils have fairly high Ir and Au contents. The parent rocks also have relatively Au contents compared to average crustal values (see, e.g., Koeberl et al. 1998 and Table 3), possibly because of mineralization in the area (e.g., Jones 1985).

5.3.5
Rare Earth Element (REE) Distribution

The chondrite-normalized (Taylor and McLennan 1985) REE distribution patterns are shown in Fig. 5 for soils collected in February 2000, and in Fig. 6 for pit soils collected in 1999. Four distinct patterns can be distinguished in Fig. 5:

1. patterns with positive Ce anomalies (Ce/Ce* from 1.25 to 2.21) and Gd_N/Yb_N values, <1.0 (from 0.57 to 0.97);
2. patterns with positive Ce anomalies (Ce/Ce* from 1.12 to 1.85) and Gd_N/Yb_N values, >1.0;
3. patterns with slightly negative or positive Ce anomalies and Gd_N/Yb_N values, <1.0;

4. patterns with slightly negative or positive Ce anomalies and positive Gd_N/Yb_N values, >1.0.

Similar positive Ce anomalies are noted for in pit samples T3 (Ce/Ce* = 2.26), B4 (1.65), T7 (2.74), T8 (1.64) and T10 (1.67). The country rocks, however, have no positive Ce anomalies. Variable REE distribution patterns with slightly positive Ce anomalies for samples from the lower zone (100 cm depth) to pronounced, positive Ce anomalies for samples from the upper zone (30 cm depth) were noted for the pit samples. Similarly, graywacke/phyllite-normalized REE distribution patterns of the soils show positive Ce anomalies, with the anomalies being more pronounced in the 30 cm depth samples (Fig. 7). Positive Ce anomalies are noted for both soils with high and low SiO_2 contents, but are more pronounced in the soils with high Fe_2O_3 contents.

All samples are enriched in both light rare earth elements (LREE) and heavy rare earth elements (HREE) relative to chondritic values, with about 75 % of the samples having La_N/Yb_N ratios greater than 4. Most of the samples have small negative Eu anomalies (slightly depleted in Eu relative to nearby REE), with Eu/Eu* values of about 0.69 to 1.03.

Silt and sand samples have lower overall REE abundances in comparison to those samples with abundant clay-sized material, although the shapes of the REE distribution patterns are similar. The influence of high quartz contents on the absolute REE contents in the soils is seen in samples S19 and S24, which have the highest concentrations of SiO_2 (93.6 and 92.8 wt%, respectively) and low REE contents.

Figure 8 shows a plot of Eu/Eu* versus Gd_N/Yb_N, in comparison to the values for upper continental crust (Taylor and McLennan 1995). Most of the soils plot into the field of post-Archean sedimentary rocks (Taylor and McLennan 1995). The majority of the samples show a slight relative depletion of Eu, with Eu/Eu* values in the range of 0.70 to 0.95 and Gd_N/Yb_N values of less than 2.0.

6 Discussion

Major and trace element concentrations in sediments result from the competing influences of provenance (source rocks), weathering, diagenesis, sediment sorting and the aqueous geochemistry of individual elements (Rollinson 1993).

6.1 Provenance (Source Rocks)

Rare earth elements (REE), Th, Sc, and, to a lesser extent, Cr and Co, are the most useful elements for provenance characterization, because they are among the least soluble trace elements. These elements are believed to be transported exclusively in the terrigenous component of a sediment and, therefore, reflect the composition of the source materials (McLennan et al. 1980; Rollinson 1993).

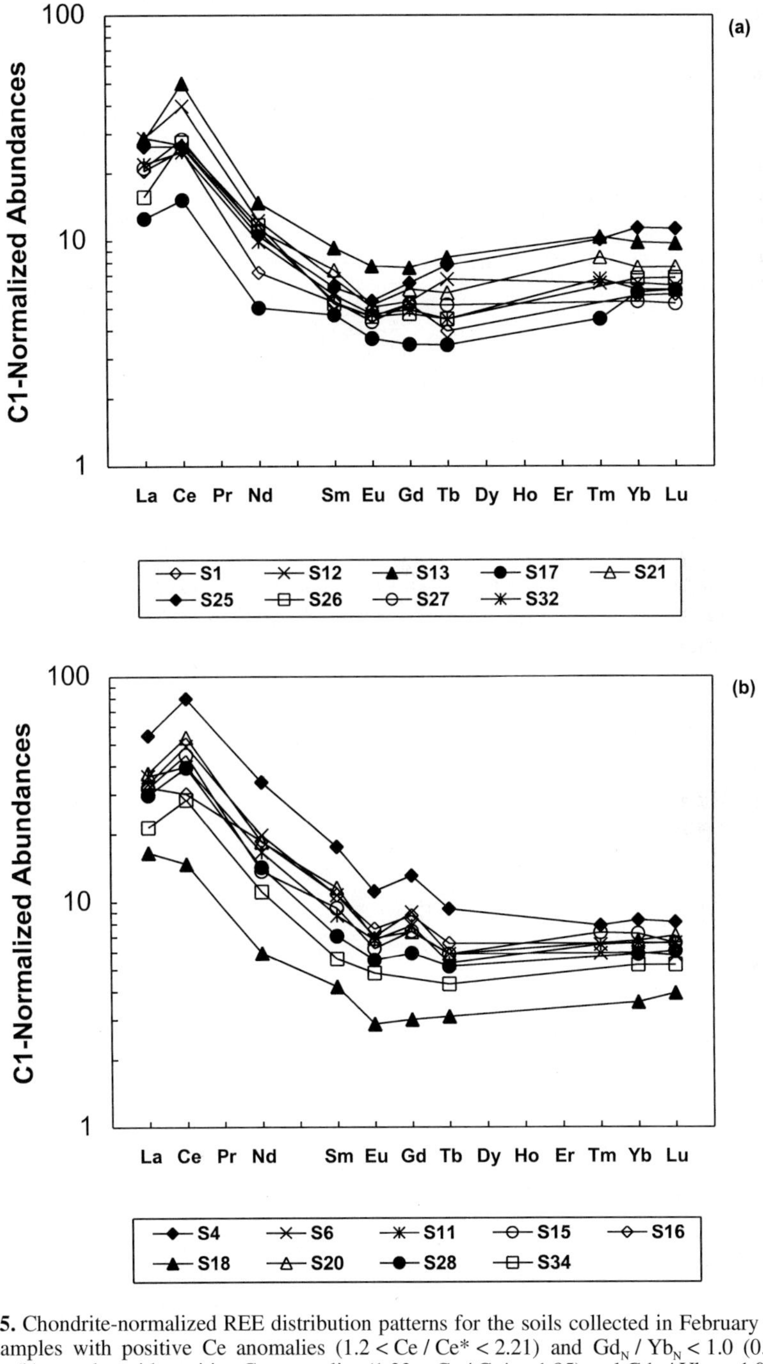

Fig. 5. Chondrite-normalized REE distribution patterns for the soils collected in February 2000: (a) samples with positive Ce anomalies ($1.2 < Ce / Ce^* < 2.21$) and $Gd_N / Yb_N < 1.0$ (0.57 to 0.97); (b) samples with positive Ce anomalies ($1.22 < Ce / Ce^* < 1.85$) and $Gd_N / Yb_N > 1.0$ (1.03 to 1.57). Normalization factors after Taylor and McLennan (1985).

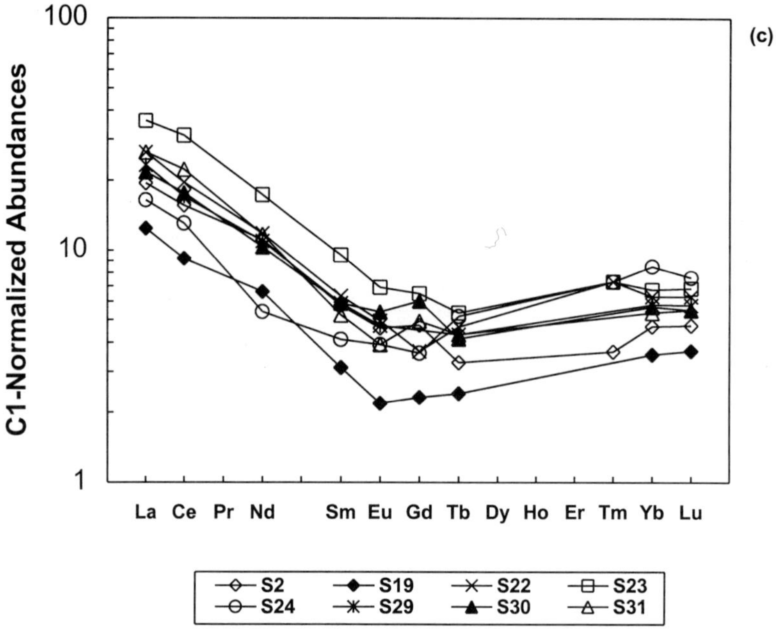

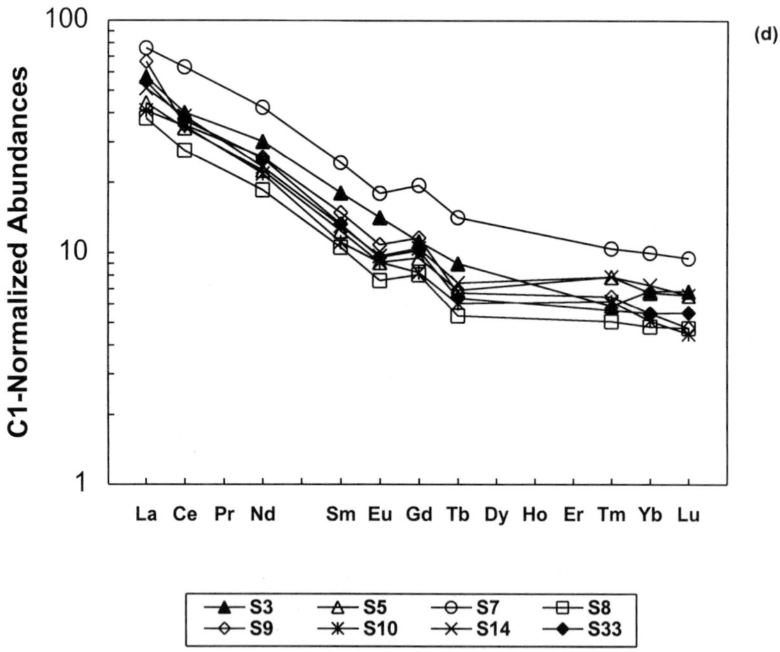

Fig. 5. (cont.) (c) samples with slight positive or negative Ce anomalies and $Gd_N / Yb_N < 1.0$; (d) samples with slight positive or negative Ce anomalies and $Gd_N / Yb_N > 1.0$. Normalization factors after Taylor and McLennan (1985).

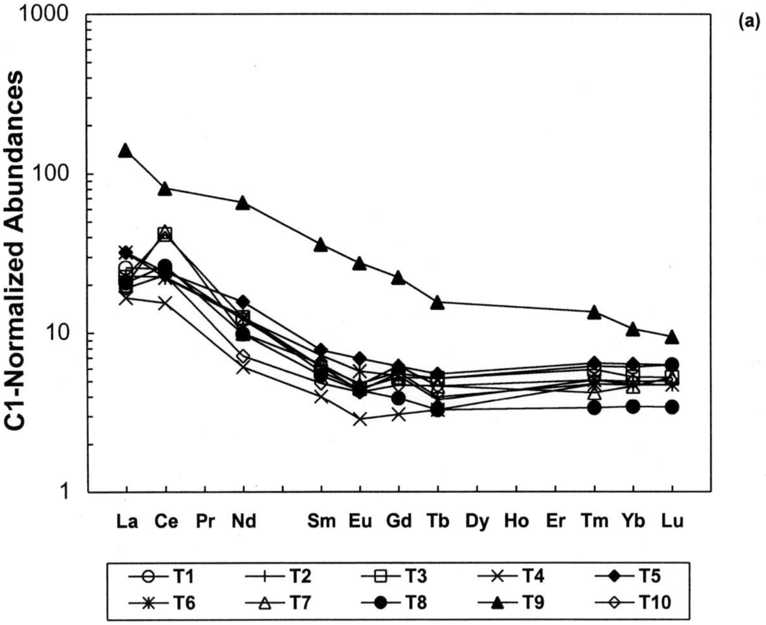

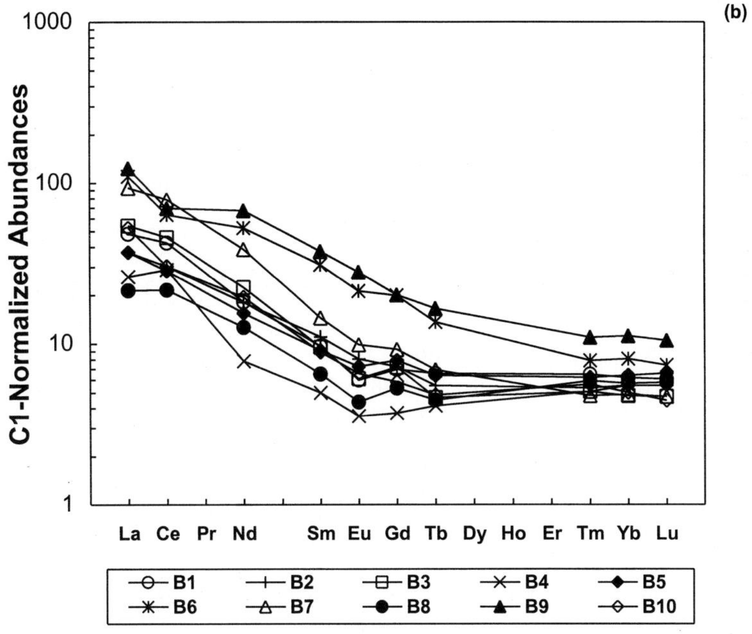

Fig. 6. Chondrite-normalized REE distribution patterns for the pit soils investigated: (a) 30 cm depth (samples denoted T); (b) 100 cm depth (samples denoted B). Normalization factors after Taylor and McLennan (1985).

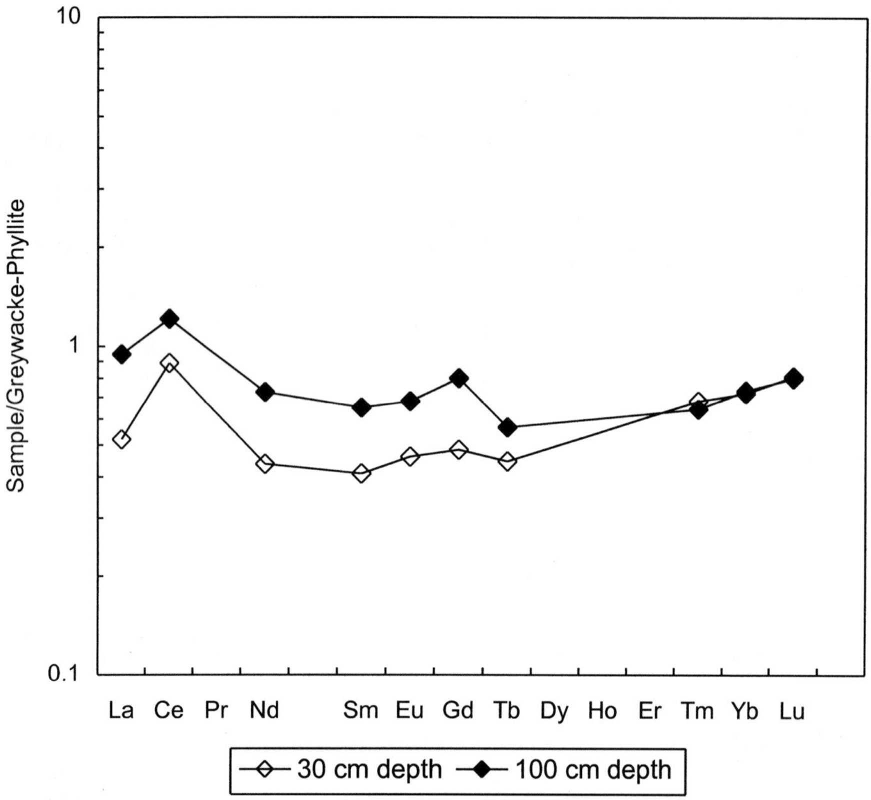

Fig. 7. Graywacke/phyllite-normalized REE distribution patterns for the soils investigated; open diamonds represent average of 30 cm depth samples; solid diamonds represent average 100 cm depth samples.

The immobile elements La and Th are more abundant in felsic than in basic rocks, whereas Sc and Co are more concentrated in basic rocks than in felsic rocks (Taylor and McLennan 1985; Wronkiewicz and Condie 1987). Thus, the concentration of these elements and the corresponding elemental ratios (e.g., La/Sc, Th/Sc, Th/Co and La/Co) in soils may be useful for provenance determination. Basic end members (e.g., basalt) should have high Co/Th and low La/Sc, whereas felsic end members (e.g., granite) should have low Co/Th and high La/Sc. Condie and Wronkiewicz (1990) and McLennan and Taylor (1991) used plots such as Th/Sc vs Sc and Cr/Th vs Sc/Th as indicators of contrasting felsic and mafic provenance.

The concentrations of Th, Hf, Co, Zr, La, and Sc are compared in Th-Hf-Co, Th-Co-Zr/10, and La-Th-Sc ternary diagrams (Fig. 9). Both strongly incompatible (Th) and strongly compatible (Sc and Co) elements, as well as those related to heavy minerals (Hf and Zr), are represented. The La-Th-Sc plot for the Bosumtwi soils (Fig. 9a) shows a relatively limited dispersion of data. The data plot parallel

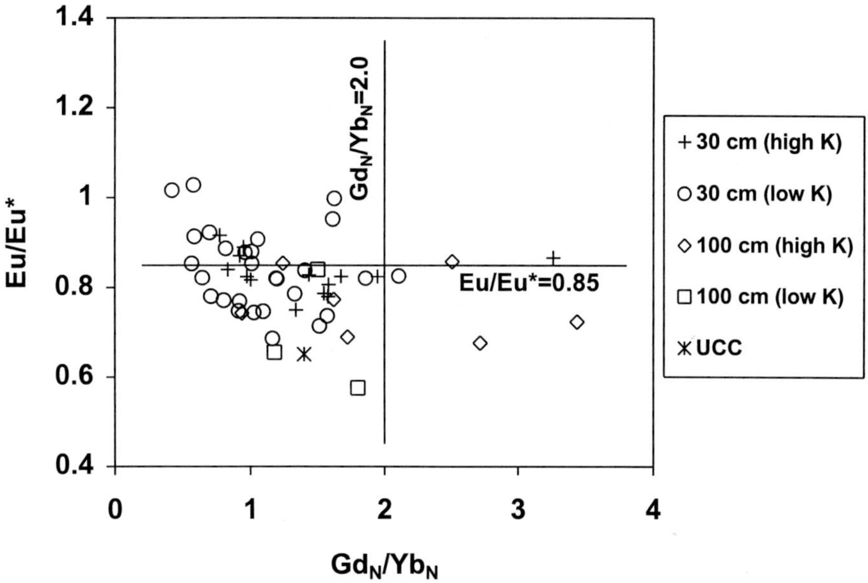

Fig. 8. Plot of Eu/Eu* versus Gd_N / Yb_N for the soil samples investigated in this study. The data point for upper continental crust (Taylor and McLennan 1985) is included for comparison.

to the La-Sc tie-line, away from the Th corner. The plot indicates significant contributions to these soils from both basic and felsic rocks. In the Th-Hf-Co plot (Fig. 9b), there is a cluster of data near the shale, graywacke-phyllite and the granite dike compositions and a general trend of the data towards the Co corner, away from the Th corner. This suggests a strong contribution of mafic rocks to the soils, and a slight enrichment of Hf, reflecting the presence of zircon in the soils. In the Th-Co-Zr/10 diagram (Fig. 9c), most of the data cluster near the Zr/10 corner, away from the Pepiakese Granite point. This shows the contribution of heavy minerals (e.g., zircon) to the soils. There is a slight enrichment in Co and a general depletion in Th, indicative of a significant contribution from a basic source. The high contents of Zr in the soils are associated with high contents of Hf and may reflect the presence of zircon in these soils. According to Taylor and McLennan (1995), an influence of zircon on the rare earth element (REE) distribution is expected, if the Zr content exceeds 300 ppm. The negative correlation between Zr and Gd_N/Yb_N ($r = -0.6$) may indicate that the heavy rare earth element (HREE) content is not controlled by zircon, a finding which is also in agreement with the lack of a correlation between Zr and HREE. However, the relative increase of heavy REE content in the samples with the highest Zr contents is documented by low La_N/Yb_N ratios of <2 and $Gd_N/Yb_N < 1.0$.

The relative contributions from felsic and mafic-ultramafic sources should be reflected in the abundances of Zr and Cr, as these elements reflect zircon and

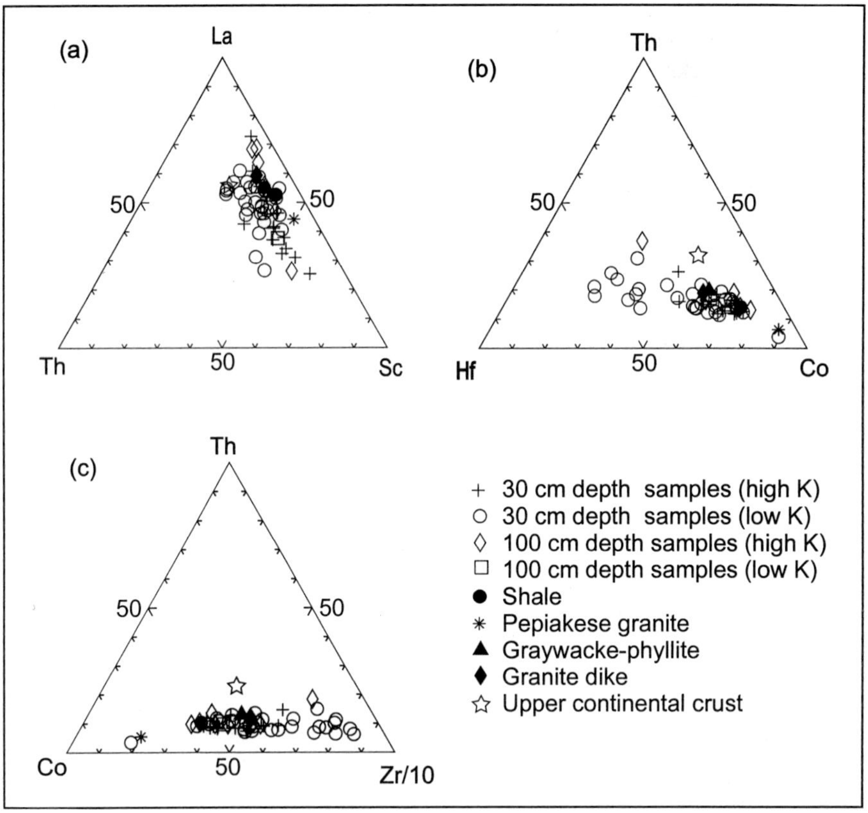

Fig. 9. Ternary diagrams showing relationships between compatible elements (Sc, Co) and incompatible elements (Th, La), as well as the zircon contribution effect (Hf, Zr). (a) La-Th-Sc, (b) Th-Hf-Co (c) Th-Co-Zr / 10.

chromite content (Wronkiewicz and Condie 1987), respectively. Most of the soil samples have high concentrations in both elements, indicating significant input of both felsic and mafic-ultramafic components, and the variable Cr/Zr ratios (0.17 to 18.2) reflect varying proportions of different source rock types. The low Cr/Zr values of, especially, samples S18, S19, S21, S22, S23, S24 and S25 may reflect larger proportions of felsic relative to basic rocks in the source area of these soils or could be the result of sorting. Probably all Cr/Zr values less than 1.0 indicate the dominance of felsic over basic source material. It is noteworthy that Cr/Zr reaches a maximum of 18.2 in sample S20, as do Cr, Co, and Ni contents, suggesting a particularly high proportion of mafic-ultramafic component. The high contents of Fe_2O_3 (14.8 wt%) and MgO (2.46 wt%) in this sample also suggest a significant contribution of mafic rocks to this soil.

Erosion of the Pepiakese Granite, with high Ni, Cr and Co contents (Koeberl et al. 1998; Jones 1985), could possibly have contributed to these high contents.

There are also dolerite and amphibolite dikes around the crater, which could have provided Cr, Ni, and MgO. Similar elevated Ni and Cr contents in Precambrian sedimentary rocks have been reported by Gibbs et al. (1986). Taylor and McLennan (1985) also reported relatively high levels of Cr and Ni in sedimentary rocks from an Archean sedimentary sequence, which they attributed to a combination of provenance and secondary enrichment effects. The high concentrations of Cr, Ni, and Co in the gravels with ferruginous nodules suggest they could be due to a combination of provenance and secondary enrichment effects.

The enrichment of Ir and Au in the soils, in comparison with upper continental crust, is not surprising, because of the close proximity of the crater to a gold mining region. The Konongo gold mine is located only about 25 km to the northeast, and the Obuom gold prospect is situated about 10 km south of the lake shore. An area rich in gold might also be rich in the geochemically related Ir (Jones 1985). Iridium contents for the parent rocks are not available for comparison, but it is quite likely that the Bosumtwi target rocks provided the Ir in the soil. The parent rocks are known to be rich in Au (see, e.g., Koeberl et al. 1998 and Table 3). The possibility of meteoritic material incorporated into the soils, however, cannot be ruled out, due to erosion and incorporation of impactites, which could have carried a meteoritic component.

The slightly different REE distribution patterns of the soils may be due to the mixture of various country rocks (granites, phyllites, shales, greywacke, etc.) at different proportions (e.g., Koeberl et al. 1998) and locally different weathering processes. The physicochemical processes leading to the formation of soils and the release and mobilization of REE during weathering are well documented (e.g., Nesbitt 1979; Humphris 1984; Taylor and McLennan 1985; Cullers et al. 1988; Middelburg et al. 1988; Braun et al. 1990; Condie et al. 1995; Aubert et al. 2001). From these studies, it has become clear that REE mobility is mainly controlled by the climatic weathering conditions and the stability of the primary REE-carrying minerals. Results from studies of weathering profiles (e.g., Nesbitt 1979; Duddy 1980; and Topp et al. 1987) have shown that when REE are fractionated during weathering processes, residuals become enriched in LREE (from La to Eu) and depleted in HREE (from Gd to Lu).

Moderate correlations between the REEs (LREE and HREE) and Al_2O_3 suggest that phyllosilicates and clay are important hosts of REEs. Thus, the REE abundance is, at least in part, a direct effect of alteration processes and not of primary component mixing. The lack of correlations between La and Zr, and between Yb and Zr suggest that the control of zircon on REE abundance is probably less important than that of phyllosilicates, both for light and heavy REEs. Silt and sand samples have lower overall REE abundances in comparison to those samples with a lot of clay-sized material, although the shapes of the REE distribution patterns are similar, suggesting that the major control is a dilution effect from quartz, a mineral with very low REE abundance (Taylor and McLennan 1985; McLennan 1989). High quartz contents dilute the absolute REE contents in the soils, as seen in samples S19 and S24. According to Cullers et al. (1975) and McLennan (1989), the reasons for differences in absolute REE abundances between sediments and parent rocks are attributed primarily to the REE content of the source rocks, as modified by weathering processes, and to exchange reactions in the depositional environment, and not to differences in REE contents between various phyllosilicate minerals.

The soils are generally characterized by high La/Th and low Th/U ratios (La/Th = 4.14 and Th/U = 2.6 respectively). In contrast, UCC has high Th/U and low La/Th (Th/U = 3.8 and La/Th = 2.8 respectively). According to Bhatia and Taylor (1981), high Th/U and low La/Th ratios possibly indicate a substantial contribution from volcanic or magmatic sources.

6.2 Source Area Weathering

The intensity of chemical weathering in the soils may be quantified using a chemical index of alteration (CIA) (Nesbitt and Young 1982), which is calculated, using molecular proportions, as $CIA = [Al_2O_3/(Al_2O_3 + K_2O + Na_2O + CaO^*)] \times 100$ (CaO* represents CaO associated with the silicate fraction of the sample). The range (57 - 98) and average (83) CIA values for the Bosumtwi soils are higher than values for average shale (CIA = 73) (Nesbitt and Young 1984) or deep sea mud (CIA = 69) (Kronberg et al. 1986). The country rock values range from 47 for the Pepiakese Granite to 80 for shale. Taylor and McLennan (1985) reported that average upper continental crust has a CIA value of 50, residual clays range from 85 to 100, and pure kaolinite has a value of 100. This places most of the soils from Bosumtwi into the residual clay category. According to Nesbitt and Young (1982), such a high degree of chemical alteration is caused by weathering under humid tropical conditions. Those soils with high CIA values (>90) also have the lowest K_2O and Na_2O contents (Table 2). The clear depletion in Na_2O, CaO, K_2O and MgO of the soils, along with their high CIA values, indicates that their source area underwent intense chemical weathering. There appear to be no distinct differences between the CIA values for high SiO_2 and low SiO_2 soils (Table 2).

Weathering trends on Al_2O_3-CaO + Na_2O-K_2O (A-CN-K) and Al_2O_3-CaO + Na_2O + K_2O-Fe_2O_3 + MgO (A-CNK-FM) plots (Nesbitt and Young 1984, 1989), together with a SiO_2/10-CaO + MgO-Na_2O + K_2O (S/10-CM-NK) plot, are shown in Fig. 10. A scaling factor for SiO_2 is used to bring the data more to the center of the diagram without altering their relative positions (Bhatia and Crook 1986). As seen in the A-CN-K diagram (Fig. 10a), all the soils and even the parent rocks show a marked depletion of K, Na and Ca, compared with upper continental crust (UCC). They plot towards the Al_2O_3 corner. Some of the soil samples are so weathered that they plot on this apex. In the A-CNK-FM diagram (Fig. 10b), a shift towards the FM apex suggests an influence of the ferruginous nodules in the soils and some possible concentration of heavy minerals, possibly ilmenite and hematite, which are common in the soils developed over Birimian rocks. Similar observations were noted by Vital and Stattegger (2000) for stream sediments from the Amazon river. In the S/10-CM-NK diagram (Fig. 10c), most of the soils plot towards the S/10 apex, reflecting an abundance of quartz (XRD analysis) and SiO_2 content up to 93.6 wt%, as well as the modification of soil composition in the intense chemical weathering environment.

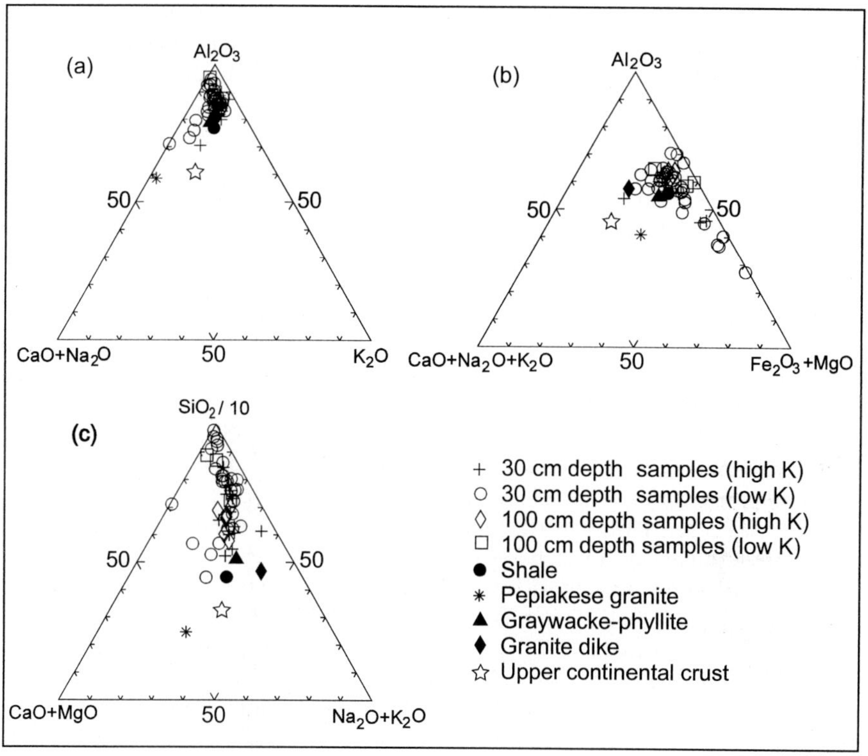

Fig. 10. Soil weathering on ternary plots of (a) A-CN-K (b) A-CNK-FM (c) S-CM-NK. *A*, Al_2O_3; *C*, CaO; *N*, Na_2O; *K*, K_2O; *M*, MgO; *S*, SiO_2 / 10; *F*, Fe_2O_3. Average composition of parent rocks (Koeberl et al. 1998) and upper continental crust (Taylor and McLennan, 1985) are included for comparison.

Compared to upper continental crust composition, these soils are more siliceous and ferruginous, with considerably lower levels of other major cations. Cations with relatively large ionic radii, such as Cs, Rb, and Ba, are often fixed in weathering profiles, whereas smaller cations (Na, Ca, and Sr) are more readily leached (Nesbitt et al. 1980; Wronkiewicz and Condie 1987; Bauluz et al. 2000). This is noted especially for Na and Ca in the Bosumtwi soils, where in some of the samples, the Na_2O and CaO values are below the detection limit of XRF spectrometry. The Bosumtwi rocks, similar to other metamorphosed rocks from Ghana, contain relatively low percentages of the alkali elements (Junner 1937).

These low values in the soils can be related to chemical weathering and/or provenance (source rocks), and, thus, are difficult to interpret. High LOI (average of 6 wt%) values are probably due to high contents of organic materials in the soils. Chemical composition is closely dependent on grain size, with Al_2O_3 generally higher in samples with a lot of clay-sized materials, and SiO_2 higher in the sandy or coarser materials. This strongly reflects mineralogical composition, whereby phyllosilicates dominate the fine fraction, and quartz the coarse fraction.

The somewhat higher concentration of most trace elements (TTE, REE, and LILE) in soils with higher Al_2O_3 compared to those with higher SiO_2 suggests that their enrichments may be associated with the presence of phyllosilicates or clay-sized materials. The positive correlations of Zr, Hf, Ta, and Nb with SiO_2 seem to reflect their association with minerals preferentially concentrated in quartz-rich rocks. This shows that all these elements are concentrated more in the sandy and quartz-rich soils, as they have negative correlations with Al_2O_3 and their concentrations are generally low in the soils containing abundant clay-sized material.

6.3 Cerium Anomalies

Most Bosumtwi soils have moderate to pronounced positive Ce anomalies in chondrite-normalized REE distribution patterns. Similarly, graywacke/phyllite-normalized REE distribution patterns of the soils show positive Ce anomalies, with the anomalies being more pronounced in the 30 cm depth samples (Fig. 7). Oxidation converts Ce^{3+} to Ce^{4+}, which is possibly stabilized as insoluble CeO_2, whereas the other REEs in the more soluble 3+ ionic state are leached by circulating groundwater (Marsh 1991). This leads to positive Ce anomalies in top soils. Similar pronounced, positive Ce anomalies have been reported from other lateritic terrains (see, e.g., Braun et al. 1990; Mongelli 1993; and Prudêncio et al. 1995). Rankin and Childs (1976) observed for some soils from New Zealand that positive Ce anomalies can be related to the presence of Fe-Mn concretions, whereas Braun et al. (1990) noted for lateritic profiles from Cameroon that Fe and Ce can behave differently. Our study shows that, in the Bosumtwi soils, most of the positive Ce anomalies are associated with high Fe_2O_3 contents.

6.4 Pit Samples

There is no obvious change in absolute concentrations of the various elements between the upper and lower zones for any of the pits (Table 4; Fig. 11). This may be due to the fact that both zones are within the same soil horizon (loose clay). However, there is a generally higher concentration of REEs in the lower zone, which is phyllosilicate-rich and contains more clay-sized material. This is reflected in the positive correlation between Al_2O_3 and the REEs. The lower REE content of the upper zone is probably due to the diluting effect of quartz. Comparison of various trace element abundances is shown in the multi-element diagram of Fig. 12, where selected trace elements in the soils have been normalized to upper continental crust values (Taylor and McLennan 1985). The figure shows that the soils are enriched in Cr, Co, and Ni, and depleted in the other elements compared to upper continental crust. These compositional differences imply that the source area of the soils is enriched in mafic or ultramafic components relative to UCC. Apart from Zr, Hf, and Ta, which show positive

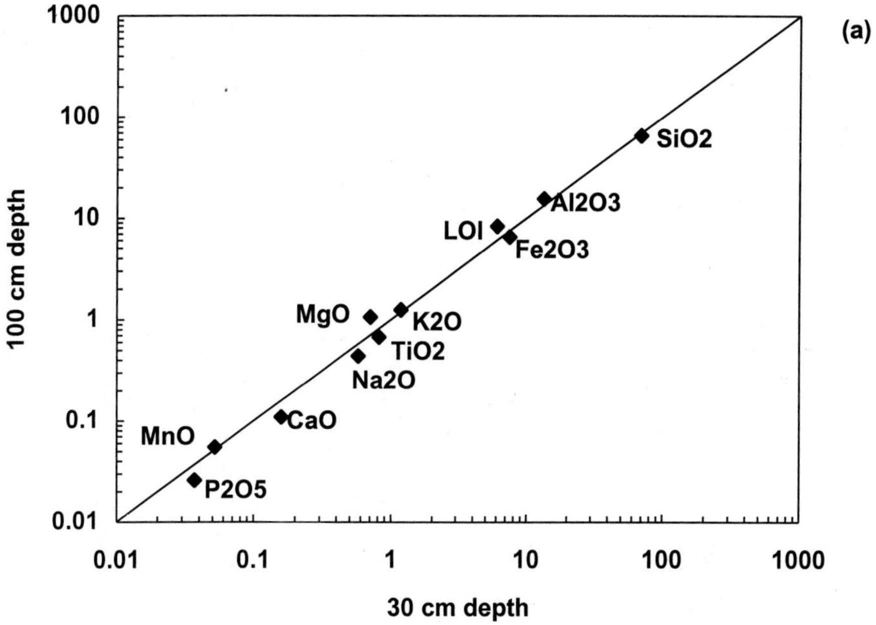

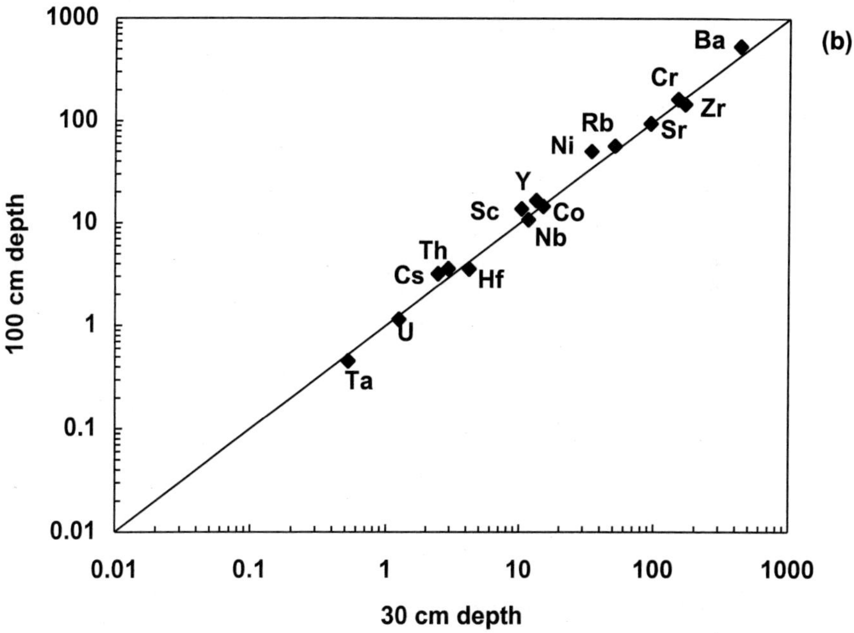

Fig. 11. Plots of (a) major element and (b) selected trace element contents of pit soil samples from 100 cm depth versus contents in top layer soil (30 cm depth).

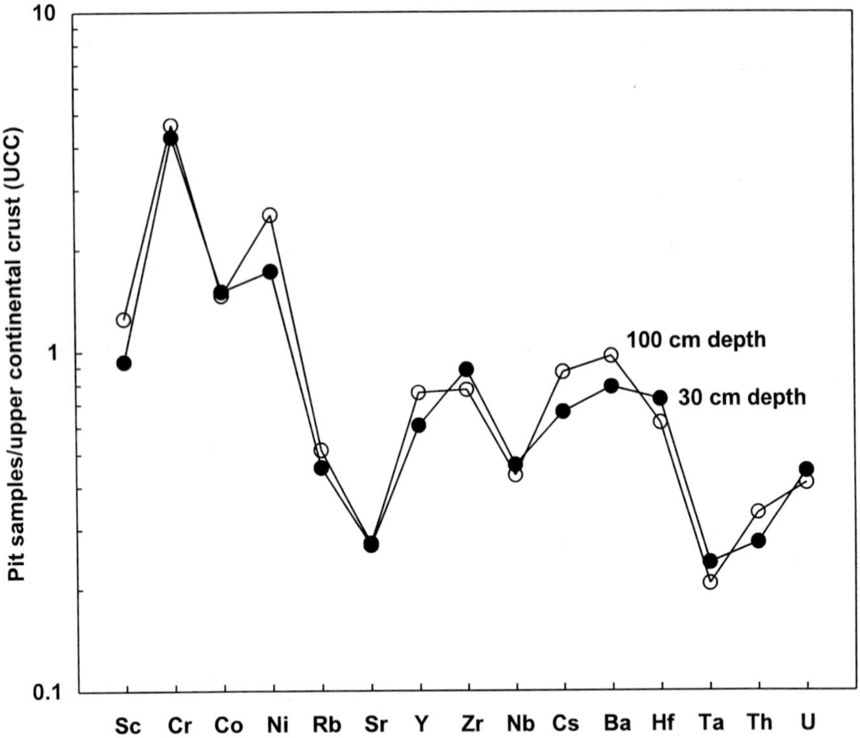

Fig. 12. Plots of average concentrations for selected trace elements calculated for all top (30 cm depth) and bottom (100 cm depth) soil samples from the pits, normalized to upper continental crust (UCC). Normalization factors after Taylor and McLennan (1985).

correlations with SiO_2 and are, thus, concentrated in the upper zone, all the other elements show positive correlations with Al_2O_3 and are generally higher in the more phyllosilicate-rich or clay-rich lower zone.

6.5 Relationship of Soil Composition with Airborne Radiometric Data (Equivalent Concentration of Potassium)

The sampling was planned to cover all the ranges of equivalent concentration of potassium (K) indicated by the airborne survey. Samples S1, S7, S8, S12, S13, S14, S27, S34, P3, P5, P6, P7, P8, P9, and P10 were collected from areas of apparent high K in the radiometric data, and samples S16, S17, S18, S19, B20, S21 and S22 from the low K radiometry areas. The soil results clearly show relatively higher concentrations of K_2O corresponding with areas of high K indicated on the radiometric map. This may reflect K mobilization due to the impact event or source rocks with variable concentrations of K. Soils from high K radiometry areas also have slightly higher concentrations of Al_2O_3, MgO, and Ba, and lower SiO_2 contents. Apart from these elements, no other elements seem to distinguish the various terrains as distinguished on the airborne radiometric map, although Th and U radiometry distribution maps have features similar to the K radiometry map (Pesonen et al. 1998). The reconnaissance mineralogical studies of soils from areas of high and low K radiometry, and increasing distance from the crater rim produced similar mineralogical compositions (quartz, muscovite, clinochlore, and albite). This suggests that the K radiometry data are not influenced significantly by the local geology or that the soils are transported mixtures.

7 Summary and Conclusions

Our main observations and conclusions are summarized in the following paragraphs.

1. The soils from the Bosumtwi impact structure area are characterized by an enrichment of quartz (decrease in the Al_2O_3/SiO_2 ratio) and a depletion of chemically unstable minerals in the soils, compared to the likely precursor rocks, indicating intense chemical weathering. Compared to upper continental crust composition, these soils are more siliceous and ferruginous, with considerably lower levels of major cations. They have reached the highest degree of chemical alteration (average CIA = 83) and have a rather simple mineralogical composition dominated by quartz, muscovite, and kaolinite, with a few refractory minerals, such as zircon.
2. Compared to average upper continental crust, the soils from the Bosumtwi crater area appear to have been derived from continental sources depleted in LILE, HFSE and REE and enriched in compatible elements (Si, Fe, Co, Ni, Cr, V, As, Sb) and Ti, Au, and Ir.
3. Differences in REE, major, and trace element concentrations, and trace elemental ratios between the soils and the likely precursor rocks may be due to small amounts of some elements added or lost to an aqueous phase during intense chemical weathering, thus modifying elemental ratios and REE patterns. Compared to the parent rocks (i.e., graywacke, phyllite, shale), the soils are relatively enriched in SiO_2, Zr, Hf, Nb, Sb, Ta, and W, reflecting the association with minerals preferentially concentrated in quartz-rich rock. Hydraulic sorting of minerals during the formation of the soils may have concentrated some elements and diluted others.
4. There is a slight difference in both the chondrite-normalized and graywacke/phyllite-normalized REE distribution patterns of the soils. These soils are from an area of about 90 km^2, with different types of underlying rocks. Chondrite-normalized REE distribution patterns for all the soils show enrichments in the light rare earth elements (LREE) and exhibit small to moderate negative Eu anomalies, with about 75 % of samples having La_N/Yb_N ratios greater than 4. The soils show much variation in La_N/Yb_N (1.21 to 12.1),

suggesting a derivation from a mixture of basic and silicic rocks. Pronounced positive Ce anomalies in chondrite-normalized REE distribution patterns associated with high Fe_2O_3 are noted in the soils from 30 cm depth. Similar positive Ce anomalies are shown in the graywacke/phyllite-normalized distribution pattern. The slight differences among the REE distribution patterns of the soils probably do not reflect changes in source area composition, but instead suggest variations in mineral sorting, chemical weathering and secondary enrichment and depletion effects. A positive correlation between REE and Al_2O_3 contents in the soils suggests that the REE are contained principally in phyllosilicates. The increase of the HREE in the samples with the highest Zr contents is documented by low La_N/Yb_N ratios of less than 2.0 and Gd_N/Yb_N values less than 1.0, suggesting a significant zircon abundance in the soils.

5. The sandy and coarse-grained soil samples have lower REE abundances compared to those with abundant clay-sized materials, although the shapes of the chondrite-normalized and greywacke/phyllite-normalized REE distribution patterns are similar, suggesting that the major control is a dilution effect from quartz.
6. The abundances of Si and Al in soils do not discriminate provenance. Rather, the concentrations of Co, Ni, Sc, and Cr distinguish between different mixtures of felsic and basic precursor rocks. A better provenance discrimination is obtained via the ratios La/Sc, Th/Sc, Th/Co, Cr/Zr, and systems Th-La-Sc and Th-Co-Hf. These discriminations indicate importance of basic components in the source-area composition.
7. The concentrations of both major and trace elements in the soil samples show considerable variation, as is expected from a chemical weathering environment, but are within the range obtained for the source or parent rocks. The wide range in trace element characteristics of these soils argues for inefficient mixing of the various source lithologies during their formation. The high trace element concentrations in samples with low SiO_2 and high Al_2O_3 suggest that elemental enrichments may be associated with the presence of phyllosilicates.
8. The major elements of the soils can be accounted for as SiO_2-Al_2O_3-Fe_2O_3-H_2O, with minor TiO_2 and K_2O. Among the factors affecting the distribution of elements in lateritic soils are the abundances of these elements in the parent rocks and the nature of the weathering and laterite-forming processes operating on these rocks. Because of the significant rim topography and extensive surface drainage at Bosumtwi, mechanical dispersion is believed to play a significant part in the distribution of elements.
9. The soil results clearly show relatively higher concentrations of K_2O (up to 2.88 wt%) corresponding to areas of high K indicated on the airborne radiometric map. This may reflect K mobilization due to the impact event or source rocks or these sediments with variable concentrations of K. Soils from high K radiometry areas also have slightly higher concentrations of Al_2O_3, MgO, and Ba, and lower SiO_2 contents. Apart from these elements, no other elements seem to characterize the various terrains that are distinguished on the airborne radiometric map. It should be noted that samples from both high and

low K radiometry areas have similar mineralogical composition, dominated by quartz, muscovite, clinochlore, and albite.

Acknowledgements

The authors thank the Geological Survey Department of Ghana, especially the Director, Mr. C. E. Oduro, for field and logistic support. D.B. thanks the Austrian Academic Exchange Service (ÖAD) for a Ph.D stipend and partial financial support for field work. The authors are grateful to W.U. Reimold (University of Witwatersrand, Johannesburg) for the help with the XRF analyses. We are furthermore grateful to G. Taylor and, especially, W.U. Reimold for very detailed reviews of this manuscript. The work was supported by the Austrian Fonds zur Förderung der wissenschaftlichen Forschung, project Y58-GEO (to C.K.).

References

Aubert D, Stille P, Probst A (2001) REE fractionation during granite weathering and removal by waters and suspended loads: Sr and Nd isotopic evidence. Geochimica et Cosmochimica Acta 65: 387-406

Bauluz B, Mayayo JM, Fernandez-Nieto C, Gonzalez Lopez JM (2000) Geochemistry of Precambrian and Paleozoic siliciclastic rocks from the Iberian Range (NE Spain): implications for source-area weathering, sorting, provenance, and tectonic setting. Chemical Geology 168: 135-150

Bhatia MR, Crook KAW (1986) Trace element characteristics of graywackes and tectonic setting discrimination of sedimentary basins. Contributions to Mineralogy and Petrology 92: 181-193

Bhatia MR, Taylor SR (1981) Trace element geochemistry and sedimentary provinces: a study from the Tasman Geosyncline, Australia. Chemical Geology 33: 115-125

Boamah D, Koeberl C (1999) Shallow drilling around the Bosumtwi crater, Ghana: Preliminary results [abs]. Meteoritics & Planetary Science 34: A13-A14

Bowell RJ (1993) Mineralogy and geochemistry of tropical rainforest soils: Ashanti, Ghana. Chemical Geology 106: 345-358

Braun JJ, Pagel M, Muller JP, Bilong P, Michard A, Guillet B (1990) Cerium anomalies in lateritic profiles. Geochimica et Cosmochimica Acta 54: 781-795

Chepik A, Baranov V, Kurimo M, Multala J (1998) Joint calibration of airborne geophysical instruments in test areas in Finland and Russia. Geological Survey of Finland, Report of Investigation 144, pp 1-20

Condie KC (1991) Another look at rare earth elements in shales. Geochimica et Cosmochimica Acta 55: 2527-2531

Condie KC, Wronkiewicz DJ (1990) A new look at the Archean-Proterozoic boundary: Sediments and tectonic setting constraint. In: Naqvi SM (ed) Precambrian Continental Crust and its Economic Resources. Elsevier, pp 61-84

Condie KC, Dengate J, Cullers RL (1995) Behavior of rare earth elements in a paleoweathering profile on granodiorite in the Front Range, Colorado, USA. Geochimica et Cosmochimica Acta 59: 279-294

Cullers RL, Chaudhuri S, Arnold B, Lee M, Wolf CW (1975) Rare earth distributions in clay minerals and the clay-sized fractions of the Lower Permian Havensville and Eskridge shales of Kansas and Oklahoma. Geochimica et Cosmochimica Acta 39: 1691-1703

Cullers RL, Basu A, Suttner LJ (1988) Geochemical signature of provenance in sand-sized material in soils and stream sediments near the Tobacco Root Batholith, Montana, U.S.A. Chemical Geology 70: 335-348

Davis DW, Hirdes W, Schaltegger U, Nunoo EA (1994) U-Pb age constraints on deposition and provenance of Birimian and gold-bearing Tarkwaian sediments in Ghana, West Africa. Precambrian Research 67: 89-107

Duddy JR (1980) Redistribution and fractionation of rare earth and other elements in a weathering profile. Chemical Geology 30: 363-381

Gibbs AK, Montgomery CW, O'Day PA, Erslev EA (1986) The Archean-Proterozoic transition: Evidence from the geochemistry of metasedimentary rocks of Guyana and Montana. Geochimica et Cosmochimica Acta 50: 2125-2141

Govindaraju K (1989) 1989 compilation of working values and sample description for 272 geostandards. Geostandards Newsletter 13: 1-113

Götze J, Lewis R (1994) Distribution of REE and trace elements in size and mineral fractions of high-purity quartz sands. Chemical Geology 114: 43-57

Grasty RL (1987) The design, construction and application of airborne gamma-ray spectrometer calibration pads – Thailand. Geological Survey of Canada Paper 87 – 10, 34 pp

Gregory AF, Horwood JL (1961) A laboratory study of gamma-ray spectra at the surface of rocks. Ottawa, Mines Branch Research Report R85, 52 pp

Hirdes W, Davis DW, Lüdtke G, Konan G (1996) Two generations of Birimian (Paleoproterozoic) volcanic belts in northeastern Cote d'Ivoire (West Africa): Consequences for the "Birimian controversy". Precambrian Research 80: 173-191

Humphris SE (1984) The mobility of the rare earth elements in the crust. In: Henderson P (ed) Rare Earth Element Geochemistry. Elsevier, pp 317-340

Jones WB (1985) Chemical analyses of Bosumtwi crater target rocks compared with Ivory Coast tektites. Geochimica et Cosmochimica Acta 49: 2569-2576

Jones WB, Bacon M, Hastings DA (1981) The Lake Bosumtwi impact crater, Ghana. Geological Society of America Bulletin 92: 342-349

Junner NR (1937) The geology of the Bosumtwi caldera and surrounding country. Gold Coast Geological Survey Bulletin 8: 1-38

Koeberl C (1993) Instrumental neutron activation analysis of geochemical and cosmochemical samples: A fast and proven method for small sample analysis. Journal of Radioanalytical and Nuclear Chemistry 168: 47-60

Koeberl C, Bottomley RJ, Glass BP, Storzer D (1997a) Geochemistry and age of Ivory Coast tektites and microtektites. Geochimica et Cosmochimica Acta 61: 1745-1772

Koeberl C, Reimold WU, Pesonen LJ, Brandt D (1997b) New Studies of the Bosumtwi Impact Structure, Ghana: The 1997 Field Season [abs]. Meteoritics & Planetary Science 32: A72-A73

Koeberl C, Reimold WU, Blum JD, Chamberlain CP (1998) Petrology and geochemistry of target rocks from the Bosumtwi impact structure, Ghana, and comparison with Ivory Coast tektites. Geochimica et Cosmochimica Acta 62: 2179-2196

Kolbe P, Pinson WH, Saul JM, Miller EW (1967) Rb-Sr study on country rocks of the Bosumtwi crater, Ghana. Geochimica et Cosmochimica Acta 31: 869-875

Krongberg BI, Nesbitt HW, Lam WW (1986) Upper Pleistocene Amazon deep-sea fan muds reflect intense chemical weathering of their mountainous source lands. Chemical Geology 54: 283-294

Lecomte P (1988) Stone line profiles: Importance in geochemical exploration. Journal of Geochemical Exploration 30: 35-61

Leube A, Hirdes W, Mauer R, Kesse GO (1990) The early Proterozoic Birimian Supergroup of Ghana and some aspects of its associated gold mineralization. Precambrian Research 46: 139-165

Marsh JS (1991) REE fractionation and Ce anomalies in weathered Karoo dolerite. Chemical Geology 90: 189-194

McLennan SM (1989) Rare earth elements in sedimentary rocks: Influence of provenance and sedimentary processes. In: Lipin BR, McKay GA (eds) Geochemistry and mineralogy of rare earth elements. Reviews in Mineralogy 21: 169-200

McLennan SM, Taylor SR (1991) Sedimentary rocks and crustal evolution: Tectonic setting and secular trends. Journal of Geology 99: 1-21

McLennan SM, Nancy WB, Taylor SR (1980) Rare earth element-thorium correlations in sedimentary rocks, and the composition of the continental crust. Geochimica et Cosmochimica Acta 44: 1833-1839

Middelburg JJ, Van der Weijden CH, Woittiez JRW (1988) Chemical processes affecting the mobility of major, minor and trace elements during weathering of granitic rocks. Chemical Geology 68: 253-273

Mongelli G (1993) REE and other trace elements in granitic weathering profile from "Serre", southern Italy. Chemical Geology 103: 17-25

Moon PA, Mason D (1967) The geology of ¼° field sheets 129 and 131, Bompata SW and NW. Ghana Geological Survey Bulletin 31: 1-51

Multala J (1981) The construction of gamma-ray spectrometer calibration pads. Geoexploration 19: 33-46

Murali AV, Parthasarathy R, Mahadevan TM, Sankar Das M (1983) Trace element characteristics, REE patterns and partition coefficients of zircons from different geological environments – A case study on Indian zircons. Geochimica et Cosmochimica Acta 47: 2047-2052

Nance WB, Taylor SR (1977) Rare earth element patterns and crustal evolution – II. Archean sedimentary rocks from Kalgoorlie, Australia. Geochimica et Cosmochimica Acta 41: 225-231

Nesbitt HW (1979) Mobility and fractionation of rare earth elements during weathering of a granodiorite. Nature 279: 206-210

Nesbitt HW, Young GM (1982) Early Proterozoic climates and plate motions inferred from major element chemistry of lutites. Nature 299: 715-717

Nesbitt HW, Young GM (1984) Prediction of some weathering trends of plutonic and volcanic rocks based on thermodynamic and kinetic considerations. Geochimica et Cosmochimica Acta 48: 1523-1534

Nesbitt HW, Young GM (1989) Formation and diagenesis of weathering profiles. Journal of Geology 97: 129-147

Nesbitt HW, Markovics G, Price RC (1980) Chemical processes affecting alkalines and alkaline earths during continental weathering. Geochimica et Cosmochimica Acta 44: 1659-1666

Ojamo H, Pesonen LJ, Elo S, Hautaniemi H, Koeberl C, Reimold WU, Plado J (1997) The Bosumtwi Impact Structure, Ghana: International geophysical co-operation at the best [abs].

In: Kaikkonen P, Komminaho K, Salmirinne H (eds) Sovelletun geofysiikan XI neuvottelupäivät. Oulun yliopisto, Oulu, pp 10-11

Pesonen LJ, Koeberl C, Ojamo H, Hautaniemi H, Elo S, Plado J (1998) Aerogeophysical studies of the Bosumtwi impact structure, Ghana [abs]. Geological Society of America, Abstracts with Programs 30/7: A190

Pesonen LJ, Plado J, Koeberl C, Elo S (1999) The Lake Bosumtwi meteorite impact structure, Ghana: Magnetic modeling [abs]. Meteoritics & Planetary Science 34: A91-A92

Plado J, Pesonen LJ, Koeberl C, Elo S (2000) The Bosumtwi meteorite impact structure, Ghana: A magnetic model. Meteoritics & Planetary Science 35: 723-732

Prudêncio MI, Gouveia MA, Sequeira Braga MA (1995) REE distribution in present-day and ancient surface environments of basaltic rocks (Central Portugal). Clay Minerals 30: 239-248

Rankin PC, Childs CW (1976) Rare earth elements in iron-manganese concretions from some New Zealand soils. Chemical Geology 18: 54-64

Reimold WU, Koeberl C, Bishop J (1994) Roter Kamm impact crater, Namibia: Geochemistry of basement rocks and breccias. Geochimica et Cosmochimica Acta 58: 2689-2710

Reimold WU, Brandt D, Koeberl C (1998) Detailed structural analysis of the rim of a large, complex impact crater: Bosumtwi crater, Ghana. Geology 26: 543-546

Rollinson HR (1993) Using Geochemical Data: Evaluation, Presentation, Interpretation. Longman, 352 pp

Taylor SR, McLennan SM (1985) The Continental Crust: Its Composition and Evolution. Blackwell, 312 pp

Taylor SR, McLennan SM (1995) The geochemical evolution of the continental crust. Reviews of Geophysics 33: 241-265

Taylor PN, Moorbath S, Leube A, Hirdes W (1992) Early Proterozoic crustal evolution in the Birimian of Ghana: Constraints from geochronology and isotope geochemistry. Precambrian Research 56: 97-111

Topp SE, Salbu B, Roaldset E, Jorgensen P (1984) Vertical distribution of trace elements in lateritic soil (Suriname). Chemical Geology 47: 159-174

Vital H, Stattegger K (2000) Major and trace elements of stream sediments from the lowermost Amazon River. Chemical Geology 168: 151-168

Wagner R, Reimold WU, Brandt D (2001) Bosumtwi impact crater, Ghana: A remote sensing investigation. This volume

Woodfield PD (1966) The geology of the ¼° field sheet 91, Fumso NW. Ghana Geological Survey Bulletin 30: 1-66

Wright JB, Hastings DA, Jones WB, Williams HR (1985) Geology and Mineral Resources of West Africa. Allen and Unwin, London, pp 38-45

Wronkiewicz DJ, Condie KC (1987) Geochemistry of Archean shales from the Witwatersrand Supergroup, South Africa: Source-area weathering and provenance. Geochimica et Cosmochimica Acta 51: 2401-2416

Appendix

Table A. Soil samples from the Bosumtwi impact structure collected in February 2000; all from 30 - 40 cm depths. %K from aero-radiometry data (Ojamo et al. 1997), GPS locations

Sample No.	Location	%K	Description
S1	06°32.882'N 001°22.561'W	1.2-2.0	Reddish-brown clay with weathered rock fragments
S2	06°32.943'N 001°22.520'W	0.4-0.8	Yellowish-brown, sandy clay
S3	06°33.034'N 001°22.417'W	0.4-0.8	Brownish clay
S4	06°33.133'N 001°22.344'W	0.4-0.8	Yellowish-brown, sandy clay with weathered rock fragments
S5	06°33.331'N 001°22.367'W	0.4-0.8	Yellowish-brown, silty clay
S6	06°33.475'N 001°22.328'W	0.4-0.8	Yellowish brown, silty clay
S7	06°33.559'N 001°22.235'W	0.8-1.2	Grayish-brown, silty clay with some gravels/ferruginous nodules
S8	06°33.644'N 001°22.178'W	1.2-2.0	Brownish, silty clay
S9	06°33.585'N 001°22.100'W	0.4-0.8	Yellowish-brown, silty clay with few gravels
S10	06°33.528'N 001°22.092'W	0.4-0.8	Brownish clay with some rock fragments
S11	06°33.082'N 001°22.048'W	0.4-0.8	Brownish clay
S12	06°33.432'N 001°21.099'W	0.8-1.2	Brownish clay
S13	06°33.546'N 001°21.126'W	0.8-1.2	Brownish clay with weathered fragments
S14	06°33.861'N 001°20.989'W	0.8-1.2	Grayish-brown, silty clay
S15	06°34.161'N 001°20.910'W	0.4-0.8	Yellowish-brown, silty clay
S16	06°34.430'N 001°20.749'W	0.0-0.4	Yellowish-brown, silty clay
S17	06°35.676'N 001°23.093'W	0.0-0.4	Chocolate brown, silty sand
S18	06°36.090'N 001°22.647'W	0.0-0.4	Light brown, silty sand
S19	06°35.765'N 001°22.137'W	0.0-0.4	Light brown, silty sand
S20	06°35.887'N 001°21.449'W	0.0-0.4	Brown, silty clay with some ferruginous nodules
S21	06°36.491'N 001°20.653'W	0.0-0.4	Brownish, silty clay
S22	06°36.735'N 001°27.449'W	0.0-0.4	Brownish-gray sand
S23	06°36.141'N 001°27.400'W	0.4-0.8	Light brown, silty clay

Table A. (cont.)

Sample No.	Location	%K	Description
S24	06°35.768'N 001°27.308'W	0.4-0.8	Light brown, silty sand
S25	06°35.481'N 001°27.161'W	0.4-0.8	Light brown silt
S26	06°35.277'N 001°27.084'W	0.4-0.8	Light brown gravel
S27	06°34.692'N 001°27.025'W	0-8-1.2	Reddish-brown gravel
S28	06°34.385'N 001°26.887'W	0.4-0.8	Brownish gravel
S29	06°34.353'N 001°26.593'W	0.4-0.8	Grayish-brown, silty sand
S30	06°34.055'N 001°26.251'W	0.4-0.8	Light brown, silty clay
S31	06°33.722'N 001°25.836'W	0.4-0.8	Reddish-brown, silty sand
S32	06°33.382'N 001°25.837'W	0.4-0.8	Light brownish, silty sand
S33	06°33.064'N 001°25.859'W	0.4-0.8	Brownish gravel
S34	06°32.969'N 001°25.801'W	0.8-1.2	Light brown, silty sand

Table B. Pit samples from the Bosumtwi impact structure collected during the shallow drilling in 1999. %K from aero-radiometry data (Ojamo et al. 1997), GPS locations

Pit No.	Sample No.	Location	Depth (cm)	%K	Description
1	T1	06°33.976'N 001°24.097'W	30-35	0.4-0.8	Reddish-brown clay with a few quartz grains
"	B1	"	115-125	0.4-0.8	Weathered suevite with brown clay
2	T2	06°33.794'N 001°23.755'W	30-35	0.4-0.8	Yellowish-brown, sandy clay with grayish patches
"	B2	"	100-110	0.4-0.8	Mixture of yellowish and grayish, sandy clay
3	T3	06°33.496'N 001°23.178'W	30-35	0.8-1.2	Yellowish-brown, sandy clay with light grayish patches
"	B3	"	100-110	0.8-1.2	Yellowish-brown, weathered graywacke with clay
4	T4	06°34.116'N 001°24.170'W	30-35	0.4-0.8	Brownish gravel with clay
"	B4	"	90-100	0.4-0.8	Weathered suevite with brownish clay
5	T5	06°33.698'N 001°22.933'W	30-35	1.2-2.0	Yellowish-brown, sandy clay
"	B5	"	100-110	1.2-2.0	Yellowish-brown, sandy clay with light grayish and reddish clay
6	T6	06°33.904'N 001°22.811'W	30-35	1.2-2.0	Light yellowish-brown, sandy clay with few quartz grains

Table B. (cont.)

Pit No.	Sample No.	Location	Depth (cm)	%K	Description
"	B6	"	100-105	1.2-2.0	Yellowish-brown clay with clasts of granite
7	T7	06°34.511'N 001°22.946'W	30-35	1.2-2.0	Brownish gravel with some clay
"	B7	"	100-110	1.2-2.0	Reddish-brown sand with some clay
8	T8	06°34.703'N 001°23.003'W	30-40	1.2-2.0	Light brown, silt clay with a few quartz grains
"	B8	"	110-120	1.2-2.0	Weathered graywacke with some brownish clay
9	T9	06°33.940'N 001°24.190'W	30-40	0.8-1.2	Mixture of reddish and grayish sand
"	B9	"	150-170	0.8-1.2	Mixture of reddish and grayish sand
10	T10	06°35.666'N 001°23.332'W	30-35	0.8-1.2	Light brown, sandy clay with some gravels and weathered graywacke
"	B10	"	130-140	0.8-1.2	Weathered graywacke with brownish clay

Tektite Origin in Oblique Impacts: Numerical Modeling of the Initial Stage

Natalia Artemieva

Institute for Dynamics of Geospheres, Leninsky Prospect 38-6, 117334 Moscow, Russia. (nata_art@mtu-net.ru)

Abstract. Numerical modeling of the initial stage of the growth of a Bosumtwi-size crater was performed to reproduce the formation of impact melt from the upper 50 m of the target surface. Such melt, fragmented into cm-sized particles, could produce the Ivory Coast tektite strewn field. Melt production from the surface layer of the target due to an asteroidal impact is studied with 3D numerical simulations for various impact angles (15° - 60° to horizon) and velocities (11 - 40 kms^{-1}). For this range in impact angles and velocities, projectile sizes are within the range of 400 to 1100 m, according to the scaling law, and produce the same transient cavity of 9 km in diameter (final crater diameter is 10.5 km). The mass of the near-surface melt (tektite material) is almost constant for all impact velocities, but high velocity impact (>20 kms^{-1}) accelerates the melted ejecta, whereas in the case of low-velocity impact (11 kms^{-1}) all the melt is contained within the crater. The amount of tektite material increases with decreasing impact angle (with a maximum at 15°), but low angle impacts produce melt which is strongly contaminated by projectile material. Pure target melt arises during an impact with an "intermediate" impact angle of 30° - 50°. This melt is ejected from the growing crater with high velocity and may produce tektites after disruption into cm-sized particles. The process of disruption is not considered in this paper, but it is shown that solid particles may be transported to distances of hundreds of km in the post-impact atmospheric flow.

1 Introduction

Tektites are naturally occurring glasses, generally up to several centimeters in diameter, found in four distinct strewn fields (see Table 1). Microtektites (less than 1 mm in diameter) have been found in deep-sea cores of three of the four strewn fields (Glass 1972). Tektites of a given strewn field are related to each other by their chemistry, age, petrologic, and physical characteristics (see reviews by Glass 1990; Koeberl 1990, 1994). Compositional data (chemical and isotopic) strongly suggest that tektites are formed as a result of melting of terrestrial sedimentary rocks during the impact cratering process. The European and the Ivory Coast strewn fields are located several hundred kilometers away from their probable source craters - Ries and Bosumtwi, respectively. The North American strewn field has been linked to the Chesapeake Bay impact structure (Poag et al. 1994; Koeberl et al. 1996). The parent crater for the Australasian tektites has not yet been found, but variations in microtektite concentrations (Glass and Pizzuto 1994) and ^{10}Be contents (Tera et al. 1983; Blum et al. 1992) point strongly towards a location in Indochina. The meteoritical component in tektites is very low (with maximum of about 0.06 wt% for Ivory Coast tektites, Koeberl and Shirey 1993). At the same time tektites differ from all other types of natural glasses, such as obsidians and impact melts, by having very low water contents of 0.002–0.02 wt% (Gilchrist et al. 1969; Koeberl and Beran 1988) compared to several percent for normal impact glasses. A high degree of chemical homogeneity for regions above 0.1 mm in size (see, for example, Glass 1990; Koeberl 1990, and references therein) and physical homogeneity (density, refractive index - Friedman 1963) are also typical for tektites. These facts, as well as petrographic properties (Chao 1963), indicate that tektites originate from a high-temperature melt. The only exception is the group of Muong Nong-type tektites, which are found within the Australasian strewn field and differ from regular tektites by being of larger size, being heterogeneous in composition, and by having a layered structure (Chapman and Scheiber 1969; Schnetzler 1992). The noble gas contents (Matsuda et al. 1993) and the low ferric to ferrous iron ratios (Fudali et al. 1987; Jakes et al. 1992) indicate tektite solidification in the upper Earth atmosphere with low oxygen fugacity. The study of cosmogenic radionuclides (^{10}Be and ^{26}Al) provides further proof of the terrestrial origin of tektites (Pal et al. 1982): if tektites were exposed to cosmic radiation in space, then the ^{26}Al/^{10}Be ratio must be between 2.7 and 5.4, but Middleton and Klein (1987) found that it is about 0.07 in only one sample and even lower in others. This means that the ^{10}Be in the tektites has been introduced from sediments that have absorbed ^{10}Be that was produced in the terrestrial upper atmosphere. Moreover, the concentration of ^{10}Be in the environment is a strict function of the depth from the surface - most of it is concentrated in the upper 20 m (Valette-Silver et al. 1983). This limit may vary as a function of different rock types, rainfall rates, and porosity, but probably by less than a factor of 10. It means that tektites were formed exclusively of the top few hundred meters of the target.

One more fact should be mentioned - all the strewn fields are on one side of the known (or inferred) location of the parent crater, but this direction changes for different strewn fields: the European strewn field is to the east from the Ries crater, the Ivory Coast one is to the west from the Bosumtwi, the North American one - to the south from the Chesapeake Bay, and all the Australasian tektites are to the south of Indochina. This position of the strewn fields indicates that tektites formed after oblique impacts, otherwise the ejecta deposition would have been more or less symmetrical.

Table 1. Characteristics of the four strewn fields (based on Koeberl 1994; Montanari and Koeberl 2000).

	North American	Central European	Ivory Coast	Australasian
Age (Ma)	35.4	15	1.1	0.77
Area (10^6 km)	10–30	0.3	4	50
Total mass (10^6 t)	300–42,000	5?	20	2,000?
Source crater	Chesapeake Bay	Ries	Bosumtwi	
Source crater diameter (km)	85	24	10.5	
Distance from source crater (km)	≥1300	250–420	300–1000	

Geochemical and geophysical studies provide important data, but do not answer two principal and closely linked questions: 1) how do tektites form during the impact and are then transported for hundreds kilometers, and 2) are they really rare or are they a normal type of distal ejecta? Numerical modeling, based on the analysis of the available geochemical data summarized shortly in the previous paragraphs may be a useful tool to investigate the tektite origin and then to discuss how rare they are.

It seems reasonable to divide the process of tektite origin into three stages:

- near-surface melt production during crater growth, and its ejection from the crater
- melt transport to higher altitudes and its disruption into a set of smaller particles of various sizes
- particle motion through the atmosphere disturbed by the post-impact gas dynamic flow.

In this paper I model the first stage of tektite formation, i.e., I calculate the amount of high-temperature melt of the near-surface material that is produced in an oblique impact. I also try to answer the question - which factors are critical for an effective melt production from the upper layer of the target. For this purpose, impacts with various impact velocities and trajectory inclinations are modeled, and the influence of projectile and target materials is discussed. It is probable that some exotic coincidence of impact conditions causes tektite production, but the first approximation is to check the most basic variables (impact angle, impact velocity, and nature of target material).

2 Hydrocode and Equation of State (EOS)

Hydrocode modeling is a fundamental tool to study melt production in planetary impact events. Most of these studies have been limited to vertical impacts (O'Keefe and Ahrens 1977; Pierazzo et al. 1997). The first results of melt production in oblique impacts were presented by Pierazzo and Melosh (2000). These simulations are carried out using the CTH code keeping all parameters of the impact constant except the impact angle. The purpose of this study differs from that presented by Pierazzo and Melosh in two respects: (i) it models melt production for an impact crater of constant size (i.e., the case of Bosumtwi crater, whose rim-to-rim diameter is about 10 km), not for a constant projectile size, and (ii) only the melt of the upper surface layer is considered. The thickness of this layer doesn't exceed 50 - 200 m on the basis of tektite geochemical studies.

The 3D simulations of oblique impacts are carried out using the SOVA hydrocode (Shuvalov 1999; Shuvalov et al. 1999). This code allows to model multidimensional, multimaterial, large deformation, and strong shock wave flows. The Analytical Equation of State (ANEOS) package (Thompson and Lauson 1972) is used to prepare equation of state tables, describing pressure and temperature dependencies on the density and internal energy of the projectile and target materials over a wide range of both parameters (from cold solid state to hot plasma). These tables are then used by SOVA.

The relatively low spatial resolution of 10 cells per projectile radius (cppr) is used. Study of the effect of spatial resolution on melt production has shown that the estimated amount of melt at 10 cppr is about 20 % lower than the melt estimated at the resolution of 20 cppr (for the impact velocity of 20 kms^{-1}). This melt deficiency depends only slightly on the impact angle. This result of the decrease of melt with decreasing spatial resolution differs from that of Pierazzo et al. (1997), where the computational losses are much higher (about 40 %). The reason for this discrepancy requires further research, but it may be due to much larger number of tracer particles (about 100,000) in the present simulations, compared to only 100 particles used by Pierazzo et al. (1997).

Constant spatial resolution creates one more serious problem - when 20 cells per projectile diameter (400 - 1000 m) are used, only 1 - 2 computational cells describe the upper layer of the surface. This is comparable to the thickness of the shock wave in numerical modeling by Eulerian codes. Moreover, shock waves strongly decay near the free boundary (surface-atmosphere) - see, for example, Melosh (1989, p. 64). It means that the amount of surficial melt strongly depends on the resolution of the computation. An additional study of impacts with high spatial resolution in the upper layer shows that the value of the impact melt volume has artificial computational maximum for 2 cells in the upper layer and this maximum is 20 % higher than the estimated melt for 8 cells in that layer.

Granitic material (Pierazzo et al. 1997) or wet tuff (Zamyshliayev and Yevterev 1990) are used for both the projectile and the Earth's surface. Whereas the EOS parameters are identical, the projectile and the target are treated as different materials and an accurate boundary is constructed in each mixed cell at each time step. Air fills the space above the target to simulate the Earth's atmosphere. The presence of the atmosphere does not influence shock melting, but is of great importance for the development of the atmospheric expansion plume and ejecta motion. Initially each cell of the target is marked with a tracer particle. These particles move through the grid, recording the maximum pressure in a given Lagrangian volume. A set of specific tracer particles marks the upper surface layer between zero and 50 m deep.

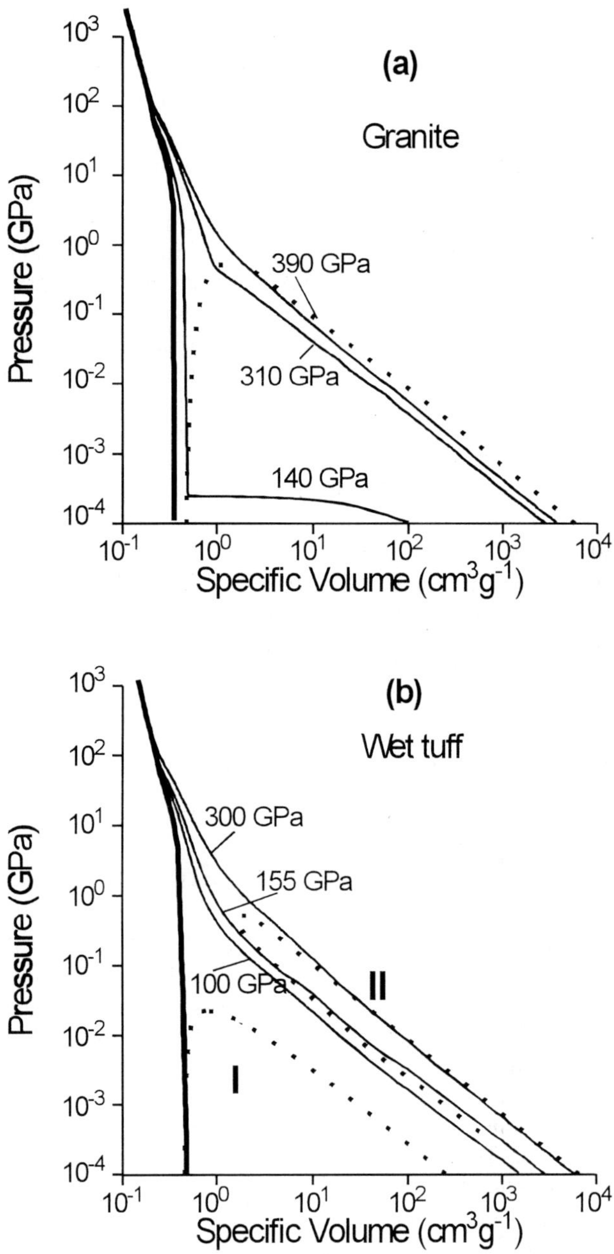

Fig. 1. Hugoniot curve (thick black line), isentropes (thin black lines) and the boundary of the two-phase region (dashed lines) for a) granite and b) wet tuff with a water content of 15 %. High-temperature may arise after shock compression above 100 GPa. Numbers near the isentropes show the value of maximum shock compression (initial point at the Hugoniot curve).

Shock pressures needed for total or incipient melting are defined by the proper equation of state for granite and wet tuff. Figure 1 (a, b) shows the Hugoniot curve (possible shock-compressed states), some isentropes (release paths from the shocked state), and the two-phases (liquid-vapor) boundary, for granite and wet tuff, respectively. It is widely accepted that isentropes crossing the two-phase curve on the right of the critical point correspond to the states "vapor + liquid droplets", while isentropes crossing the two-phase curve on the left of the critical point describe "liquid + gas bubbles". Thus, one can conclude that for the granite EOS in use the release from the shock pressures in the range from 100 to 300 GPa gives a high-temperature melt needed for tektite production. Granite compressed above 300 GPa is totally vaporized after decompression. In the case of wet tuff target (15 % water) the two-phase region is more complex. Curve I in Fig. 1b corresponds to the "water-steam" transition (tuff remains in solid state), while curve II describes transformation of silica into vapor. Between these curves there is a mixture of water vapor with solid particles. Phase transitions in real target materials are even more complex. But estimates for the high-temperature tuff melt are practically the same as for pure granite - initial shock compression should range roughly between 100 - 300 GPa (higher accuracy of this estimate requires much more sophisticated EOS models).

Summarizing geochemical data and EOS-based estimates, one can conclude that the target region no deeper than 50 m and compressed up to 100 - 300 GPa during the impact should be considered as the main source of tektites. For brevity, material from this region will be referred to as the "tektite" material further in this paper and its volume and mass designations will be V_{tek} and m_{tek}, respectively. It is not identical to real tektites, but an appropriate material for tektite production.

3 Bosumtwi Crater - Estimates for the Projectile

The 1.07 Ma Bosumtwi Crater in Ghana (06°32'N, 01°25'W), modeled here, is a 10.5-km-diameter complex well-preserved impact structure associated with the Ivory Coast tektite strewn field (Barnes 1961; Jones et al. 1981; Koeberl et al. 1997, 1998, and references therein). The crater is almost completely filled with Lake Bosumtwi, which measures 8 km in diameter and has a current maximum depth of about 80 meters. The rim rises about 250 - 300 m above the lake level.

Hydrocode simulations of cratering impacts rarely extend to the end of the crater excavation stage. To estimate parameters of the projectile that are able to form the Bosumtwi crater I use well-known scaling laws. To find the transient cavity diameter D_{tr} associated with the observed crater rim diameter D. I apply the Croft (1985) model: $D_{tr} = D_*^{0.15} \cdot D^{0.85}$, where D_* is a critical parameter defining the crater diameter when the collapse begins and it depends on the target material, its strength and planet gravity. For the Earth D_* is about 4 km for crystalline rocks (Neukum and Ivanov 1994). Hence, the transient crater diameter for Bosumtwi is estimated as about 9 km. Assuming a vertical impact one can use the scaling law

Table 2. Projectile diameter estimates (in km) for different impact velocities and angles, resulting in a ~10 km diameter crater.

	Impact angle				
Impact velocity (kms^{-1})	90°	60°	45°	30°	15°
40	0.39	0.42	0.47	0.57	0.82
20	0.57	0.62	0.69	0.83	1.2
11.2	0.78	0.84	0.94	1.14	1.64

connecting projectile and crater parameters, developed by Schmidt and Housen (1987):

$$D_{tr} = 1.16 \cdot (\rho_{pr} / \rho_t)^{1/3} \cdot D_{pr}^{0.78} V^{0.43} g^{-0.22}, \quad (1)$$

where ρ_{pr} and ρ_t are the densities of projectile and target, respectively, D_{pr} is the projectile diameter, g is the acceleration of gravity, V is the impact velocity. Even in the case of same target and impactor materials the same transient cavity may be produced by a high-velocity small projectile or by a low-velocity larger body. On the other hand, the impact may be substantially different from the vertical, with the most probable impact angle being 45° (Gilbert, 1893; Shoemaker, 1962; Melosh, 1989). The influence of impact angle θ on crater size is weakly investigated numerically, but the results of experimental work (Gault and Wedekind 1978; Schultz and Gault 1990) indicate that for particulate targets the cratering efficiency decreases as $\sin \theta$. Therefore, the impact velocity V in (3.1) should be substituted with its vertical component $V \sin \theta$ (Chapman and McKinnon 1986).

Table 2 summarizes estimates of the projectile diameter (in km) for various impact velocities and various impact angles. These data are used in my simulations to produce a Bosumtwi-like crater in each run. Usually the simulations are stopped much earlier, before the final stage (collapse) of the crater formation begins (but much later than the duration of the compression stage); thus, our assumptions are not the results of strict numerical modeling, but are based on the scaling law (1). A recent study of transient cavity growth in oblique impact (Ivanov and Artemieva 2001) shows that high-velocity natural oblique impacts may be as efficient as vertical ones unlike the low-velocity laboratory impacts. Thus, the scaling law (1) should be rewritten for oblique impacts using a more sophisticated way than simple correction by $\sin \theta$.

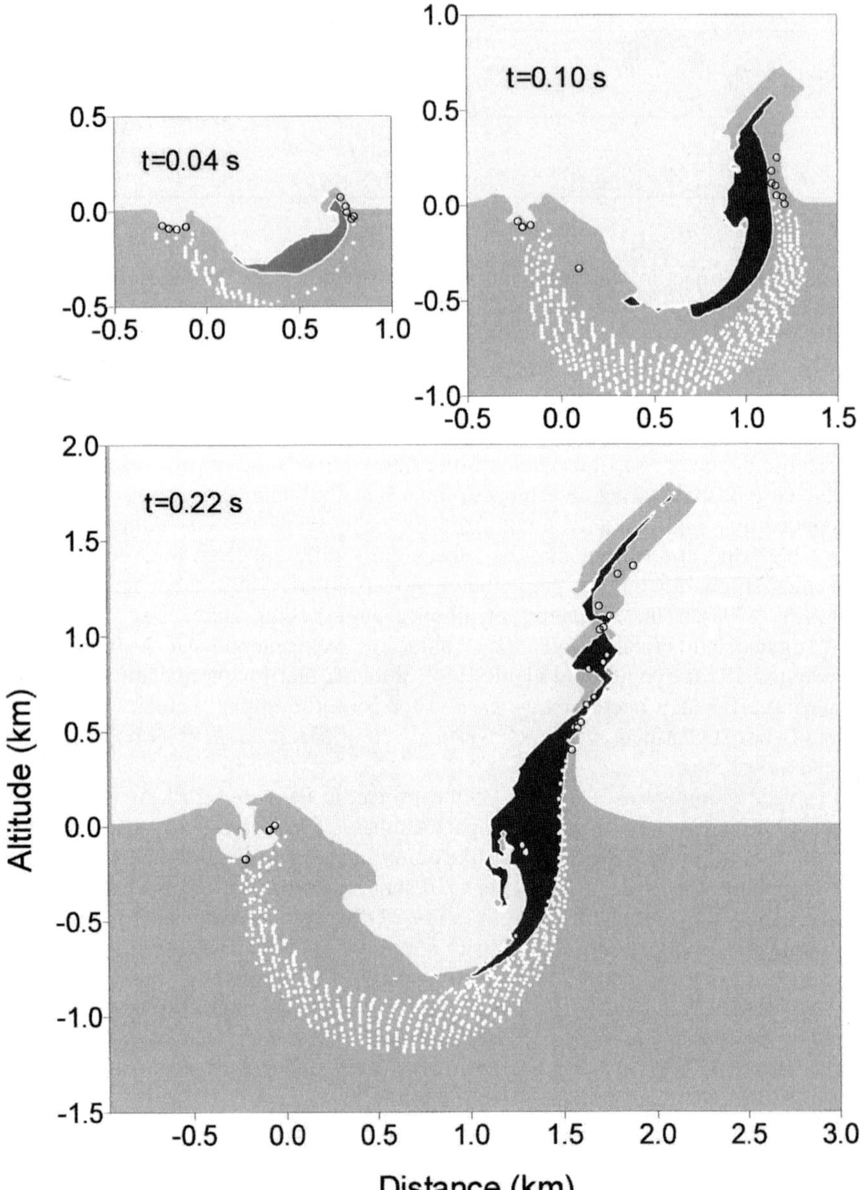

Fig. 2. Initial stage of a 40 kms^{-1} impact with impact angle of 45° for time moments t = 0.04, 0.1, and 0.22 s. Light grey shading corresponds to atmosphere, grey shading shows the target material, and black represents the projectile. Solid white circles mark material compressed in the range 100 - 300 GPa, open black circles mark upper surface material in the same pressure range.

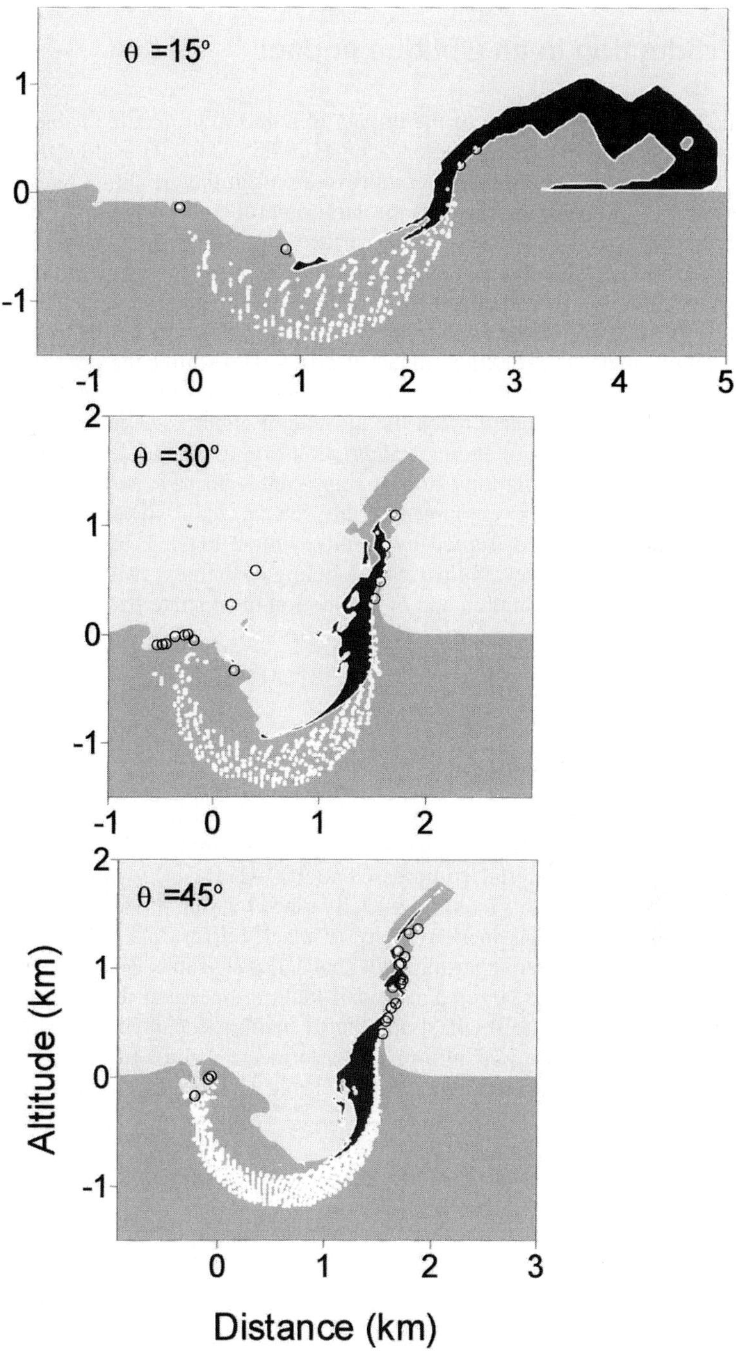

Fig. 3. Post-impact flow for impact angles 15°, 30°, and 45°. Legend is the same as in Fig. 2.

4
Melt Production in an Oblique Impact

Figure 2 shows the initial stage of the impact of a 400-m projectile (Table 2) with the velocity of 40 kms^{-1} and trajectory angle of 45°. At 0.02 s after the initial contact, the projectile is completely compressed within the growing crater, strong shock waves are propagating through the target and the atmosphere. At this time no material has been ejected above the surface yet. Some portion of the target material behind the projectile is compressed to tektite-producing pressures (open black circles), but it is characterized by rather low initial ejection velocity, with no possibility to reach the upper atmosphere. Therefore, it should not be treated as a possible tektite melt. The "tektite" material arises 0.04 s after the contact just in front of the expanding projectile (approximately 1 km away from the impact point). After 0.26 s these ejecta reach the altitude of about 1 - 2 km and continues to move with high velocity of about 10 kms^{-1}, conserving a rather high temperature (above 1800 K). The simulations stop at this point - no new melt from the top layer arises later (the shock pressure near the free surface is lower than 50 GPa). The existing melt will be disrupted by aerodynamical loading and inertia forces into particles of various sizes. Numerical modeling of the next two stages requires description of melt fragmentation and of the motion of separate particles in the gas flow. This part of the calculations is still in progress.

5
Impact Angle Variations

Figure 3 shows the central cross section of the post-impact flow for different impact angles. Target material compressed to 100 - 300 GPa is marked by solid white circles. The part of this material initially placed within the upper 50 m of the target ("tektite" material) is marked by open black circles. The total amount of near surface melted material strongly depends on θ. Table 3 summarizes the computational results. The largest amount of melting occurs at the lowest impact angle modeled, 15°, and the smallest amount of melting occurs at 60°. This is not surprising: in the case of a grazing impact the projectile, sliding over the surface, produces the melt primarily from the upper layer of the target, while the main

Table 3. Near surface (upper 50 m) melt production versus impact angle for a Bosumtwi-like impact

	Angle			
	15°	30°	45°	60°
Mass (Mt)	51	34	23	14
V_{tek}/V_{pr}	0.1	0.25	0.24	0.23

V_{tek}, tektite material volume; V_{pr}, projectile volume component of melt in vertical (or near vertical) impacts is produced from deeper layers of the target, as the shock wave decays strongly near the free surface. The ratio of tektite material volume to projectile volume, V_{tek}/V_{pr}, is practically constant for all impact angles under consideration and sharply decreases to 0.1 for the lowest angle of 15°. It should be mentioned that the total amount of melt also increases as the impact angle decreases, if the same crater size, not the same projectile size, is applied. Rescaling the results by Pierazzo and Melosh (2000) to a constant crater volume (not to a constant projectile volume as in their paper), the maximum melt production is estimated for a 30° impact angle (a factor of 1.7 higher than for the vertical impact). Certainly, tektite material is a rather small part of the total amount of melt produced by the impacts. The average ratio of melt volume to projectile volume for 10 - 30 kms^{-1} impacts is 20 - 30 for a vertical impact (Pierazzo et al. 1997) and drops ten times for a 15° impact, while the ratio of tektite material volume to projectile volume never exceeds 0.25 (see Table 3).

In the numerical simulations reported here the strength of the material is not taken into account. This is clearly an oversimplification, which could, in principle, lead to incorrect results, because in the case of very oblique impacts shear heating may play an important role as reported by Schultz (1996). Shear heating has also been implicated in the production of pseudotachylite veins and sheets in the target rocks of large impact craters (Reimold 1995; Spray and Thompson 1995). The influence of shear heating on melt production has been discussed in details by Pierazzo and Melosh (2000) and they concluded that, most probable, it may not be important if it is confined to narrow shear zones. The influence of shear heating on tektite production is even weaker, because in this case we need not just target melt, but extremely high-velocity melt, while in the shear heating zones velocities are low (pseudotachylite has been usually found within impact structures).

The idea that tektites may be produced by low-angle oblique impacts is not a new one. Vickery (1993) applied the theory of jetting (the phenomenon occurs at the interface of two obliquely impacting surfaces) to propose that tektites originate in high-velocity post-impact jets. However, the results of that work are not encouraging, because high-velocity jets are strongly contaminated with projectile material while the real tektites are not. Although jetting is not considered here, a similar conclusion may be made for the materials ejected from the growing crater after an impact at a low impact angle (<30°). This melt is a mixture of projectile and target materials. In the present modeling no mechanism of mixing is included, but, if two materials (target and projectile) are close to each other in the simulations, then they would be mixed in nature. Only well separated melts (as for 45°) can produce pure tektites without being contaminated with projectile material. One more observation can be used as evidence against a shallow impact: ejecta are moving along the surface in a rather dense atmosphere and are not capable of transporting the melt up to hundreds of kilometers. On the other hand, high impact angles (>60°) result in a practically vertical column of near-surface melt ejecta. The melt may reach the upper atmosphere only above the impact site - and it is not clear how to deliver the melted material a long distance away. Thus, only impacts with angles ranging from 30° to 50° are the best candidates for the "tektite parent" impact event.

Table 4. Near-surface melt production versus velocity variations (impact angle 45°)

	Velocity (kms^{-1})		
	40	20	11.2
Mass (Mt)	23	21	28
V_{tek}/V_{pr}	0.24	0.068	0.046

V_{tek}, tektite material volume; V_{pr}, projectile volume

6 Velocity Variations

Table 4 shows the effects of impact velocity variations. The total mass of "tektite" material depends only slightly on the impact velocity, whereas the ratio of this material volume to the projectile volume decreases by a factor of five for the lowest impact velocity of 11.2 kms^{-1}. Our modeling reveals that the impact melt created in a low-velocity impact stays totally inside the crater and mixes with the projectile melt. At the same time, high-velocity impacts (40 kms^{-1}) produce high-

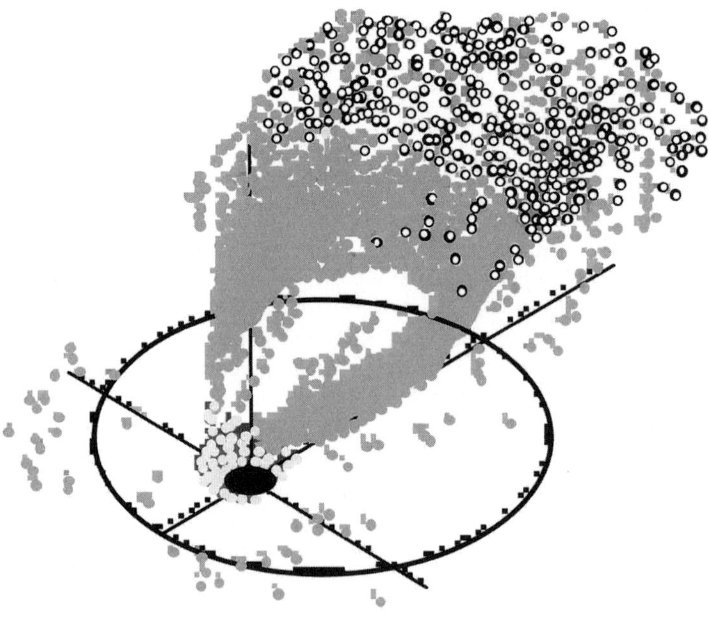

Fig. 4. Spatial distribution (in 3D) of "tektite" material for a 40 kms^{-1} impact 0.26 s after the contact. Light-grey circles mark low-velocity material (<2 kms^{-1}), dark-grey - moderate velocity (2 - 10 kms^{-1}) and white circles - high speed material (>10 kms^{-1}). The full black oval in the center corresponds to the initial position and size of the projectile, and the black line marks current crater diameter.

velocity melt ejecta with better chances to transport the near-surface melt to large distances.

The probability of a high-velocity asteroidal impact is rather low. Ivanov et al. (2001) used the list of osculating orbits "astorb.dat" (as presented on http://asteroid.lowell.edu) and applied the Öpik method (Öpik 1951; Wetherill 1967) to calculate probabilities and impact velocities of Near Earth Asteroids (NEA). The probability of impacts at velocities below 20 kms^{-1} exceeds 70 - 80 %, depending on the accuracy of trajectory definition, whereas the probability of high-velocity impacts (>35 kms^{-1}) drops below 1 %. As our simulations showed, at least a 20 kms^{-1} impact is needed to produce tektite material and to eject it from the growing crater. It means that only one third of the total asteroidal flux seems to be able produce craters with tektites.

High-velocity cometary impacts are not discussed here. Their contribution to the total flux of cosmic objects is not known properly: we have no evidence of this type of impacts, neither any serious objections against a possibility these impacts.

The 3D spatial distribution of "tektite" material for the 45°, 40 kms^{-1} impact is shown in Fig. 4. The black oval represents the initial projectile size and position, the solid curve corresponds to the current crater size (0.26 s after the impact it is approximately one-half of the final size). Tektite material is ejected in a rather narrow angle in the direction of the projectile flight path. The upper portions of these ejecta have high velocity (above 10 kms^{-1}) and may escape from the atmosphere.

7
Target and Projectile Material Variations

It is obvious that the upper layer (50 - 100 m) of the surface has fractures, cavities (large-scale porosity), as well as real pores, filled with air or water. Much lower shock pressures are needed to melt porous material rather than solid targets. Estimates show that the value of shock pressure needed to melt granite drops from 56 GPa for solid granite to only 20 GPa for granite with 30 % porosity. Modeling with a porous upper layer revealed a twofold increase in the mass of tektites compared to the case of dry (no water) and high porosity targets.

Practically all results presented here have been obtained for a dry granite target (which it is not quite correct, as some type of sedimentary rocks should be used, but no proper EOS are available so far for such rocks). In several simulations with wet tuff target (15 % of water) an increase of vapor production and an increase of the ejection velocity are observed. While the velocity increase is favourable for tektite production, abundant water in the plume leads to fast disruption of the target melt, as demonstrated by Kieffer and Simonds (1980). Interaction of melt with water results in the formation of fine dust, not in the formation of melted droplets several centimeters in diameter. We should remember that tektites are impact glasses of high quality (O'Keefe 1994), i.e., they have very low water contents and a high degree of homogeneity. The problem of tektite devolatilization (upper layer of the target contains about 10 wt% of water, while tektites contain less than 0.02 wt%) is the most important problem for the next stage of this study.

8
Trajectory of Small Particles in the Atmosphere

One of the principal objections against the impact origin of tektites concerns the motion of small particles through the atmosphere. The experimental study by Chapman and Larson (1963) concluded that tektites solidified from the melt in vacuum and entered the Earth's atmosphere at high velocities and low angles, which would be incompatible with an impact origin. A study of liquid jets passing through the atmosphere (Adams 1963) concluded that the jets would break up into fine mist before traversing hundreds of kilometers. The study of a solid particle flight through Earth's atmosphere similarly concludes that a tektite-size sphere could not traverse the distance required for strewn field formation. The idea of a temporary atmospheric hole, created by a cometary impact, which could support the tektite flight (Lin 1966), has been also been rejected by Chapman and Gault

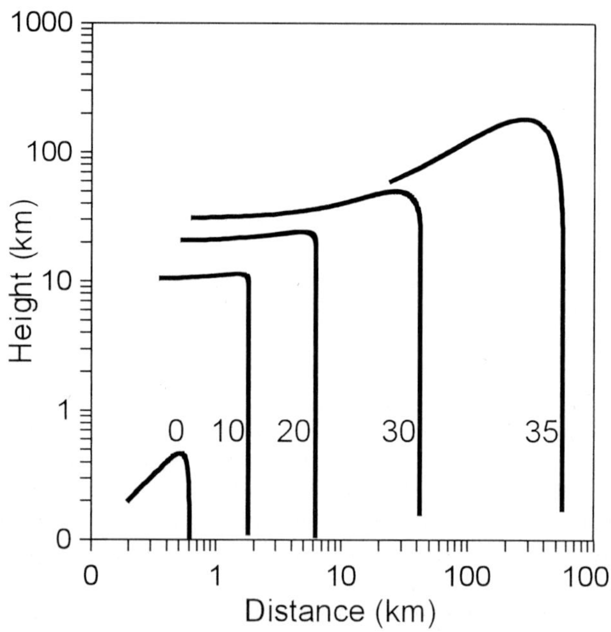

Fig. 5. Trajectories of particles launched from various initial altitudes (values near the curves) and subjected to gravity and atmospheric drag. Particles diameter is 3 cm, initial velocity - 10 kms^{-1}, initial angle - 45°. The gas is immobile (undisturbed stationary atmosphere).

(1967) because of the excessive energy needed for the required atmospheric disturbance. A comet of the required energy could create a huge 300-km-diameter crater on the Earth, whereas tektite parent craters Bosumtwi and Ries are an order of magnitude smaller and Chesapeake Bay is four times smaller. Nevertheless, these 30-years-old ideas should be re-evaluated via numerical modeling, including the joint solution of gas-dynamic equations and differential equations for solid particle motion.

Figure 5 shows the trajectory of a 3-cm-diameter particle launched with an initial velocity of 10 kms^{-1} at various altitudes in the Earth's atmosphere with standard stratification and no gas motion. If the particle is launched from the surface ($h = 0$), the length of the trajectory does not exceed 600 m. Only particles that start their trajectory above 30 km (or reach this altitude at this velocity in the gas flow) can traverse distances of hundreds or thousands of kilometers. But the particles' motion in the post-impact flow differs substantially from the motion in an undisturbed atmosphere: particles follow the gas flow up to the altitudes of 30 - 50 km and the total trajectory length may easily exceed several hundreds km. One more effect should be taken into account - the motion of a swarm of particles differs from the motion of an isolated particle. This effect was investigated in experiments (Schultz and Sugita 1994) and was confirmed by numerical modeling (Artemieva and Shuvalov 1996): a "leading" fragment with the highest velocity moves ahead, while all other fragments fly in its wake ("aerodynamic shadow"). As the leading fragment decelerates enough, another fragment becomes the leader and the resulted swarm of the particles looks like a jet.

9 Conclusions and Future Study

Numerical simulations presented here investigate the formation and fate of high-temperature melts from the upper (near surface) layer of the target during an asteroid impact, and allow to make preliminary estimates of the melt mass and its velocity and spatial distribution. The most suitable conditions for the origin of tektites arise in the case of a high-velocity impact (above 20 kms^{-1}), with the impact angle ranging from 30° to 50°. It is important to note that a very oblique impact is not needed to produce tektites. Approximately 30 % of the impacts have velocities exceeding 20 kms^{-1} and ~50 % of the impacts have impact angles in the desirable interval. It means that about 15 % of the Earth's craters should have tektites strewn fields. The percent might be smaller, because in addition to high impact velocity and specific (but not far from average) impact angle we need a rather large projectile (to produce a crater of at least 10 km) and silica-rich target (but silica is a very abundant material on the Earth's surface).

The results described in this paper should be considered as preliminary. Many important questions (influence of water content on melt production, cometary impact, more realistic EOSs) should be considered in the future study. The modeling should be continued to describe later stages of melt evolution - its disruption into fragments (droplets, bombs) and the consequent particle's motion through the atmosphere, disturbed by the intensive post-impact flow.

10
Discussion – Are Tektites Really Rare?

The preliminary modeling described above shows that no special conditions are needed to produce tektite material - the impact velocity should be slightly above average impact velocity for the Earth (~18 kms^{-1}), impact angle should be close to the most probable impact angle of 45° (probability of an impact angle between 30°- 60° is 50 %). Certainly, the influence of target and projectile material is not yet clear, but the simulations show that tektites should not be a rare phenomenon in impact cratering. What are the reasons for tektites "deficiency" on the Earth? The first one is obvious - only craters exceeded a certain diameter may produce distal ejecta. The estimate for this critical size ranges from 3 km to ~10 km (Melosh 1989). Excluding small craters we reduce the amount of possible parent craters 3 times - from about 150 to 49. The second restriction is crater age. Small glass bodies break down to clays rather quickly. It is impossible to estimate the exact critical age for tektites and it should undoubtedly depend on the environmental conditions and tektite chemical composition, but the oldest tektite strewn field is ~35 Ma and most impact glasses are much younger (Montanari and Koeberl 2000). If now we exclude old craters from the crater list (say, older than 50 Ma), we then remain with only 13 craters which, in principle, may be connected with tektites - see Table 5. Three of these 13 craters are actually parent craters for tektites (Bosumtwi, Ries, and Chesapeake Bay). Remembering that we need a high velocity oblique impact we can conclude that indeed we do have enough tektites. Certainly, these estimates are rather crude (probabilities really

Table 5. Possible parent craters for tektites

Name	Diameter (km)	Age (Ma)	Method	References
Bosumtwi, Ghana	10.5	1.07±0.05	^{40}Ar-^{39}Ar	Koeberl et al. 1997
Chesapeake Bay, USA	85	35.3±0.2	stratigraphy	Poag and Aubry 1995
El'Gygytgyn, Russia	18	3.58±0.04	^{40}Ar-^{39}Ar	Layer 2000
Haughton, Canada	20.5	23.4±1.0	^{40}Ar-^{39}Ar	Jessberger 1988
Kara-Kul, Tajikistan	52	<25	stratigraphy	Gurov et al. 1993
Karla, Russia	12	10	stratigraphy	Masaitis 1999
Logancha, Russia	20	25±20	stratigraphy	Fel'dman et al. 1985
Logoisk, Belarus	17	40±5	stratigraphy	Masaitis 1999
Mistastin, Canada	28	38±4	^{40}Ar-^{39}Ar	Mak et al. 1976
Montagnais, Canada	45	50.5±0.8	^{40}Ar-^{39}Ar	Bottomley and York 1988
Popigai, Russia	100	35.7±0.2	^{40}Ar-^{39}Ar	Bottomley et al. 1997
Ries, Germany	24	15.1±1.0	^{40}Ar-^{39}Ar	Stauchacher et al. 1982
Zhamanshin, Kazakhstan	13.5	0.9±0.1	^{40}Ar-^{39}Ar	Deino and Becker 1990

apply only to really large numbers, larger than 150) and the list of known impact structures on the Earth is not complete (for example, the parent crater for the Australasian tektites has not been identified yet). Nevertheless, the widely accepted theory of the rarity of tektites is not quite correct and should be revised.

Acknowledgments

I thank E. Pierazzo, C. Koeberl, and J. Plado for their helpful comments and useful questions.

References

Adams E (1963) Aerodynamic analysis of tektites and their hypothetical parent bodies. In: O'Keefe JA (ed) Tektites. University of Chicago Press, Chicago & London, pp 150-166

Artemieva NA, Shuvalov VV (1996) Interaction of shock waves during the passage of a disrupted meteoroid through the atmosphere. Shock Waves 5: 359-367

Barnes VE (1961) Tektites. Scientific American 205: 58-65

Blum JD, Papanastassiou DA, Koeberl C, Wasserburg GJ (1992) Nd and Sr isotopic study of Australian tektites: New constrains on the provenance and age of target materials. Geochimica et Cosmochimica Acta 56: 483-492

Bottomley R, York D (1988) Age measurement of the submarine Montagnais impact crater. Geophysical Research Letters 15: 1409-1412

Bottomley R, Grieve RAF, York D, Masaitis V (1997) The age of the Popigai impact event and its relation to events at the Eocene/Oligocene boundary. Nature 388: 365-368

Chao ECT (1963) The petrographic and chemical characteristics of tektites. In: O'Keefe JA (ed) Tektites. University of Chicago Press, Chicago & London, pp 51-94

Chapman CR, McKinnon WB (1986) Cratering of planetary satellites. In: Burns JA, Matthews MS (eds) Satellites. University of Arizona Press, Tucson, pp 492-580

Chapman DR, Gault DE (1967) Critique of "Cometary impact and the origin of tektites". Journal of Geophysical Research 72: 2695-2699

Chapman DR, Larson HK (1963) On the lunar origin of tektites. Journal of Geophysical Research 68: 4305-4358

Chapman DR, Scheiber LC (1969) Chemical investigation of Australasian tektites. Journal of Geophysical Research 74: 6737-6776

Croft SK (1985) The modification stage of basin formation: conditions of ring formation. In: Schultz PH, Merrill RB (eds) Multi-ring basins, Proceedings of Lunar and Planetary Science 12A. Pergamon Press, NewYork, pp 227-257

Deino AL, Becker TA (1990) Laser-fusion $^{40}Ar/^{39}Ar$ ages of acid Zhamanshinite [abs]. Lunar and Planetary Science XXI: 271-272

Fel'dman VI, Mironov YuV, Melikhov BA, Ivanov BA, Basilevsky AT (1985) Astroblemes on trap rock: Structural features and differences from impact structures on other targets (in Russian). Meteoritika 44: 139-145

Friedman I (1963) The physical properties and gas content of tektites. In: O'Keefe JA (ed) Tektites. University of Chicago Press, Chicago & London, pp 130-136

Fudali RF, Dyar MD, Griscom DL, Schreiber HD (1987) The oxidation state of iron in tektite glass. Geochimica et Cosmochimica Acta 51: 2749-2756

Gault DE, Wedekind JA (1978) Experimental studies of oblique impacts. Proceedings of the Lunar and Planetary Science Conference 9: 3843-3875

Gilbert GK (1893) The moon's face, a study of the origin of its features. Bulletin of the Philosophical Society of Washington (DC) 12: 241-292

Gilchrist J, Thorpe AN, Senftle FE (1969) Infrared analysis of water in tektites and other glasses. Journal of Geophysical Research 74: 1475-1483

Glass BP (1972) Bottle green microtektites. Journal of Geophysical Research 77: 7057-7064

Glass BP (1990) Tektites and microtektites: key facts and inferences. Tectonophysics 171: 393-404

Glass BP, Pizzuto JE (1994) Geographic variation in Australasian microtektite concentrations: Implications concerning the location and size of the source crater. Journal of Geophysical Research 99: 19,075-19,081

Gurov EP, Gurova EP (1980) Shock metamorphosed rocks of the Elgygytgyn meteorite crater in Chukotka (in Russian). Meteoritika 39: 102-109

Gurov EP, Gurova EP, Rakitskaya RB, Yamnichenko AYu (1993) The Karakul depression in Pamirs - the first impact structure in central Asia [abs]. Lunar and Planetary Science XXIV: 591-592

Ivanov BA, Artemieva NA (2001) Transient cavity scaling for oblique impacts [abs]. Lunar and Planetary Science XXXII: # 1327 (CD-ROM)

Ivanov BA, Neukum G, Wagner R (2001) Size-frequency distribution of planetary impact craters and asteroids. In: Marov M, Rickman H (eds) Collisional Processes in the Solar System, Astrophysics and Space Science Library, Kluwer Academic Publishers, in press

Jakes P, Sen S, Matsuishi K, Reid AM, King EA, Casaniva I (1992) Silicate melts at super liquidus temperatures: Reduction and volatilization [abs]. Lunar and Planetary Science XXIII: 599-600

Jessberger EK (1988) ^{40}Ar-^{39}Ar dating of the Haughton impact structure. Meteoritics 23: 233-234

Jones WB, Bacon M, Hastings DA (1981) The Lake Bosumtwi impact crater, Ghana. Geological Society of America Bulletin 92: 342-349

Kieffer SW, Simonds CH (1980) The role of volatiles and lithology in the impact crating process. Reviews of Geophysics and Space Physics 18: 143-181

Koeberl C (1990) The geochemistry of tektites: an overview. Tectonophysics 171: 405-422

Koeberl C (1994) Tektite origin by hypervelocity asteroidal or cometary impact: target rocks, source craters, and mechanisms. In: Dressler BO, Grieve RAF, Sharpton VL (eds) Large Meteorite Impacts and Planetary Evolution. Geological Society of America Special Paper 293: 133-151

Koeberl C, Beran A (1988) Water content of tektites and impact glasses and related chemical studies. Proceedings of Lunar and Planetary Science Conference 18: 403-408

Koeberl C, Shirey SB (1993) Detection of a meteoritic component in Ivory Coast tektites with rhenium-osmium isotopes. Science 261: 595-598

Koeberl C, Bottomley R, Glass BP, Storzer D (1997) Geochemistry and age of Ivory Coast tektites and microtektites. Geochimica et Cosmochimica Acta 61: 1745-1772

Koeberl C, Poag CW, Reimold WU and Brandt D (1996) Impact origin of the Chesapeake Bay structure and the source of the North American tektites. Science 271: 1263-1266

Koeberl C, Reimold WU, Blum JD, Chamberlain CP (1998) Petrology and geochemistry rocks from the Bosumtwi impact structure, Ghana, and comparison with Ivory Coast tektites. Geochimica et Cosmochimica Acta 62: 2179-2196

Layer PW (2000) Argon-40/argon-39 age of the El'gygytgyn impact event, Chukotka, Russia. Meteoritics & Planetary Science 35: 591-599

Lin SC (1966) Cometary impact and the origin of tektites. Journal of Geophysical Research 71: 2427-2437

Mak EK, York D, Grieve RAF, Dence MR (1976) The age of the Mistastin Lake crater, Labrador, Canada. Earth and Planetary Science Letters 31: 51-57

Masaitis VL (1999) Impact structures of northeastern Eurasia: the territories of Russian and adjacent countries. Meteoritics & Planetary Science 34, 691-711

Matsuda JI, Matsubara K, Koeberl C (1993) Origin of tektites: Constrains from heavy noble gas concentrations. Meteoritics 28: 586–589

Melosh HJ (1989) Impact cratering: A Geologic Process. Oxford University Press, Oxford, 246 pp

Middleton R, Klein J (1987) ^{26}Al: measurement and applications. Philosophical Transactions of the Royal Society London A323: 121-143

Montanari A, Koeberl C (2000) Impact stratigraphy: The Italian record. Springer Verlag, Lecture Notes in Earth Sciences 93, Berlin - Heidelberg, 364 pp

Neukum G, Ivanov BA (1994) Crater size distributions and impact probabilities on Earth from lunar, terrestrial-planet, and asteroid cratering data. In: Gehrels T (ed) Hazards due to Comets and Asteroids. University of Arizona Press, Tucson, pp 359-416

O'Keefe JA (1994) The origin of tektites. Meteoritics 29: 73-78

O'Keefe JD, Ahrens TJ (1977) Impact-induced energy partitioning, melting, and vaporization on terrestrial planets. Proceedings of the Lunar Science Conference 8: 3357-3374

Öpik EJ (1951) Collision probabilities with the planets and the distribution of interplanetary matter. Proceedings of Royal Irish Academy 54A: 165-199

Pal DK, Tuniz C, Moniot RK, Kruse TH, Herzog GF (1982) Berillium-10 in Australasian tektites: Evidence for a sedimentary precursor. Science 218: 787-789

Pierazzo E, Melosh HJ (2000) Melt production in oblique impacts. Icarus 144: 252-261

Pierazzo E, Vickery AM, Melosh HJ (1997) A reevaluation of impact melt production. Icarus 127: 408-422

Poag CW, Aubry M-P (1995) Upper Eocene impactites of the U.S. East Coast: Depositional origins, biostratigrphic framework, and correlations. Palaios 10: 16-43

Poag CW, Powars DS, Poppe L, Mixon RB (1994) Meteoroid mayhem in Ole Virginny: Source of the North American tektite strewn field. Geology 22: 691-694

Reimold WU (1995) Pseudotachylite in impact structures – Generation by friction melting and shock brecciation? A review and discussion. Earth-Science Reviews 39: 247-265

Schmidt RM, Housen KR (1987) Some recent advances in the scaling of the impact and explosion cratering. International Journal of Impact Engineering 5: 543-560

Schnetzler CC (1992) Mechanism of Muong Nong-type tektite formation and speculation on the source of Australasian tektites. Meteoritics 27: 154-165

Schultz PH (1996) Effect of impact angle on vaporization. Journal of Geophysical Research 100: 21,117-21,135

Schultz PH, Gault DE (1990) Prolonged global catastrophes from oblique impacts. In: Sharpton VL, Ward PD (eds) Global Catastrophes in Earth History. Geological Society of America Special Paper 247: 239-261

Schultz PH, Sugita S (1994) Penetrating and escaping the atmospheres of Venus and Earth [abs]. Lunar and Planetary Science XXV: 1215-1216

Shoemaker EM (1962) Interpretation of lunar craters. In: Kopal Z (ed) Physics and Astronomy of the Moon. New York/London, Academic: 283-359

Shuvalov VV (1999) Multi-dimensional hydrodynamic code SOVA for interfacial flows, application to thermal layer effect. Shock Waves 9: 381-390

Shuvalov VV, Artemieva NA, Kosarev IB (1999) 3D hydrodynamic code SOVA for multimaterial flows, application to Shoemaker-Levy 9 comet impact problem. International Journal of Impact Engineering 23: 847-858

Spray JG, Thompson LM (1995) Friction melt distribution in a multi-ring impact basin. Nature 373: 130-132

Stauchacher T, Jessberger EK, Dominik B, Kirsten T, Schaeffer OA (1982) ^{40}Ar-^{39}Ar ages of rocks and glasses from the Nördlinger Ries crater and the temperature history of impact breccias. Journal of Geophysics 51: 1-11

Tera F, Middleton R, Klein J, Brown L (1983) Berillium-10 in tektites. EOS (Transactions American Geophysical Union) 64: 284

Thompson SL, Lauson HS (1972) Improvements in the Chart D radiation-hydrodynamic CODE III: Revised analytic equations of state. Sandia National Laboratory Report SC-RR-71 0714, Albuquerque, 119 pp

Valette-Silver N, Brown L, Klein J, Middleton R, Pavic MJ, Tera F (1983) Berillium-10: Vertical Distribution in solis and sediments. In: Yearbook of the Carnegie Institution'82, Carnegie Institution, Washington, pp 462-463

Vickery AM (1993) The theory of jetting; application to the origin of tektites. Icarus 105: 441-453

Wetherill GW (1967) Collisions in the asteroid belt. Journal of Geophysical Research 72: 2429-2444

Zamyshliayev BV, Yevterev LS (1990) Models of Dynamic Deforming and Failure for Ground Media (in Russian). Nauka, Moscow, 215 pp

Geology and Magnetic Signatures of the Neugrund Impact Structure, Estonia

Sten Suuroja[1,2], Tarmo All[1,3], Jüri Plado[3,4], and Kalle Suuroja[1]

[1]Geological Survey of Estonia, 12618 Tallinn, Estonia. (s.suuroja@egk.ee; t.all@egk.ee)
[2]Institute of Mining, Tallinn Technical University, Estonia.
[3]Institute of Geology, University of Tartu, 51014 Tartu, Estonia.
[4]Geological Survey of Finland, FIN-02151 Espoo, Finland.

Abstract. The Early Cambrian Neugrund crater, Estonia, has a rim-to-rim diameter of 7 km. In addition to the structural elements common in impact structures (crater cavity, rim wall, crater filling, ring fault), there occurs a specific, up to 70 m deep and 200 - 500 m wide circular canyon between the rim wall and the crater. Theoretically, Neugrund should also have a central uplift, but there is neither geological nor geophysical evidence as yet. The seismic reflection data show that the disturbed strata around the structure are spatially delimited by the ring fault, which has a diameter of about 21 km. The ring fault is interpreted as the outer limit of the impact structure. The true depth of the crater, as well as the distribution of post-impact sediments and impactites within the crater, is still unknown. Magnetic modelling and comparison with other impact structures of the Fennoscandian shield (e.g., Kärdla, Lappajärvi, Lockne) suggest that the maximum depth of the interface between autochthonous and allochthonous breccias is approximately 700 m. A distinct, slightly elongated negative magnetic anomaly about 5 km in diameter marks the crater cavity. This magnetic low is caused mainly by weakly magnetic post-impact sedimentary infill and its screening effect, but possibly also by allochthonous breccias and post-impact oxidation processes. The absence of high-amplitude local anomalies within the minima suggests a lack of highly magnetic melt bodies. The magnetic data show no evidence of a central uplift.

1 Introduction

The Early Cambrian (~540 Ma; Suuroja and Suuroja 2000) Neugrund impact structure (59°20´N; 23°31´E) is located in the Gulf of Finland north-east of Osmussaar Island (see Abels et al., this volume; Fig. 1). The well-preserved

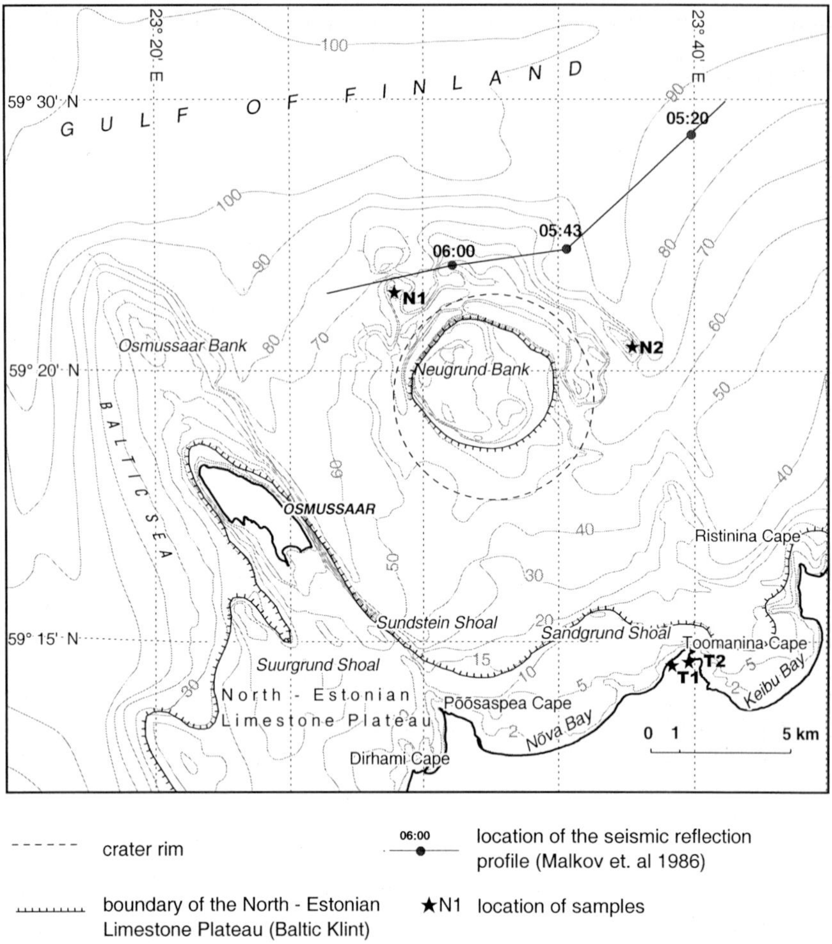

Fig. 1. Bathymetric map of the Neugrund structure area.

circular crater has a rim-to-rim diameter of 7 km and is surrounded by a 21-km-wide area of deformed sedimentary target rocks. The crater cavity is filled with post-impact Cambrian and Ordovician sedimentary rocks. The Ordovician layers consist of hard, erosion-resistant limestone, forming a nearly circular 4.5-km-wide central plateau inside the crater, the Neugrund Bank (Fig. 2). A 200 - 500 m wide and up to 70 m deep semicircular canyon surrounds the central plateau (Suuroja and Suuroja 2000) along the inner side of the crater wall.

The Neugrund impact structure was discovered in the course of integrated geological and geophysical mapping at the NW of Estonia. The first medium-scale geological mapping of the mainland was carried out in 1965 - 1969 (Kala and Eltermann 1969). Deformations in the Vendian and Early Cambrian terrigenous rocks of a drill core from Osmussaar Island (Fig. 2a) were investigated. The geological mapping of the basement of the onshore area was conducted in 1984 - 1987 (Suuroja et al. 1987). An about 20 m thick layer of brecciated sedimentary rocks was discovered at the top of the Lontova Formation (Early Cambrian) in one core (F-331; Figs. 2a and 3). Simultaneously, marine geological mapping of the

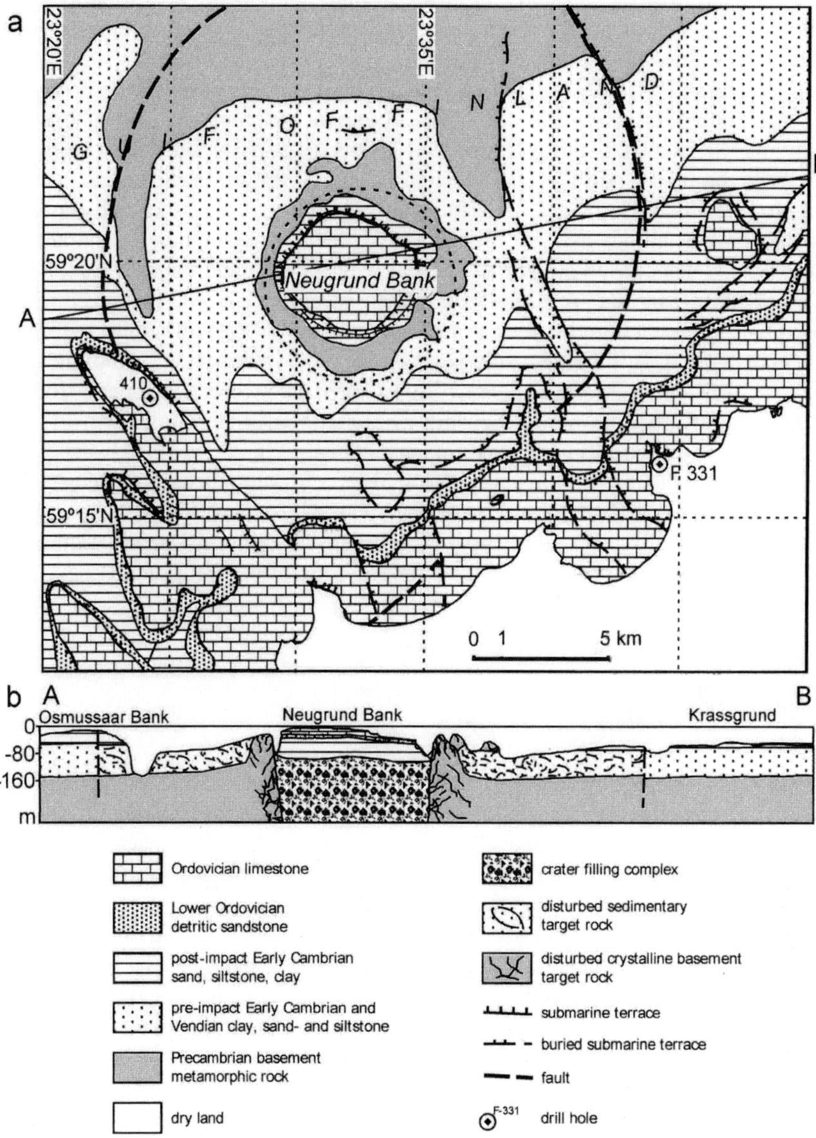

Fig. 2. (a) Generalised bedrock map and (b) cross-section of the Neugrund structure area.

offshore area was carried out (Malkov et al. 1986). An east–west seismic profile north of the Neugrund Bank (Figs. 1 and 4) revealed small hills and short ridges, which were first interpreted as being of glacial origin (moraine walls). During subsequent marine geological mapping across the Neugrund structure, seismoacoustic reflection recording was performed along five N - S profiles and bottom sediments were sampled with a cavity-corer (Talpas et al. 1993). A comprehensive geological study of the Estonian shelf (Lutt and Raukas 1993) gave only slight indication of the Neugrund structure. The interpretations once again relied on the glacial theory.

In the years 1994 - 1995 numerous brecciated erratic boulders were discovered in the course of large-scale (1 : 50 000) geological mapping of the onshore area (including the islands) (Suuroja et al. 1998). Suuroja and Saadre (1995) suggested that these boulders might be associated with an impact structure. This idea prompted seismic reflection and marine magnetic study of the Neugrund Bank and its nearest surroundings in 1996 (Suuroja et al. 1999b). Aeromagnetic mapping of this area had been carried out already earlier (Metlitckaya and Papko 1992), but no traces of the Neugrund impact structure had been recognised. In the course of subsequent marine geological research (Suuroja et al. 1999a), the impact structure was investigated by diving, side-scan sonar and submarine sampling. The discovery of shock metamorphosed rocks provided direct evidence for the hypotheses derived from remote sensing data, and confirmed the impact origin of the Neugrund structure (Suuroja and Suuroja 2000).

The aim of the present paper is to analyse the geological and geophysical characteristics of the Neugrund structure. The physical properties of impact-influenced rocks of the structure and erratic boulders are discussed and a magnetic model composed to describe the subsurface structure of the Neugrund crater.

2 Bedrock Geology

The Neugrund impact structure is well expressed on the marine geological map (Fig. 2a) at the edge of the erosional boundary with the Precambrian basement. The North-Estonian Limestone Plateau, consisting of post-impact Ordovician limestones, is located to the south of the structure. The shallow northern edge of the plateau is partially submerged. The plateau borders on a well-jointed escarpment - the Ordovician cliff of the Baltic Klint (Fig. 1). The submarine cliff is well traceable, as Quaternary deposits only rarely cover it. The isolated and circular submerged cliff-island, about 4.5 km in diameter, the so-called Neugrund Bank, constitutes the central sedimentary infill of the Neugrund crater.

In the northern part of the area, above the rim wall and in the bottoms of buried valleys west and north-east of the structure, the crystalline basement crops out. It is composed of Precambrian (Paleoproterozoic) metamorphic rocks formed by the Svecofennian Orogeny. The data obtained from drill holes (Fig. 2a), five submarine outcrops, and more than 350 erratic boulders derived from the

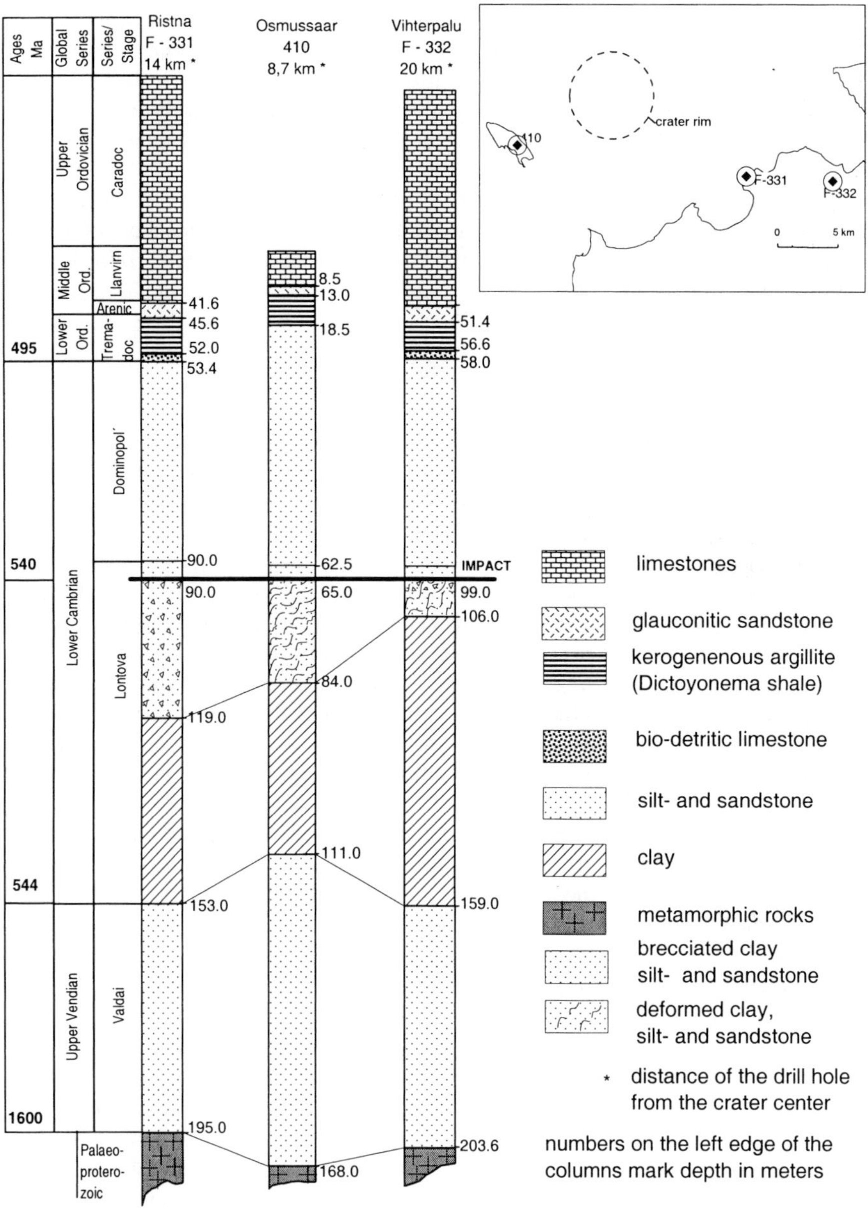

Fig. 3. Stratigraphic columns of the drill cores outside the area of Neugrund structure.

Neugrund impact structure, show that the crystalline basement consists of migmatized gneisses (54 %) and amphibolites (46 %). Brecciated crystalline basement rocks crop out around the Neugrund Bank, on the crater rim wall. Small outcrops of crystalline basement rocks behind the rim consist in large (up to several hundred metres in diameter) blocks ejected from the crater (Suuroja and Suuroja 2000). The Neugrund crater is filled (from base to top) with: (i) impact deposits of unknown thickness; (ii) post-impact Early Cambrian silt- and sandstones with a thickness of over 40 m; (iii) Ordovician sandstone, shale, and limestone of more than 40 m thickness (Fig. 2).

The ring fault (Fig. 2) is expressed in the bedrock relief by a 10 - 80 m high escarpment. Inside the ring fault the sedimentary cover and the crystalline basement are strongly deformed, but almost intact outside (see Fig. 7b in Suuroja and Suuroja 2000).

3
Quaternary Deposits

A map of Quaternary deposits (Fig. 4a) was compiled through integrated geological mapping of the on- and offshore areas of NW Estonia and combining the results with the bedrock map (Kala and Eltermann 1969; Malkov et al. 1986; Talpas et al. 1993; Suuroja et al. 1998; Suuroja et al. 1999b). The distribution of the Quaternary deposits is clearly influenced by the pre-Quaternary bedrock relief, which reflects the structures of the partially exposed Neugrund impact crater.

The Quaternary deposits of the impact structure area can be roughly divided into the ones distributed to the north or to the south of the submerged Baltic Klint (Figs. 1 and 4). In the areas southward of the Baltic Klint (on the North Estonian Limestone Plateau), where the water depth is less than 15 m, the thickness of the Quaternary deposits is minor (0 - 2 m). The coastal deposits (sand, gravel, pebbles) of the present Baltic Sea (Limnea Sea) are widespread in submerged areas with a very thin (less than 0.5 m) Quaternary cover.

In the northern part of the area, where the water depth generally exceeds 15 m, the sea floor is covered with 1 - 10 m of mud deposited during the post-glacial Limnea Sea stage. Upper Pleistocene, glaciolacustrine varved clays of the Baltic Ice Lake crop out around the rim wall, in especially great thickness of 30 - 60 m in the circular canyon.

Nowadays, the Neugrund Bank appears as an erosional area, partly covered with coarse-grained sediments (cobbles, pebbles, coarse-grained sand). Mud is deposited in most regions with water depths greater than 50 m. The coastal area is mostly non-depositional or erosional, except the areas of clastic sedimentation. However, the sedimentation north of the Neugrund Bank is minimal due to strong seawater currents (Suuroja et al. 1999a).

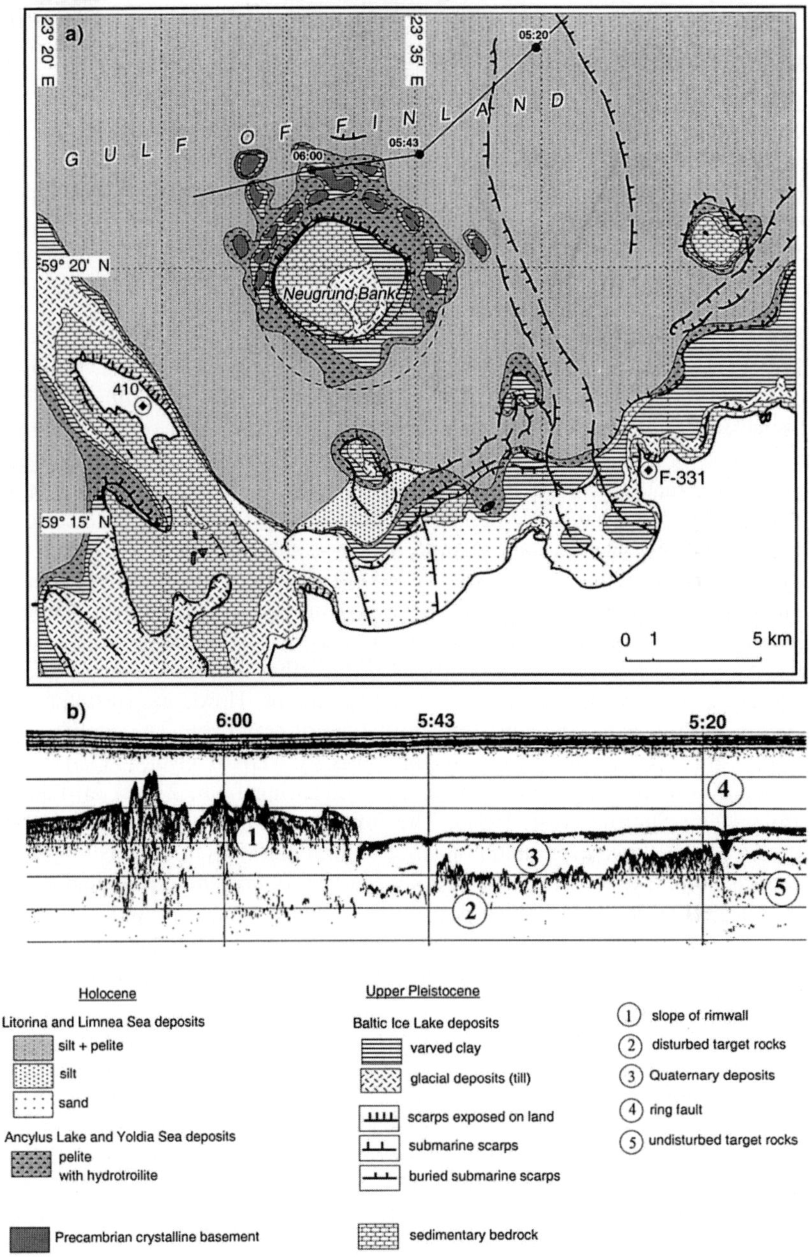

Fig. 4. (a) Quaternary deposits map of the Neugrund area and (b) seismic reflection record across the northern part of the structure. Recording frequency was 0 - 425 Hz; distance between horizontal lines is about 32 m in water (modified after Malkov et al. 1986).

4 Geophysical Signatures

The subsurface of the Neugrund structure has been investigated by seismic and magnetic methods, the morphometry and volume of the crater cavity by magnetic modelling. The latter was supported by petrophysical data of samples collected from the crater rim and erratic boulders derived from it.

4.1 Marine Seismic Investigations

Malkov et al. (1986) recorded the first seismic signatures of the Neugrund structure during a reflection survey of the Estonian side of the Gulf of Finland. One east–west profile studied by them a few kilometres north of the Neugrund Bank is presented in this paper (Figs. 1 and 4b). The profile revealed a rugged morphology directly north of the Neugrund Bank, the sedimentary cover to the north-west, and the Quaternary deposit on an eroded Precambrian surface to the north-east (Fig. 4). The earlier, seismic surveys, carried out in the Baltic Sea (Flodén 1981), revealed the block and fault structure of the Svecofennian basement (Flodén 1980). Malkov et al (1986) interpreted the above-mentioned rugged morphology as small hills and short ridges of glacial origin. The fault observed north-east of time mark 05:20, as all the other faults, were interpreted as the regular block and fault structure of the basement. However, we interpreted these hills and ridges to constitute the outer slope of crater rim. The correlation of fault structures fixed in the others seismograms by Malkov et al. (1986) and by Talpas et al. (1993) suggests the occurrence of a ring fault, 21 km in diameter (Suuroja 1996; Suuroja et al. 1997). The impact origin of the Neugrund crater largely facilitates the interpretation of the structural elements in the seismograms.

A group of researchers from Stockholm University and the Geological Survey of Estonia headed by Tom Flodén performed a geophysical survey of the area in 1996. The new seismic and magnetic data allowed them to refine the outline of the crater and the structural patterns beyond the crater rim (Suuroja and Suuroja 1999b, 2000).

4.2 Magnetic Studies

The Geological Survey of Byelorussia conducted an aeromagnetic survey of northern and western Estonia and the surrounding shelf in 1988 - 1992 (Metlitckaya and Papko 1992), with the aim of outlining the general structural and tectonic pattern of the Precambrian basement. The terrain clearance of the survey was 300 m on the sea. The east-west profiles proceeded at 0.5 km intervals. Total field readings were taken approximately every 50 m. The accuracy of the survey was ± 3 nT.

The magnetic data of the Neugrund crater and its surroundings were analysed. The aeromagnetic map (Fig. 5) shows that the regional magnetic field around the

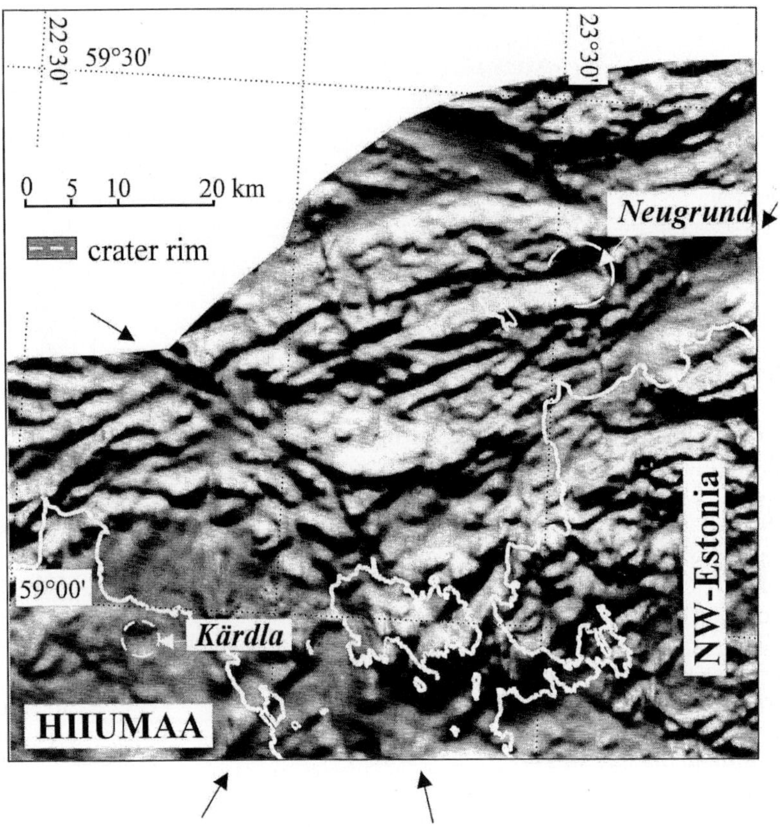

Fig. 5. Shaded relief map of aeromagnetic anomalies of NW Estonia, illuminated from northwest. Arrows indicate linear magnetic anomaly belts due to regional shear zones.

Neugrund impact structure has a complicated pattern due to variable magnetic properties of the Svecofennian metamorphic belt (Puura and Vaher 1983; Koistinen et al. 1996; Korhonen et al. 1999). The regional north-east and northwest trending faults and shear zones (Koistinen et al. 1996; Koppelmaa and Kivisilla 2000) are revealed as distinct linear magnetic anomaly belts (Fig. 5). The structure is located over a positive NEE-trending linear magnetic anomaly, but not influenced by regional shear zones. The central part is superposed on the western limit of magnetic minimum. The map of aeromagnetic anomalies (Fig. 6) demonstrates the presence of an arc of positive anomalies (up to 400 nT), which corresponds to the northern, eastern and south-eastern parts of the crater rim. The amplitudes of the anomalies are equal to those in the surroundings. Therefore, the impact-related anomalies are fairly similar to the regional field, and were not recognized as caused by an impact event in the early 1990s. However, the rim wall of the Neugrund structure produces a local bow-shaped positive anomaly pattern bordering the minimum related to the crater cavity in the east. This anomaly

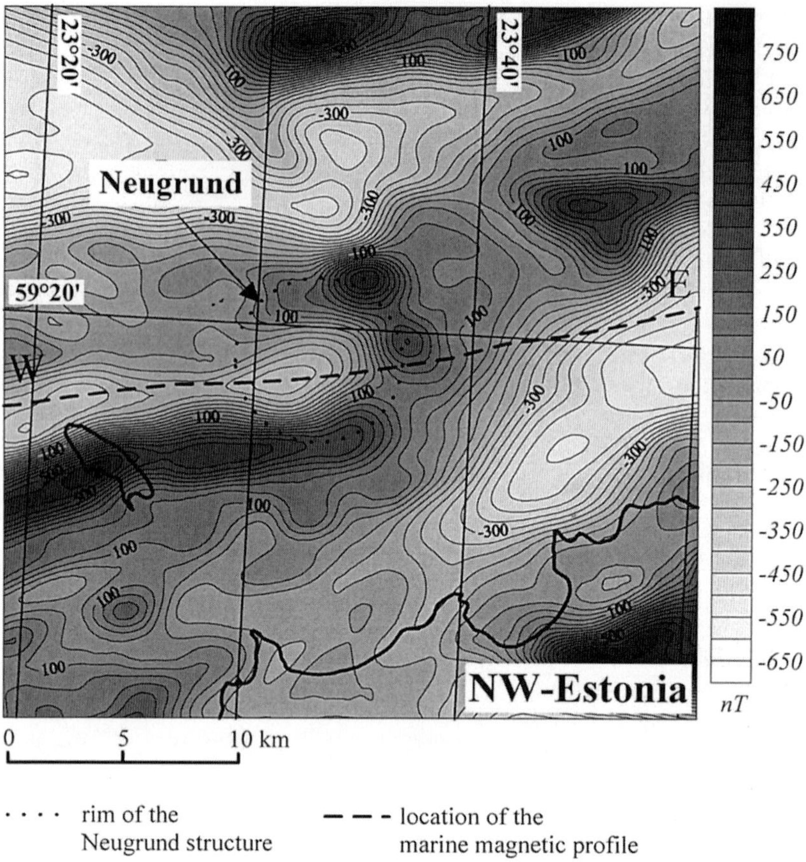

Fig. 6. Aeromagnetic anomalies map of the Neugrund structure area.

pattern can be followed mainly as a secondary structure within the prevailing linear anomaly pattern typical of the surrounding metamorphic Svecofennian rocks (Suuroja et al. 1987). Figure 5 shows the existence of a similar, but weakly expressed, anomaly pattern to the west of the structure. Altogether, these positive features form a circular anomaly pattern around the magnetic minimum.

To interpret the magnetic anomalies the forward modelling technique was used, where we calculated a magnetic response curve along the west-east magnetic profile (Fig. 6). The data were obtained during the marine study in 1996 and are first published in the present paper. By changing the shapes of polygonal prisms (for impact breccias and crater fill) and polygons (for regional bodies), we tried to match the model curve to fit the observed data by trial-and-error techniques.

For modelling the structure, we made the following assumptions, based on results of seismic and geological mappings: (i) the pre-impact deposits and the lowermost part of post-impact deposits in the surrounding area are preserved; (ii) impactites within the crater proper are preserved; (iii) the post-impact Cambrian and Ordovician rocks fill the crater cavity and (iv) the crest of the rim is partly exposed on the sea floor.

To remove the long-wavelength magnetic anomalies responding to deeper sources and unrelated to the subsurface impact structure, the data were filtered with a low-pass filter with a radius of 8 km. This transformation provided a residual magnetic anomaly curve describing the anomalies produced by the near-surface sources. For the modelling, physical properties of the impact-influenced crystalline rocks (Table 1) and data from the basement of western Estonia (Table 2; Puura and Vaher 1983) were used. The petrophysical measurements were performed at the Geological Survey of Finland with instruments and techniques described in Puranen and Sulkanen (1985). Due to the absence of oriented samples no orientation of the natural remanent magnetisation (NRM) was obtained. Declination of 326° and inclination of 60° were used to specify the direction of remanence in the target units of the magnetic model. This direction corresponds to the Early Svecofennian (1.88 Ga; Pesonen et al. 1989), i.e., to the age of the crust in this particular area. The rocks from the rim wall show relatively low susceptibilities (Table 1), probably due to weathering and increased oxidation along impact-produced fractures. Also, the NRM intensity yields Koenigsberger ratios generally below 2. However, according to Puranen (1989), the mean magnetic susceptibility of the Svecofennian metavolcanic belt in SE Finland is higher ($8{,}450 \times 10^{-6}$ SI). The same is valid for the crystalline rocks in western

Table 1. Physical properties of the crystalline rocks from the Neugrund impact structure and surroundings

Sample	Rock type	Loc.	ρ_W (kgm^{-3})	Φ (%)	χ (10^{-6} SI)	NRM (mAm^{-1})	Q
10a	Amphibolite	N1	2729	7.4	760	25	0.9
13	Amphibolite	N1	3199	0.6	12180	310	0.6
63	Brecciated amphibolite	N2	2773	0.4	740	210	7.1
66	Brecciated amphibolite	N1	2631	4.8	3290	50	0.3
10b	Migmatite granite	N1	2644	2.0	230	20	1.8
26	Migmatite	N1	2463	4.5	250	20	1.5
18	Brecciated granitoid	N1	2498	3.5	1180	15	0.4
19	Weathered biotite-gneiss	N1	2605	3.4	305	25	2.0
23	Biotite gneiss	N1	2649	3.4	520	110	5.1
3-21	Biotite gneiss	N1	2644	6.5	700	50	1.7
47	Biotite gneiss	N1	2625	1.9	250	10	1.3
T-5	Brecciated amphibolite	T1	2661	7.0	740	10	0.3
7	Brecciated amphibolite with veins of calcite	T2	2706	7.9	790	10	0.3

Loc., Location (see Fig. 1); ρ_W, density of water-saturated samples; Φ, porosity; χ, magnetic susceptibility; *NRM*, intensity of natural remanent magnetization; *Q*, Koenigsberger ratio

Table 2. Physical properties of the unshocked target rocks in western Estonia (data from Puura and Vaher 1983).

Petrology	ρ_W (kgm^{-3})	χ (mean) (10^{-6} SI)	χ (range) (10^{-6} SI)	Q
Migmatized metamorphic rocks	2760	3775	10-111,500	1.13
Metamorphic rocks	2800	4875	10-67,500	1.23
Amphibolite	2850	4000	10-62,800	1.32
Biotite-gneiss	2630	9100	65-67,500	0.8

ρ_W, density of water-saturated samples; χ, magnetic susceptibility; Q, Koenigsberger ratio

Estonia (Table 2), but the maximum values recorded there reach $110{,}000 \times 10^{-6}$ SI (Puura and Vaher 1983). Therefore, to model the short-wavelength anomalies surrounding the Neugrund structure, we used susceptibility values between $10{,}000 \times 10^{-6}$ and $90{,}000 \times 10^{-6}$ SI. According to the results of petrophysical measurements of different rock types in the Neugrund structure area (Table 1), the Q-value was set to 1.5. Also the density and porosity of impact-influenced rocks were measured, but no specific trends due to the impact were recorded. The porosity seemed to be slightly higher in intensively brecciated rocks (Table 1), but the weathering could also be the cause of this effect.

We assumed that the impact did not alter the magnetic properties of target rocks beyond the crater rim. The main source of the magnetic minimum is located inside the crater due to practically nonmagnetic sedimentary rocks and weakly magnetic allochthonous impact breccias. The model following these assumptions depicts a 700-m-deep crater cavity, filled with sedimentary rocks down to the 270 m depth, and, from 270 to 700 m, with breccias (Fig. 7). To ensure a better fit of the observed and calculated curves it was necessary to add some highly magnetic blocks simulating the variation in magnetic properties of the metamorphic belt outside the crater area. The relative elevation of the magnetic field in the central part of the crater, as well as the sharp narrow positive peaks (up to 150 nT) on both sides of the crater, reflect the geometry of the structure. In the case of the homogeneous surrounding, the raised walls will produce a 3-km-wide positive anomaly with amplitude of up to 40 nT (Fig. 7a). However, in the real field (Fig. 7b) this anomaly can be recognised only in the western part of the crater. On the eastern side anomalies from highly magnetic target rocks overprint the rim anomaly.

The central uplift of the Neugrund structure is not evident in the magnetic field. However, we added this feature into the model according to the size criteria for terrestrial impact structures (e.g., Grieve and Pesonen 1996). The diameter of the central uplift (1.5 km) was estimated from the equation of Pike (1985). The Söderfjärden impact structure, western Finland, similar to Neugrund in age and diameter (see Abels et al., this volume), reveals a 287-m-high central uplift with a diameter of ~1.2 km. (Abels et al. 1998). Based on these, and data of few other similar-sized structures (Crooked Creek, Ragozinka, Serpent Mound), we assume

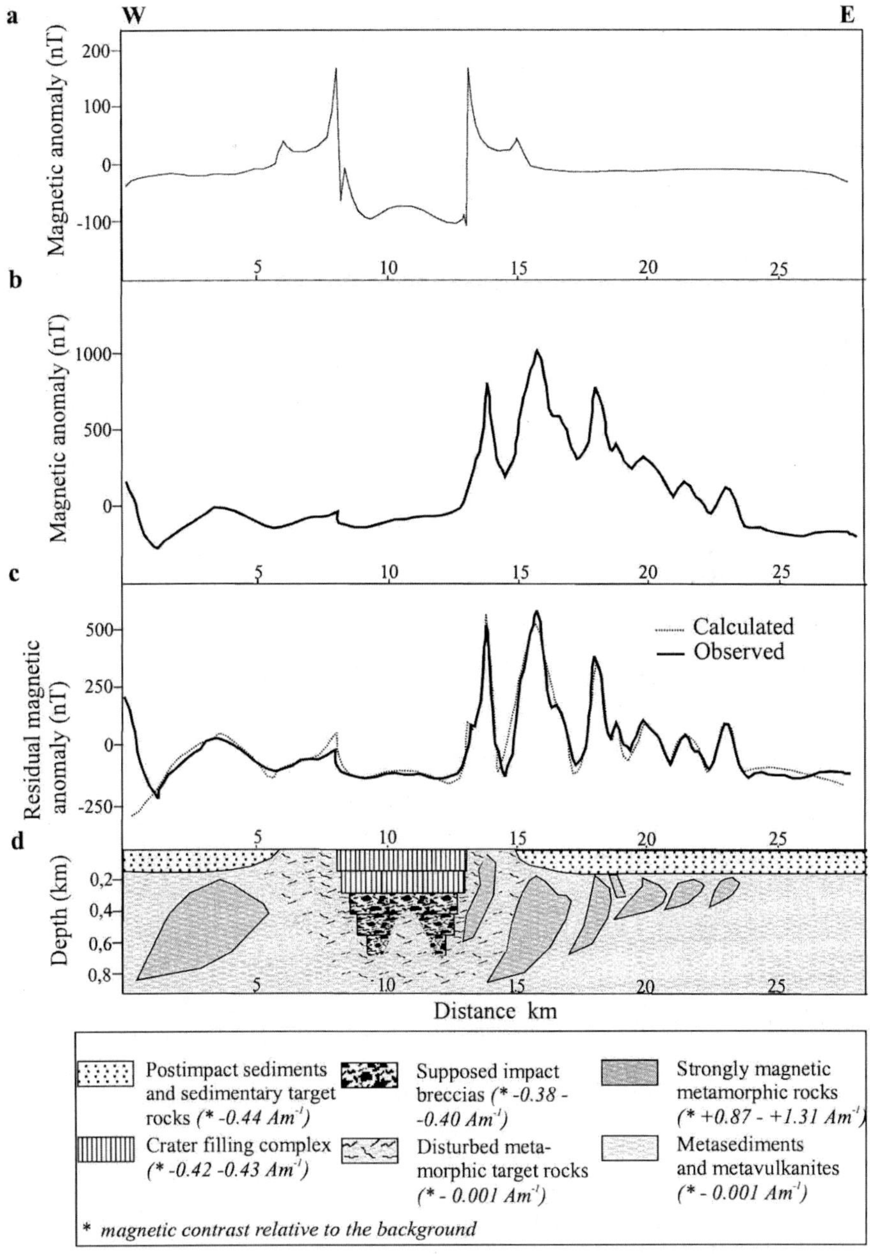

Fig. 7. Magnetic model of the Neugrund structure; (a) theoretical effect of the weakly magnetic complex of crater filling rocks and rim wall. Homogeneous background producing the level field of 0 nT; (b) observed magnetic anomaly curve (marine survey); (c) residual anomaly, and (d) calculated magnetic effect of model.

that the uplift in Neugrund is ~300 m high. This value was used in the model (Fig. 7d) as well.

5 Discussion and Conclusions

The Neugrund structure formed during Early Cambrian Lontova times of ~540 Ma (Suuroja and Suuroja 2000; Torsvik et al. 1992), in a shallow epicontinental sea as a result of the impact of an extraterrestrial body of about 400 m in diameter and possibly of iron meteorite composition (Suuroja and Suuroja 2000). The true depth of the structure, and composition of the lowermost part of crater fill, are unknown. After the impact the crater was submerged, filled with siliciclastic deposits, and buried. The structure remained buried until the Paleo- or Neogene, when it was partially exposed by water and glacier erosion.

The geological maps reveal the outlines of structural elements (crater, rim wall, ejected blocks, ring fault) of this partially buried structure. On the bedrock geological map (Fig. 2) the structure is expressed as an extraordinary circular pattern of outcrops. Deformed and uplifted crystalline basement rocks represent the rim wall. A circular plateau of the Ordovician limestones, the Neugrund Bank, overlies the crater. Small hillock-like outcrops of the crystalline basement rocks on the sea floor around the crater rim, some hundred metres in diameter, probably represent ejected megablocks. The tens of metres high escarpments in the target rocks follow the ring fault. The boundary of the North Estonian Limestone Plateau follows the southern part of the ring fault to a certain extent.

The Quaternary deposits (Fig. 4a) above the rim wall and crater proper are largely eroded. About 10 - 40 m of glaciolacustrine (varved) clays fill the 200 - 400 m wide and up to 80 - 100 m deep circular canyon between the Neugrund Bank and the crystalline rim wall. The canyon is of erosional origin and formed during pre-Quaternary times. The deeper (over 50 m) sea floor around the crater is covered with mud of the present Baltic Sea (Limnea Sea).

Magnetic lows can be often seen in impact structures, like the Lake Saint Martin crater (Grieve and Pesonen 1992), the complex Lake Lappajärvi crater (Elo et al. 1992), Kärdla and Tvären, which are very similar to Neugrund in other aspects as well (Puura and Suuroja 1992; Ormö and Blomqvist 1996), Lockne (Sturkell and Ormö 1997), Mien (Henkel 1983), Söderfjärden (Abels et al. 1998), and Dellen (Henkel 1992). The magnetic minimum is, however, most obvious where the structure is developed in highly magnetic crystalline bedrock.

The Kärdla impact crater (4 km in diameter), partly exposed in Precambrian target rocks, is located ~60 km to the south-west from Neugrund, in northern Hiiumaa, and has a distinct circular aeromagnetic minimum (Fig. 5) of -150 nT. Inland magnetic measurements have revealed a circular minimum with an amplitude of –200 nT, surrounded by a narrow maximum with an amplitude of +100 nT (Suuroja et al. 1974). Models by Plado et al. (1996) show that the magnetic contrast between the non- or weakly magnetic crater fill and unshocked target rocks plays a leading role in producing the magnetic low. The rim wall, where the Precambrian crystalline basement is uplifted up to 200 m above its regional level, is the cause of the circular pattern of the positive anomalies. We can see almost analogous effects in the potential fields of the Neugrund and Kärdla structures due to the comparable sizes, more or less similar impact conditions and target lithologies. Here the magnetic modelling suggests an approximate crater depth of 700 m. The negative magnetic anomaly is mostly due to weakly magnetic crater fill. The post-impact oxidation of magnetite to hematite (Henkel 1992) in parautochthonous breccias could also contribute to this minimum.

The central area of the magnetic low of larger craters often contains high-amplitude, short-wavelength anomalies. These are suggested (e.g., Pilkington and Grieve 1992) them to be caused by near-surface impact-melt bodies with strong remanence, and by bodies of unaltered magnetic target rocks within the brecciated region. Most of the marine-targeted Fennoscandian impact structures do not consist of coherent impact melt bodies; in agreement with study by Kieffer and Simonds (1980). Lack of short-wavelength high-amplitude magnetic anomalies observed in several melt bearing structures (e.g., Dellen, Henkel 1992; Söderfjärden, Abels et al. 1998), suggests that Neugrund does not host any coherent highly magnetic melt body. However, if any melt is present, it does not cause specific anomalies due to low magnetisation, as observed, e.g., in Lappajärvi (Elo et al. 1992).

Many terrestrial impact craters show a decrease in density of impact-related rocks relative to the target (e.g., Lappajärvi, Elo et al. 1992; Clearwater West, Plante et al. 1990). The density of Neugrund rocks is variable, but shows also a decrease with respect to the target rocks (see Table 1). However, there should exist a significant gravity contrast between the metamorphic rocks and the infill of structure (e.g., Plado et al. 1996). Thus, a composite geophysical model basing on the gravity and magnetic data should prove more realistic. Unfortunately no gravity data for the Neugrund area are currently available. According to the estimates of Pilkington and Grieve (1992), one can be expect a Bouguer anomaly of 6 - 7 mGal.

The Neugrund impact structure was discovered during integrated geological mapping. Despite the structure's age, most of the morphological elements are well preserved. The structure is easily recognisable on geological and geophysical maps.

Acknowledgements

We are grateful to T. Flodén (Stockholm University) and the crew of the r/v "Strombus" for help in organising the marine expedition and carrying out marine geophysical investigations. We also thank V. Mäss (Estonian Maritime Museum) and the crew of the r/v "Mare" for assistance in investigating submarine targets. Special thanks go to our colleague from the Geological Survey of Estonia, A. Talpas, for assistance and useful suggestions. We are grateful to E. Sturkell and H.

Dypvik for reviews of the manuscript and constructive suggestions for improvement. The Geological Survey of Estonia, under the auspices of the state geological mapping programme and the Nordic Energy Research Programme, supplied funding. The IMPACT Programme of the European Science Foundation is thanked for help as well.

References

Abels A, Pesonen LJ, Deutsch A, Bischoff L, Lehtinen M (1998) A geographic information system-based characterization of the Söderfjärden impact structure, western Finland [abs]. Meteoritics & Planetary Science 33: A8

Elo S, Jokinen T, Soininen H (1992) Geophysical investigations of the Lake Lappajärvi impact structure, western Finland. Tectonophysics 216: 99-109

Flodén T (1980) Seismic stratigraphy and bedrock geology of the Central Baltic. Stockholm Contributions in Geology 35: 1-240

Flodén T (1981) Current geophysical methods and data processing techniques for marine geological research in Sweden. Stockholm Contributions in Geology 37/5: 49-66

Grieve RAF, Pesonen LJ (1992) The terrestrial impact cratering record. Tectonophysics 216: 1-30

Grieve RAF, Pesonen LJ (1996) Terrestrial impact craters: Their spatial and temporal distribution and impact bodies. Earth, Moon and Planets 72: 357-376

Henkel H (1983) Combined gravity, magnetic and petrophysic interpretation of an impact structure [abs]. Geoexploration 21: 287

Henkel H (1992) Geophysical aspects of meteorite impact craters in eroded shield environment, with special emphasis on electric resistivity. Tectonophysics 216: 63-89

Kala E, Eltermann G (1969) The integrated geological mapping of Western Estonia (sheet O-35-XII) at a scale of 1 : 200 000. Two maps with explanatory notes. Report of Investigation (in Russian). Geological Survey of Estonian SSR, Tallinn, 311 pp

Kieffer SW, Simonds CH (1980) the role of volatiles and lithology in the impact cratering process. Reviews of Geophysics and Space Physics 18: 143-181

Koistinen T, Klein V, Koppelmaa H, Korsman K, Lahtinen R, Nironen M, Puura V, Saltykova T, Tikhomirov S, Yanovskiy A (1996) Paleoproterozoic Svecofennian orogenic belt in the surroundings of the Gulf of Finland. In Koistinen T (ed): Explanation to the Map of Precambrian basement of the Gulf of Finland and surrounding area 1 : 1 000 000. Geological Survey of Finland, Special Paper 21: 21-57

Koppelmaa H, Kivisilla J (2000) Geological Map of the Crystalline Basement of Northern Estonia. Scale 1 : 400 000. Explanation to the map. Report of Investigation. Geological Survey of Estonia, Tallinn, 33 pp

Korhonen J, Koistinen T, Elo S, Säävuori H, Kääriäinen J, Nevanlinna H, Aaro S, Haller LÅ, Skilbrei JR, Solheim D, Chepik AK, Zhdanova L, Vaher R, All T, Sildvee H (1999) Preliminary magnetic and gravity anomaly maps of the Fennoscandian shield 1 : 10 000 000. Geological Survey of Finland Special Paper 27: 173-179

Lutt J, Raukas A (1993) Geology of the Estonian Shelf (in Estonian with English summary). Academy of Sciences of Estonia, Tallinn, 192 pp

Malkov B, Kiipli T, Rennel G, Tammik P, Dulin E (1986) The regional geological-geophysical investigation of the Baltic Sea shelf area Estonian SSR at a scale of 1 : 200 000. Report of Investigation (in Russian). Geological Survey of Estonian SSR, Tallinn, 199 pp

Metlitckaya VI, Papko AM (1992) The results of aeromagnetic mapping by working group No 49 at scales of 1 : 25 000 and 1 : 50 000 on Estonian territory 1987 - 1991. Report of Investigation (in Russian). Geological Survey of Byelorussia, Minsk, 75 pp

Nironen M (1997) The Svecofennian orogen: a tectonic model. Precambrian Research 86: 21-44

Ormö J, Blomqvist G (1996) Magnetic modelling as a tool in the evaluation of impact structures, with special references to the Tvären Bay impact crater, SE Sweden. Tectonophysics 262: 291-300

Pesonen LJ, Torsvik TH, Elming S-A, Bylund G (1989) Crustal evolution of Fennoscandia - paleomagnetic constraints. Tectonophysics 162: 27–49

Pike RJ (1985) Some morphologic systematics of complex impact structures. Meteoritics 20: 49-68

Pilkington M, Grieve RAF (1992) The geophysical signature of terrestrial impact craters. Reviews of Geophysics 30: 161-181

Plado J, Pesonen L, Elo S, Puura V, Suuroja K (1996) Geophysical research on the Kärdla impact structure, Hiiumaa Island, Estonia. Meteoritics & Planetary Science 31: 289-298

Plante L, Seguin M-K, Rondot J (1990) Etude gravimétrique des astroblémes du Lac á l'Eau Clarie, Nouveau-Quebec. Geoexploration 26: 303-323

Puranen R (1989) Susceptibilities, iron and magnetite content of Precambrian rocks in Finland. Geological Survey of Finland, Report of investigation 90: 37 pp

Puranen R, Sulkanen K (1985) Technical description of microcomputer-controlled petrophysical laboratory. Open File Report Q15/27/85/1, Department of Geophysics, Geological Survey of Finland, Espoo, 252 pp

Puura V, Suuroja K (1992) Ordovician impact crater at Kärdla, Island Hiiumaa, Estonia. Tectonophysics 216: 143-156

Puura V, Vaher R (1983) Density and Magnetic Properties of the Rocks (in Russian with English summary). In: Viiding H (ed) The Crystalline Basement of Estonian Territory. Moscow, Nauka, pp. 181-194

Sturkell EFF, Ormö J (1997) Magnetometry of the marine, Ordovician Lockne impact structure, Jämtland, Sweden. Journal of Applied Geophysics 38: 195-207

Suuroja K (1996) The geological mapping as a source of geological discoveries. Geological mapping in Baltic states. Newsletter 2: 19–22

Suuroja K, Saadre T (1995) Gneiss-breccia erratic boulders from Northwestern Estonia as witnesses of an unknown impact structure (in Estonian). Bulletin of Geological Survey of Estonia 5/1: 26-28

Suuroja K, Suuroja S (2000) Neugrund structure - the newly discovered submarine early Cambrian impact crater. In: Gilmour I, Koeberl C (eds) Impacts and the Early Earth. Lecture Notes in Earth Sciences, Springer Verlag, Berlin-Heidelberg 91: 389-416

Suuroja K, Gromov O, Kala E, Gromova G (1974) Report of prospecting for crystalline basement uplift in the eastern part of Hiiumaa Island and estimation of their suitability for producing granite splinters (in Russian). Report of investigation. Geological Survey of Estonia, Tallinn, 225 pp

Suuroja K, Koppelmaa H, Kivisilla J, Niin M (1987) The deep geological mapping of the Nõva-Haapsalu area at a scale of 1 : 200 000. Three maps with explanatory note (in Russian). Geological Survey of Estonian SSR, Tallinn, 210 pp

Suuroja K, Suuroja S, Puurmann T (1997) Neugrund structure as an impact crater (in Estonian with English summary). Bulletin of Estonian Geological Society 2/96: 32–41

Suuroja K, Kadastik E, Ploom K (1998) Geological mapping of Northwestern Estonia at a scale of 1 : 50 000. Three maps with explanatory notes (in Estonian with English summary). Geological Survey of Estonia, Tallinn

Suuroja K, Saadre T, Kask J (1999a) Geology of Osmussaar Island. Estonia Maritima 4: 39-64

Suuroja K, Suuroja S, Talpas A (1999b) The marine geological investigation of the Neugrund crater area. Three maps with explanatory notes (in Estonian). Geological Survey of Estonia, Tallinn, 180 pp

Talpas A, Väling P, Kask J, Mardla A, Sakson M (1993) The geological mapping of the shelf area of the Baltic Sea at a scale of 1 : 200 000. Report of investigation (in Russian). Geological Survey of Estonia, Tallinn, 152 pp

Torsvik TH, Smethurst MA, Van der Voo R, Trench A, Abrahamsen N, Halvorsen E (1992) Baltica. A synopsis of Vendian-Permian palaeomagnetic data and their palaeotectonic implications. Earth-Science Reviews 33: 133-152

Natural Resources of the Kärdla Impact Structure, Hiiumaa Island, Estonia

Kalle Suuroja

Geological Survey of Estonia, Kadaka tee 82, 12168 Tallinn, Estonia. (k.suuroja@egk.ee)

Abstract. The about 455 Ma (Late Ordovician, Caradoc period) Kärdla impact structure on Hiiumaa Island, Estonia (58°59´N; 21°47´E), has a 4 km rim-to-rim diameter and was formed in a shallow epicontinental sea. Several natural resources are spatially related to the Kärdla structure: 1) at the rim wall crystalline building stone is found; 2) above the rim wall chemically pure micritic (cryptocrystalline) Ordovician limestones crop out; 3) in the north-eastern part of the crater, on the slopes of the crater rim, chemically pure organodetrital (cystoid) grainstones are exposed; 4) on the slopes, dolomitized limestones of the crater rim contain galena and sphalerite mineralization is discovered; 5) outside the crater, hydrocarbons in the form of diffused patches in limestones, liquid oil in pores, asphalt and asphaltite in fissures and pores are widespread; 6) within the crater, at depths of 270 - 815 m, mineral water of Cl–Na–Ca composition with a total content of dissolved salts of 2 - 6 gl^{-1} is found; 7) around and inside the crater, the Ordovician artesian basin is directly related to the structures of the buried impact crater; 8) in the crater proper and in the surrounding area, glaciolacustrine clays are distributed; 9) on the south-eastern slope of the crater rim there are considerable reserves of gravel and sand.

Except the crystalline building stone, these resources were formed after the impact. They are largely related to the morphological features caused by the impact. Sulphide precipitation and hydrocarbon impregnation were facilitated by rock jointing and brecciation caused by the impact. The processes directly related to the impact have not induced any ore mineral concentrations.

1 Introduction

The about 455 Ma (Late Ordovician, Caradoc) Kärdla impact structure (Hiiumaa Island, Estonia: 58°59´N; 21°47´E) has a rim-to-rim diameter of 4 km and formed in a shallow epicontinental sea (Puura and Suuroja 1992). At that time the Baltica Continent was situated in the Southern Hemisphere at a latitude of 30° (Torsvik et al. 1992) in a warm climate region. Bioclastic carbonate sedimentation took place in the region. The Kärdla impact caused the formation of an about 12 km wide atoll-like island, the so-called Proto-Hiiumaa (Suuroja 1996). Immediately after the impact the crater was partially filled with slump and resurge deposits. Subsequently, anomalously rapid calcareous sedimentation continued and a more than 90 m thick layer of limestone was formed in the crater (Suuroja et al. 1994), whereas, it is only 0.5 m in the surrounding area. In the surroundings, bioclastic calcareous muds were deposited, but in the crater proper more argillaceous sediments (i.e. silt, clay, marl) accumulated. At the end of Caradoc (Rakvere age), the impact structure was completely buried. However, even today the buried structure is recognizable by a subdued morphological expression.

The natural resources discovered in the Kärdla structure area are of local significance and consist of chemically pure limestones, sand and gravel deposits, clay for ceramic industry, mineral and fresh groundwater, occurrences of base metal (lead and zinc) sulphides and hydrocarbons. Sulphide deposits and groundwater resources are of specific interest. Limestone quarries near Paluküla village were operated in the last century (Viiding et al. 1969; Kala et al. 1971). The discovery of a structural uplift of the crystalline basement (in the crater rim) (Viiding et al. 1969) initiated the search for crystalline building stone (Suuroja et al. 1974; Kala et al. 1976). For the search of mineral water (Suuroja et al. 1991) deep drilling was performed. Hydrocarbons in core samples in the crater area were found in the course of integrated geological mapping (Suuroja et al. 1994; Kattai et al. 1994). Quaternary accumulation of sand, gravel and clay has been due to crater-related morphological features of the Palaeozoic bedrock.

2
Crystalline Building Stone

In Estonia there is deficiency of high-quality crystalline building stone, especially of an aggregate made from it. Therefore, the uplifted crystalline basement, consisting mainly of metamorphic and igneous rocks (58 % migmatite granites and granite gneisses, and 42 % amphibolites and biotite amphibole gneisses) and lying at 15 - 30 m from the surface, caused practical interest. In the course of searching for and exploring the areas for mining of crystalline building stone (Suuroja et al. 1974; Kala et al. 1976), 53 holes penetrating into the basement rocks were drilled on the buried crater rim wall and in surroundings. The physical properties of the rock and crushed aggregate were investigated. The average aggregate crushing value of crystalline rocks is 9.3 %. The content of flaky particles does not exceed 12 %, the average being 7 %. The preliminary estimated reserves of crystalline building rock of the Paluküla deposit, calculated in an area of 0.43 km^2 at an average thickness of the commercial bed of 58 m, are 22×10^6 m^3.

Due to the low quality of the explored building stone and unfavourable environmental conditions, the mining plans were abandoned. However, the study of the drill cores and geophysical investigations made it possible to clear up the details of the rim wall structure. The rim wall, at least in the Paluküla sequence, consists of the uplifted (up to 240 m) and fractured crystalline basement rocks. These are similar to the basement rocks in the surroundings of the crater. Huge (diameter more than 50 m) blocks of the basement rocks lie within the deformed Cambrian sand- and siltstones in the outer side of the rim.

3 Limestones

Above the buried rim wall near Paluküla village, in an area of several square kilometres, a more than 10 m thick layer of the Upper Ordovician micritic (cryptocrystalline) limestone of the Saunja Formation corresponding to the upper part of the Nabala Stage crops out (Fig. 1). From the analyses of 22 samples, the

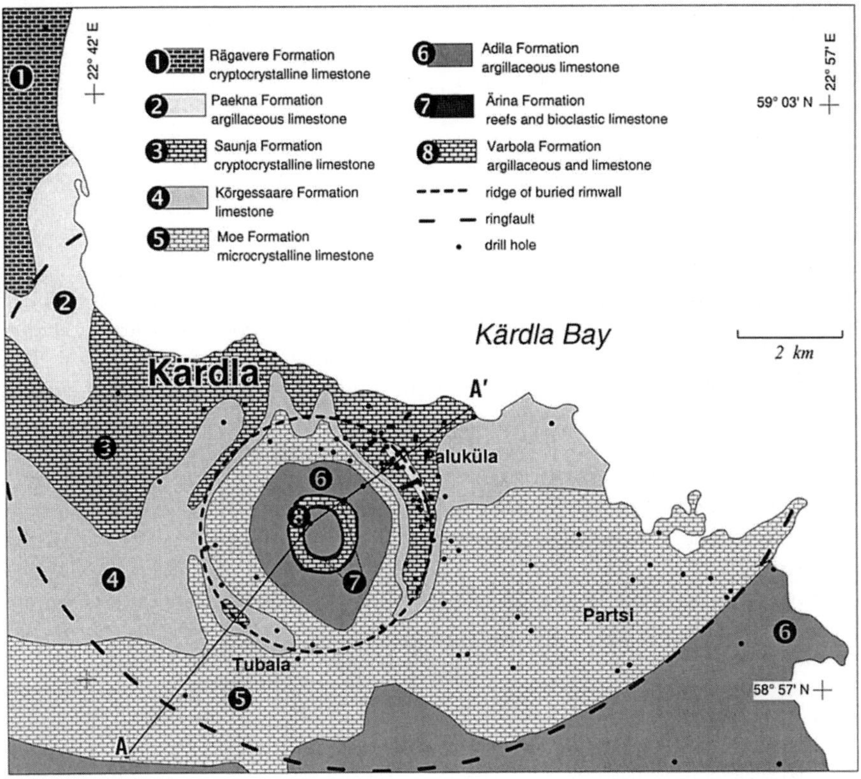

Fig. 1. Bedrock geological map of the Kärdla impact structure (modified after Suuroja et al. 1994).

average chemical composition of the limestone is: CaO - 52.8 % by weight, MgO - 1 %, Al_2O_3 - 1.45 %, Fe_2O_3 - 0.61 %, SiO_2 - 1.8 %, SO_3 - 0.6 %, P - 0.01 % (Kala et al. 1976). The preliminarily estimated reserve of the limestone is 10^6 m^3. Due to post-impact uplift and erosion of the area the limestone is available for mining.

Above the north-eastern part of the buried rim wall, on its inner and outer slopes, the oldest known reefs of the Baltica Continent, together with the surrounding facies of organodetrital (cystoid) limestones (grainstones), are distributed (Fig. 2). The thickness of the grainstones is 10 m and they are chemically pure. The average composition of 6 samples analysed is: CaO -53.4 % by weight, MgO - 0.8 %, Al_2O_3 - 1.76 %, Fe_2O_3 - 0.48 %, SiO_2 - 1.6 %, SO_3 - 0.4 %, P - 0.01 % (Kala et al. 1976). Because of their outstanding purity these grainstones could be used in the chemical, pulp and paper and metallurgical industries (Teedumäe 1997). In the past, microcrystalline organodetrital limestone was extracted in small quarries in the surroundings of the Kärdla and Paluküla villages. It was used as a building stone and for aggregate and lime production.

4 Sulphide Mineralization

On the outer slopes of the buried rim wall, near the villages of Paluküla and Tubala (Figs. 2 and 3), galena and sphalerite mineralization was discovered (Suuroja et al. 1994). Near Paluküla village, in a 4 - 5 m thick layer within the strongly dolomitized Ordovician reef limestones overlying impact-brecciated Cambrian siliciclastic target rocks, the content of lead and zinc may rise up to 10 %. Near Tubala village in drill hole F-178 (Fig. 2), within a 4 m thick limy sandstone layer covering the ejected impact breccias, the content of lead and zinc rises to 1 - 2 wt%. The spatial extent of these mineralizations is not known. The age of the formation of the sulphide mineralization is also unknown yet. However, it is post-Ordovician as the ore-bearing veins pass through the entire section of the Ordovician limestones (Fig. 4). The mineralization event is not genetically related to the impact origin of the Kärdla structure or to the post-impact hydrothermal activity possibly evoked by it, because the impact occurred at least ten million years prior to the mineralization event (Suuroja et al. 1994). The injection of the ore-bearing fluids has been a multiple process in which two separate phases can be observed. The first phase was related to extensive dolomitization of the base of the limestone section on the slopes of the rim wall, the second immediately to ore mineralization. The first phase embraced a larger space than the second one (Fig. 4). The sulphides in the buried rim were precipitated mainly in high-porosity impact fractured rocks and impact breccias and an anticline structure of the post-impact limestone on top of the uplifted rim.

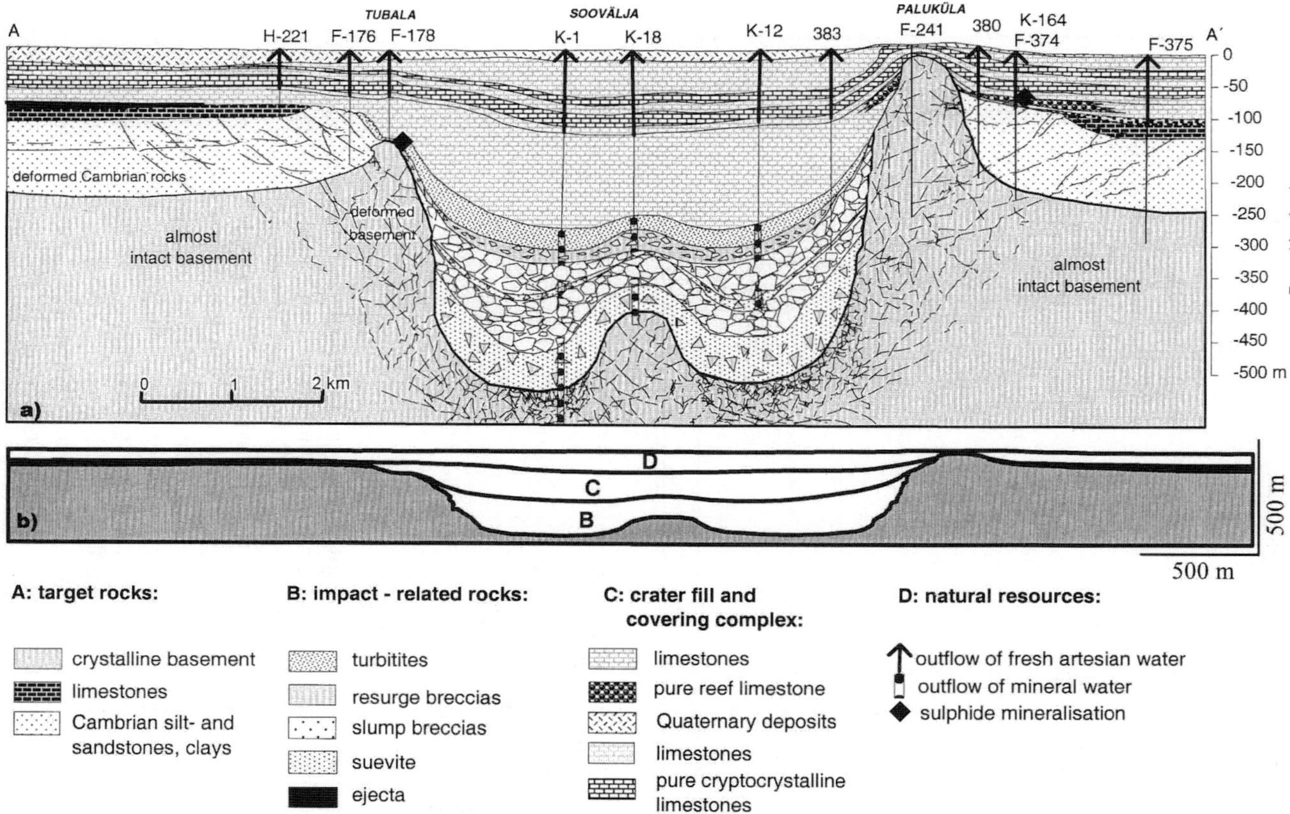

Fig. 2. a) Cross-section A–A´ of the buried Kärdla impact crater and distribution of mineral resources and groundwater. The vertical scale is exaggerated ~7 times. b) Distribution of complexes; cross-section without exaggeration. For location of A–A´ see Fig. 1.

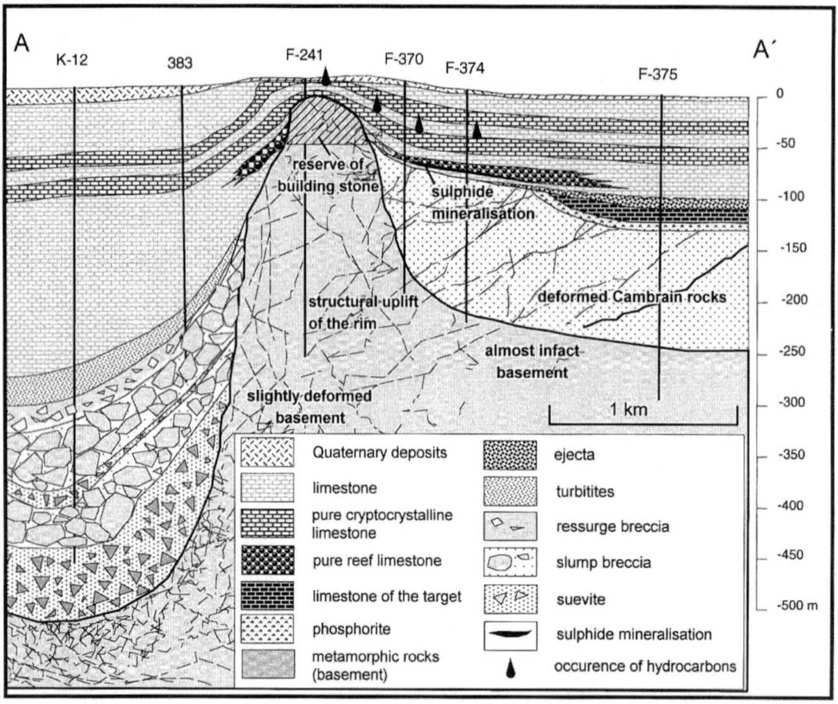

Fig. 3. Schematic geological cross-section of the rim of the Kärdla crater and distribution of the natural resources.

5 Hydrocarbons

In the uplifted section of Ordovician limestones of the crater area occurrences of hydrocarbons, such as diffused patches of oil in limestones, liquid oil in pores, asphalt and asphaltite (solid asphalt) in fissures and pores, are widespread (Suuroja et al. 1994; Kattai et al. 1994). The argillaceous limestones are commonly with patches of diffused hydrocarbons (oil), while in the pores and fissures of the purer limestone oil, asphalt and asphaltite occur (Fig. 3). As a rule, the depth of the distribution of hydrocarbons does not exceed 50 m. In deeper sedimentary and crystalline rocks hydrocarbons are very rare. The hydrocarbons of the Kärdla crater area are believed to have migrated to where they have been discovered. Their absence inside the crater proper and south-eastward of it (in the so-called shade of the Kärdla crater) seems to support that opinion (Suuroja et al. 1994). The uplifted, fractured and porous limestones above the rim wall have served as a trap for the hydrocarbon flow.

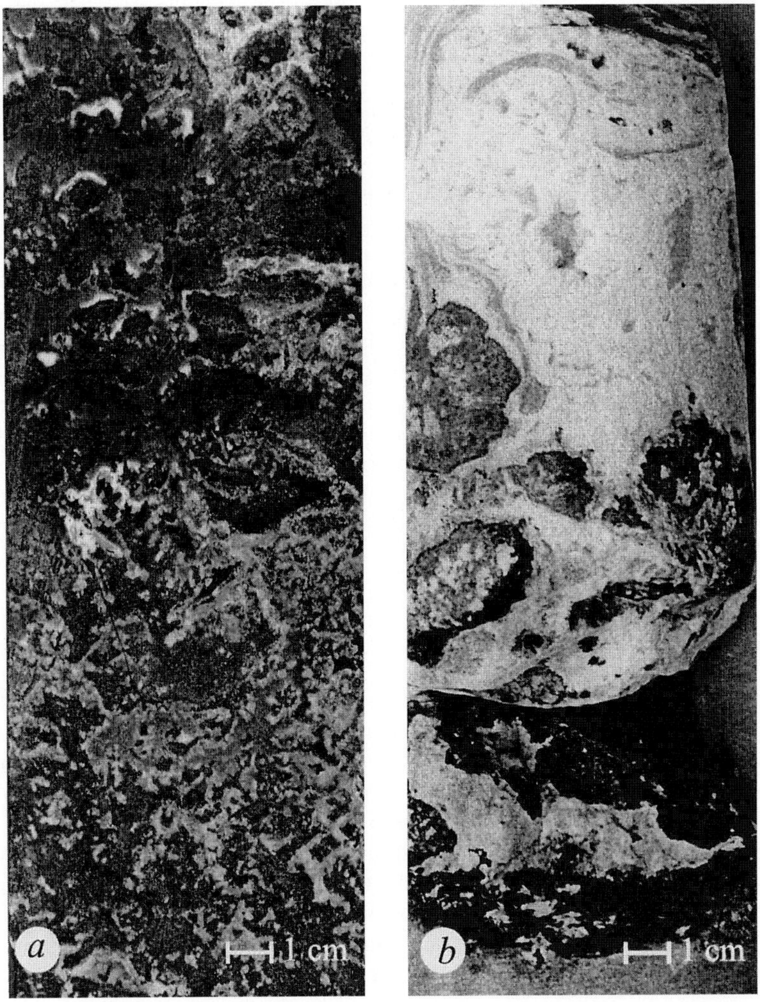

Fig. 4. Two specimens of the galena-sphalerite ore of the Kärdla crater area from drill core F-374: a) massive ore at a depth of 45.0 m; b) ore from the vein at a depth of 42.0 m. White and light grey - calcite; dark grey - sphalerite; black - galena.

6
Mineral and Fresh Groundwater

The buried Kärdla impact structure has influenced the regional groundwater flow and induced the formation of mineral and fresh groundwater basins (Figs. 2 and 3).

6.1 Mineral Water

The groundwater in which the content of total dissolved mineral salts (TDS) is 2 gl^{-1} or more is rated as mineral water in Estonia (Karise 1997). Within the Kärdla crater deep, in the section of turbidites and impact breccias at depths of 270 - 815 m, Cl-Na-Ca enriched mineral water has been discovered (Suuroja et al. 1991). TDS of the mineral water ranges from 2 - 6 gl^{-1} and grows with depth. In the surrounding area (outside the crater rim), TDS of groundwater of Cl-HCO_3-Na-Ca composition in the aquifer in Cambrian terrigenous rocks is approximately 0.5 gl^{-1}; that of HCO_3-Cl-Ca-Na water in the aquifer in Silurian–Ordovician carbonate rocks is 3 -6 gl^{-1}. Obviously the small (some km^3) groundwater basin within the crater is separated from the regional aquifer of Cambrian and Silurian-Ordovician rocks by the crater rim and does not take part in the general groundwater flow. The reserve of mineral water is estimated at about 120,000 m^3 (Suuroja et al. 1994), but the probable reserve is greater. Mineral water from the Kärdla crater (pumped from drill hole K-18, situated above the central uplift) was for many years bottled and sold as curative mineral water "Kärdla". Production ceased in 1990 for economic reasons.

6.2 Fresh Water

The buried structures of the Kärdla impact crater influence the Silurian–Ordovician artesian aquifer around and inside the impact crater. The good-quality artesian groundwater (TDS <0.3 gl^{-1}, HCO_3-Cl-Ca-Na composition) occurs in the area where the lower-lying bedrock (5 - 10 m b.s.l.) is covered by a 5 - 15 m thick layer of Quaternary glaciolacustrine deposits. The recharge of the artesian aquifer takes place 0.5 - 3 km from the discharge area, at sites where only a thin layer of water permeable soil covers the uplifted and fractured limestones above the crater rim. The velocity of groundwater movement in this aquifer, formed of fissured limestones, is up to 1,000 m per day. On the discharge area there are numerous wellsprings of natural and artificial origin.

7 Quaternary Deposits: Gravel, Sand and Clays

The deposition of Quaternary deposits has been influenced by the structures of the buried crater (Fig. 5). Above the crater rim the thickness of the Quaternary deposits is 0.5 - 2.0 m while in the crater proper it reaches up to 25 m. The Quaternary deposits contain sand and gravel and glaciolacustrine clays.

The explored reserves of the gravel and sand deposits are about 10^6 m^3. The Partsi gravel pit supplies almost the whole of Hiiumaa Island with local building materials.

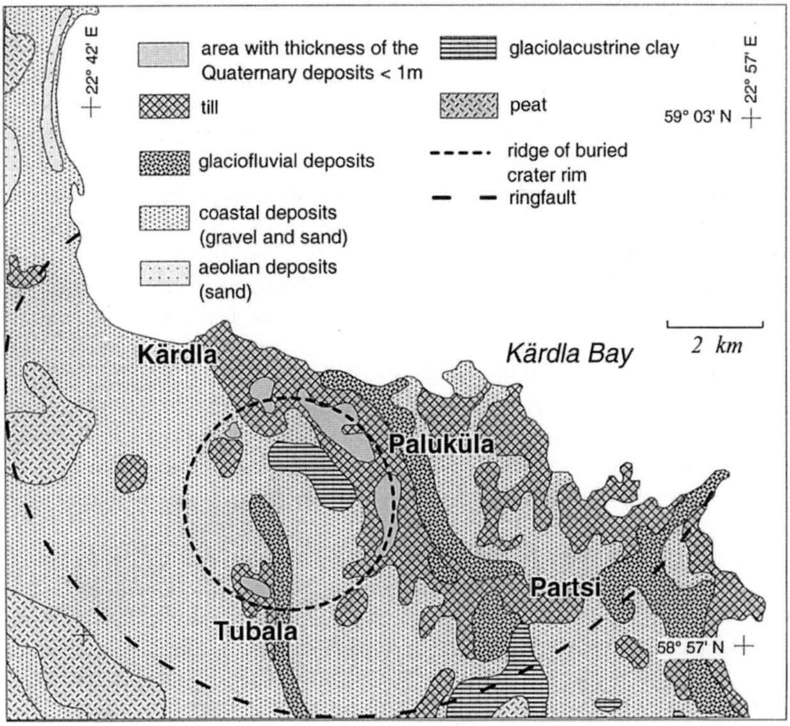

Fig. 5. Map of the Quaternary deposits of the Kärdla impact structure area (modified after Suuroja et al. 1994).

Glaciolacustrine clay deposits are widespread in the crater proper, where they are up to 15 m thick. Almost everywhere a 0.1 - 4.0 m thick layer of marine sand covers these. Rhythmic alteration of summer and winter layers (varves), which is very characteristic of glaciolacustrine clays, is practically absent in the clays of the crater area. Conditions responsible for the deposition of the clays observed in the Kärdla crater area existed in the Baltic Ice Lake about 10,000 years ago (Pirrus 1997). In the area covered with sand less than 1.0 m, the explored reserves of glaciolacustrine clays are 4×10^6 m^3. The clays are considered as a potential raw material for the ceramic industry.

8 Discussion

A considerable number of terrestrial impact structures have significant mineral resources associated with them. Prime examples are the Sudbury Structure in

Canada with its world-famous Cu-Ni-PGE deposits and the Vredefort Structure in South Africa with its Au deposits of the Witwatersrand rocks. A voluminous literature exists describing these and other deposits related to meteorite impacts (Masaitis et al. 1980; Melosh 1989; Henkel and Pesonen 1992; Deutsch and Schärer 1994; Grieve and Pesonen 1996; Koeberl 1997; Pesonen et al. 1999). In the Kärdla structure, the industrial mineral deposits, the sulphide occurrences and the water resources are not directly related to the impact processes but largely to post-impact depositional and morphological features. Sulphide minerals were deposited in fractured rocks. Post-impact hydrothermal alteration directly related to the impact has been observed in the impact-influenced target rocks and impact breccias (Puura et al. 1996), but it did not lead to the formation of the sulphide mineralization described here.

According to Korja (1995), the thickness of the crust in the Kärdla crater area is about 46 km and it probably was about the same at the time of the impact. The influence of the impact, judge by the alteration of physical properties of crystalline rocks, reached at least 500 m below the true floor (Plado et al. 1996). The distal influence, judge by the disappearing of primal fluid inclusions of the crystalline target rocks, could reach up to 10 km from the impact centre (Suuroja 1996).

Some authors (Zhukov et al. 1987) believe that the isotopic composition of sulphur of the galena of sulphide mineralization found at Kärdla suggests a crustal origin of ore-bearing fluids. Thus they have come to a conclusion favouring a cryptovolcanic origin of the Kärdla structure. However, a closely similar isotopic sulphur composition is characteristic of almost all sulphide occurrences in sedimentary rocks of Estonia. Sulphur from the post-impact sulphides may be of crustal origin but this does not confirm an endogenic origin of the Kärdla crater.

Hydrocarbon fluids observed in some other impact structures (Kieffer and Simonds 1980; Komor et al. 1988; Andersen and Burke 1996; Boer et al.1996; Grieve and Pesonen 1996; Sturkell et al. 1998) have not been recorded in the Kärdla case and no hydrocarbons have been found in fluid inclusions in quartz from the target rocks and impact breccias (Suuroja 1999). As hydrocarbons are absent in the crater and south-eastward of the crater (the shadow of the Kärdla crater), it is believed (Suuroja et al. 1994) that the hydrocarbon flow was from north-west to southeast.

The concentration of mineral resources is always connected with somewhat anomalous conditions, just like those had been in the Kärdla case. The existence of anomalous structures (crater depression, uplifted rim wall) in the basin caused anomalies in the thickness and composition of accumulated deposits. That was the case with the pure Ordovician limestones as well as with the Quaternary deposits (sand and gravel deposits, glaciolacustrine clay). At Kärdla the structural control was an important factor contributing to the formation of natural resources.

Acknowledgements

We wish to thank our colleagues at the Geological Survey of Estonia E. Kala, G. Elterman, V. Puura, V. Petersell, M. Niin, H. Koppelmaa, and E. Tammiste for assistance in carrying out our long-term investigations in the Kärdla crater area. Special thanks belong to the reviewers B. Dressler and V. Puura, whose useful comments essentially improved the manuscript. The Impact Programme of the European Science Foundation and the Geological Survey of Estonia are thanked for help.

References

Andersen T, Burke EAJ (1996) Methane inclusions in shocked quartz from the Gardnos impact breccia, south Norway. European Journal of Mineralogy 8: 927-936

Boer RH, Reimold WU, Koeberl C, Kessler SE (1996) Fluid inclusion studies on drill core samples from the Manson impact crater: Evidence for post-impact hydrothermal activity. In: Koeberl C, Anderson RR (eds) The Manson Impact Structure, Iowa: Anatomy of an Impact Crater. Geological Society of America Special Paper 302: 377-382

Deutsch A, Schärer U (1994) Dating terrestrial impact events. Meteoritics 29: 301-322

Grieve RAF, Pesonen LJ (1996) Terrestrial impact craters: their spatial and temporal distribution and impacting bodies. Earth, Moon and Planets 72: 553-591

Henkel H, Pesonen LJ (1992) Impact craters and craterform structures in Fennoscandia. Tectonophysics 216: 31-40

Kala E, Kajak K, Kajak H, Elterman G (1971) The integrated geological mapping of Island Hiiumaa at a scale of 1 : 200 000. Map with explanatory note (in Russian). Geological Survey of Estonian SSR, Keila, 624 pp

Kala E, Suuroja K, Tassa V (1976) Report of the prospecting granitic rock of Paluküla (in Russian). Geological Survey of Estonian SSR, Keila, 94 pp

Karise V (1997) Composition and properties of groundwater under natural conditions. In: Raukas A, Teedumäe A (eds) Geology and mineral resources of Estonia. Estonian Academy Publishers, Tallinn, pp 152-156

Kattai V, Lokk U, Suuroja K (1994) The distribution of natural bitumens in Estonia. Bulletin of the Geological Survey of Estonia 4: 12-16

Kieffer SW, Simonds CH (1980) The role of volatiles and lithology in the impact cratering process. Reviews of Geophysics and Space Physics 18: 143-181

Koeberl C (1997) Impact Cratering: The Mineralogical and Geochemical Evidences. Oklahoma Geological Survey Circular 100: 30-54

Komor SC, Valley JW, Brown PE (1988) Fluid inclusion evidence for impact heating at the Siljan Ring, Sweden. Geology 16: 711-715

Korja A (1995) The structure of the Svecofennian crust - growth and destruction of the Svecofennian orogen. Institute of Seismology, University of Helsinki, Helsinki, Report S-31, 36 pp

Masaitis VL, Danilin AN, Mashchak MS, Raikhlin AI, Selivanovskaja TV, Shadenov YM (1980) The Geology of Astroblemes (in Russian). Nedra, Leningrad, 231 pp

Melosh HJ (1989) Impact Cratering: A Geologic Process. Oxford University Press, New York, 245 pp

Pesonen LJ, Lahti M, Kinnunen KA (1999) Meteorite impact stuctures – their economic importance and role of geophysics [abs]. In: Extended Abstract Book, EAGE 61st Conference and Technical Exhibition, Helsinki, Finland. Paper 5-32 Science Conference, 9th, Volume 2, Pergamon, New York, 2659-2664

Pirrus E (1997) Clay. In: Raukas A, Teedumäe A (eds) Geology and mineral resources of Estonia. Estonian Academy Publishers, Tallinn, pp 360-364

Plado J, Pesonen LJ, Elo S, Puura V, Suuroja K (1996) Geophysical research on the Kärdla impact structure, Hiiumaa Island, Estonia. Meteoritics & Planetary Science 31: 289-298

Puura V, Suuroja K (1992) Ordovician impact crater at Kärdla, Hiiumaa Island, Estonia. Tectonophysics 216: 143-156

Puura V, Kirsimäe K, Kivisilla J, Plado J, Puura I, Suuroja K, (1996) Geochemical anomalies of terrestrial compounds in nonmelted impactites at Kärdla, Estonia [abs]. Metoritics & Planetary Science 31: A112-A113

Sturkell EFF, Broman C, Forsberg P, Torssander P (1998) Impact-related hydrothermal activity in the Lockne impact structure, Jämtland, Sweden. European Journal of Mineralogy 10: 589-606

Suuroja K (1996) Formation and development of the Kärdla impact structure (in Estonian). Dissertation, Institute of Geology, University of Tartu, 57 pp

Suuroja K, Gromov O, Kala E, Gromova G, Petersell V, Puura V (1974) Report of the searching uplifts of crystalline basement in the northeastern part of the Island of Hiiumaa (in Russian). Geological Survey of Estonia, Keila, 212 pp

Suuroja K, Koppelmaa H, Niin M, Kivisilla J (1991) Geological mapping of crystalline basement in Island of Hiiumaa. Two maps with explanatory note (in Russian). Geological Survey of Estonia, Keila, 210 pp

Suuroja K, Kadastik E, Mardim T (1994) The geological mapping of the Hiiumaa Island at a scale of 1 : 50 000. 6 maps with explanatory note (in Estonian with English summary). Geological Survey of Estonia, Keila, 206 pp

Suuroja S (1999) Lithologies of within-crater breccias of the Kärdla crater (in Estonian). Dissertation, Institute of Geology, University of Tartu, Tartu, 78 pp

Teedumäe A (1997) Carbonate rocks. In: Raukas A, Teedumäe A (eds) Geology and mineral resources of Estonia. Estonian Academy Publishers, Tallinn, pp 348-356

Torsvik TH, Smethurst MA, van der Voo R, Trench A, Abrahamsen N, Halvorsen E (1992) Baltica. A synopsis of Vendian-Permian palaeomagnetic data and their palaeotectonic implications. Earth-Science Reviews 33: 133-152

Viiding H, Kala E, Pobul E (1969) The Paluküla Riddle Being Solved (in Estonian with English summary). Eesti Loodus 8: 469-474

Zhukov F, Petersell V, Fomin Y (1987) Appearances of the Palaeozoic volcanity in Estonia (in Russian). Proceedings of the Academy of Sciences of Estonian SSR, Geology 36: 6-13

Seismic Correlation of the Mjølnir Marine Impact Crater to Shallow Boreholes

Filippos Tsikalas, Jan Inge Faleide, Olav Eldholm, and Henning Dypvik

Department of Geology, University of Oslo, P.O. Box 1047 Blindern, N-0316 Oslo, Norway. (filippos.tsikalas@geologi.uio.no)

Abstract. We use seismic reflection data to correlate deposits related to the 40-km-diameter Mjølnir marine impact crater to two boreholes; one near the center of the crater and another ~30 km outside the crater periphery. This correlation provides a detailed seismic stratigraphy for the impact-related deposits and post-impact strata. The biostratigraphic 142 ± 2.6 Ma age for the event determined earlier at the base of the post-impact succession should be considered a minimum impact age. This is because considerable thickening of the immediately post-impact deposits in the annular trough implies that the impact event may be slightly older, but still within the Upper Volgian.

1 Introduction

Meteorite impacts produce an array of both shallow and deep structural features in the target rocks (e.g., Melosh 1989; Grieve and Pesonen 1996). The structures from marine impacts have greater preservation potential than subaerial impact structures because the water and post-impact deposits shield the crater from erosion.

The 40-km-diameter Mjølnir impact crater in the Barents Sea (Fig. 1) is exceptionally well-preserved (Tsikalas et al. 1999). The physical impact took place on the sedimentary Bjarmeland Platform (Fig. 1), where the paleo-water depth is estimated to have been 300 - 400 m (Dypvik et al. 1996). At the time of impact, the platform contained upper Paleozoic strata, mainly carbonates and evaporites, overlain by 4-5-km-thick Mesozoic siliciclastic marine sediments (cfr. Fig. 3) (Gudlaugsson et al. 1998).

Two shallow boreholes, one near the center and another ~30 km from the crater periphery (Fig. 2), have confirmed the impact origin of the structure. The boreholes revealed brecciated sediments and a proximal ejecta layer, which contains strong iridium enrichment and shocked quartz grains (Dypvik et al. 1996;

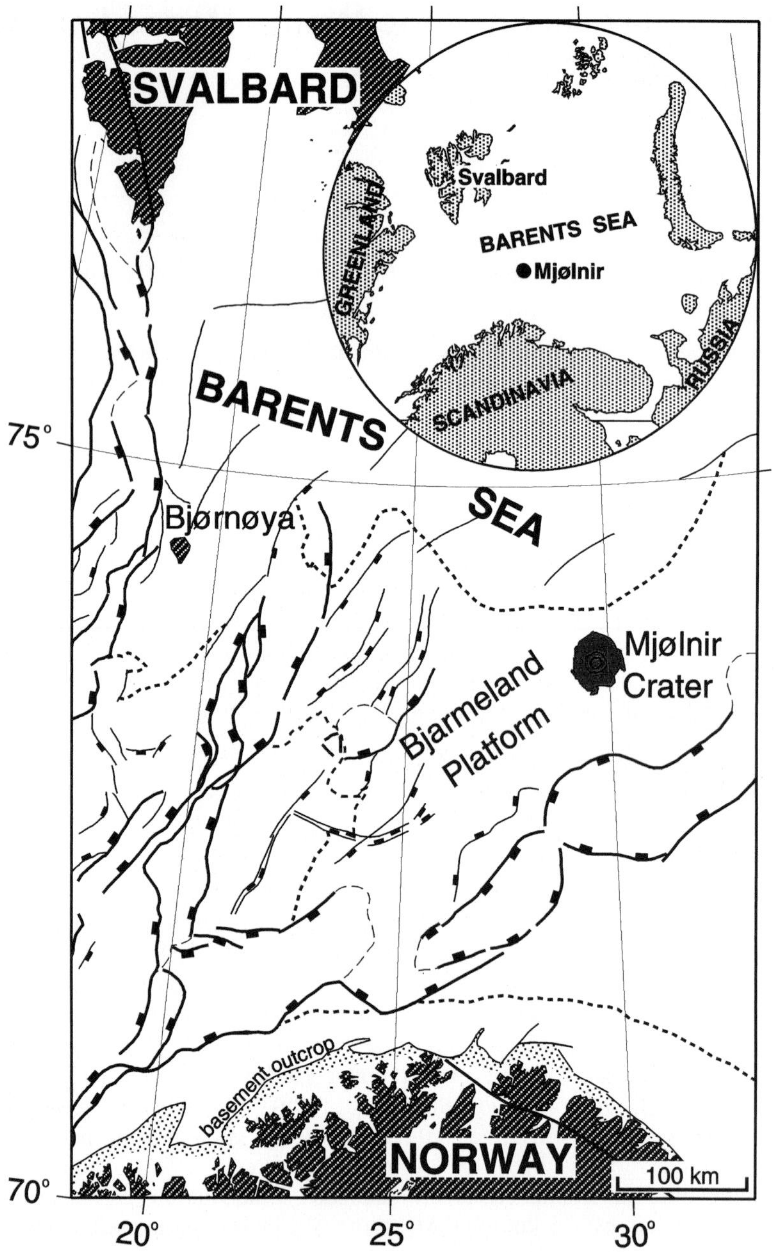

Fig. 1. Mjølnir impact crater and regional Late Paleozoic-Mesozoic structures in the southwestern Barents Sea (Faleide et al. 1993; Gudlaugsson et al. 1998). Thick dashed line denotes subcrop of base Upper Jurassic unconformity. Inset: Mjølnir impact location shown in a Late Jurassic plate reconstruction by Lawver et al. (1999).

Smelror et al. 2001). In this study, we present a detailed seismic reflection correlation to the two boreholes, and discuss the timing of the Mjølnir impact event with reference to new biostratigraphic evidence, particularly from the central borehole (Smelror et al. 2001).

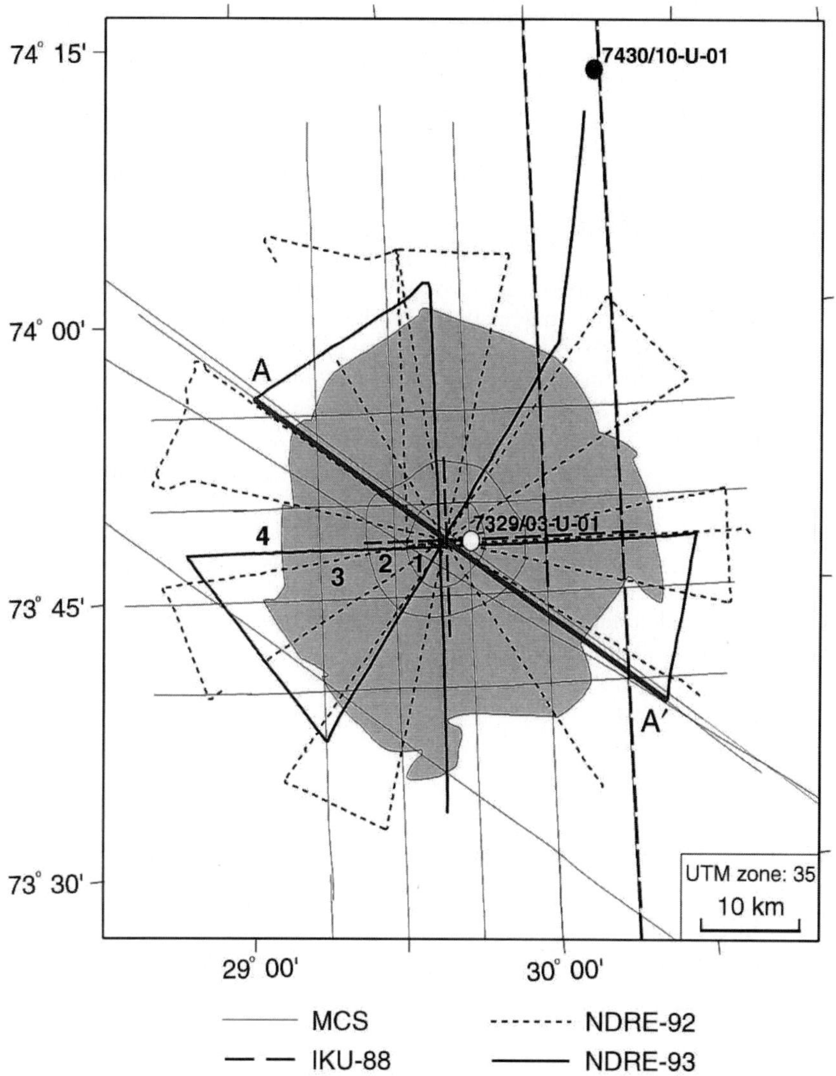

Fig. 2. Seismic reflection profile coverage, the two shallow boreholes, and the impact structure. MCS - Norwegian Petroleum Directorate conventional multi-channel profiles; IKU-88, shallow multi-channel profiles recorded by SINTEF Petroleum Research; NDRE-92/93, high-resolution single-channel profiles recorded by the Norwegian Defense Research Establishment; *1*, central high; *2*, annular basin; *3*, outer zone; *4*, platform.

2
Mjølnir Crater

The seismic investigation of the Mjølnir impact crater is based on an extensive seismic database consisting of three types of reflection profiles: high-resolution single-channel, shallow multi-channel, and conventional multi-channel (MCS) profiles (Fig. 2). The MCS profiles, acquired with a variety of recording parameters and processing sequences, image the deeper levels of the structure and outline specific features related to the cratering processes. Here, we focus on the shallower seismic structure and stratigraphy using both types of shallow reflection profiles, which retain a range of high frequencies and provide the maximum vertical resolution (Table 1).

The seismic reflection data demonstrate the presence of a broad-brimmed, bowl-shaped mass of disturbed seismic reflectivity patterns, which contrast with the uniform, horizontal stratification of the surrounding sedimentary platform (Fig. 3). Seismic mapping and analysis of the 850 - 1400 km^3 disturbed zone (Tsikalas et al. 1998a) provides insight into several cratering processes, such as brecciation, excavation, melting, gravitational collapse of the transient crater, and structural uplift (Table 2). The disturbed zone is bounded on top by a band of prominent reflectors, which define an obvious structural relief (Figs. 3 and 4). At this stratigraphic level the structure exhibits a distinct pattern of radial zonation composed of: (1) a complex outer zone; (2) an annular depression, and; (3) a central high (Figs. 3 and 4; Table 2). Steep boundary faults form a ~150-m-high near-circular rim wall around the internal structural features. Furthermore, the seismic profiles also image post-impact deformation, which is associated with an extensive overburden and is expressed by structural reactivation and differential subsidence (Fig. 4) (Tsikalas et al. 1998b).

Table 1. Mjølnir crater shallow seismic reflection database

AG	TY	SO	RF	SL (km)	RL (twt, s)	SR (ms)	FI (Hz)	VR (m)
NDRE[a] (1992)	single-channel	sparker	analog	0.15	1	1	70 - 500	4 - 8
NDRE[a] (1993)	single-channel	sleeve-gun array	digital	0.15	1	1	70 - 500	4 - 8
IKU[b] (1988)	shallow multi-channel	airgun	digital	1.5	1 - 1.5	2	8 - 180	5 - 10

AG, agency; *TY*, type; *SO*, source; *RF*, recording format; *SL*, streamer length; *RL*, record length; *twt*, two-way travel time; *SR*, sampling rate; *FI*, frequency interval; *VR*, vertical resolution
[a]NDRE, Norwegian Defense Research Establishment, Horten
[b]IKU, SINTEF Petroleum Research, Trondheim

Table 2. Mjølnir crater seismic features and dimensions, after Tsikalas et al. (1998a-c)

Morphological features	Dimensions	Character
Central high	8 km diameter	– rises ~250 m above lower Barremian platform level – top truncated by glacial erosion yielding a circular subcrop surface
Annular basin	4 km width	70 m depression
Outer zone	12 km width	complex, composed of (width, km): – peak ring (1-3) – intermediate zone (3-6.5) – marginal fault zone (2.5-7)
Cratering process features		
Excavated crater and breccia	true crater floor consists of: (1) a 4-to 6.5-km-wide trough with 1-1.3 km depth beneath annular basin; (2) a 0.05-to 0.2-km-thick unit in crater periphery filling in underlying faults blocks	– two types of breccia: *allochthonous* and *autochthonous* *allochthonous breccia*: some reflectivity locally, but mostly characterized by reflection-free zones; its top corresponds to apparent crater floor; its base defines true crater floor *autochthonous breccia*: tilted fault blocks at periphery; turns into disrupted and chaotic refractions, diffractions and reflection-free zones
Impact melts	estimate: ~10-30 km^3 volume	most probably dispersed melts in crater periphery; not yet cored
Gravitational collapse	16 km diameter and 4.5 km deep transient cavity; collapse factor, 2.5	Mjølnir exhibits much larger collapse than typical terrestrial craters (average collapse factor, 1.6)
Structural uplift	1.5-2 km	within range (although on the lower limit) of predicted structural uplift based on empirical relations from the terrestrial impact record

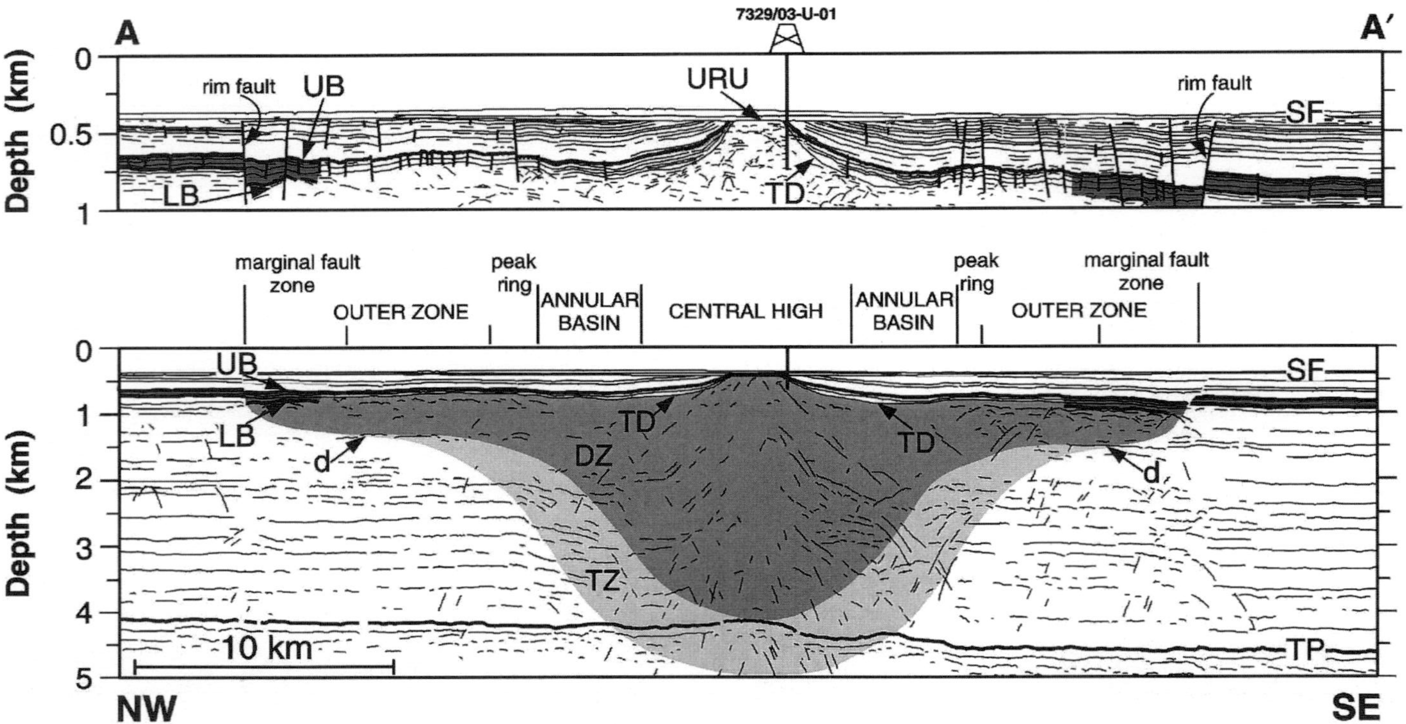

Fig. 3. Seismic type sections along profile AA´ in Fig. 2. Top - Interpreted high-resolution single-channel profile. Bottom - Interpreted multichannel profile. *DZ*, area of intensely disturbed seismic reflections; *TZ*, transitional area of less disturbance. Reflectors UB and LB bound the time of impact. *SF*, sea floor; *URU*, late Cenozoic upper regional unconformity; *UB*, lower Barremian; *TD*, the first continuous reflector above the disturbed seismic reflections; *LB*, base Upper Jurassic; *TP*, top Permian; *d*, low-angle décollement (modified from Tsikalas et al. 1998b).

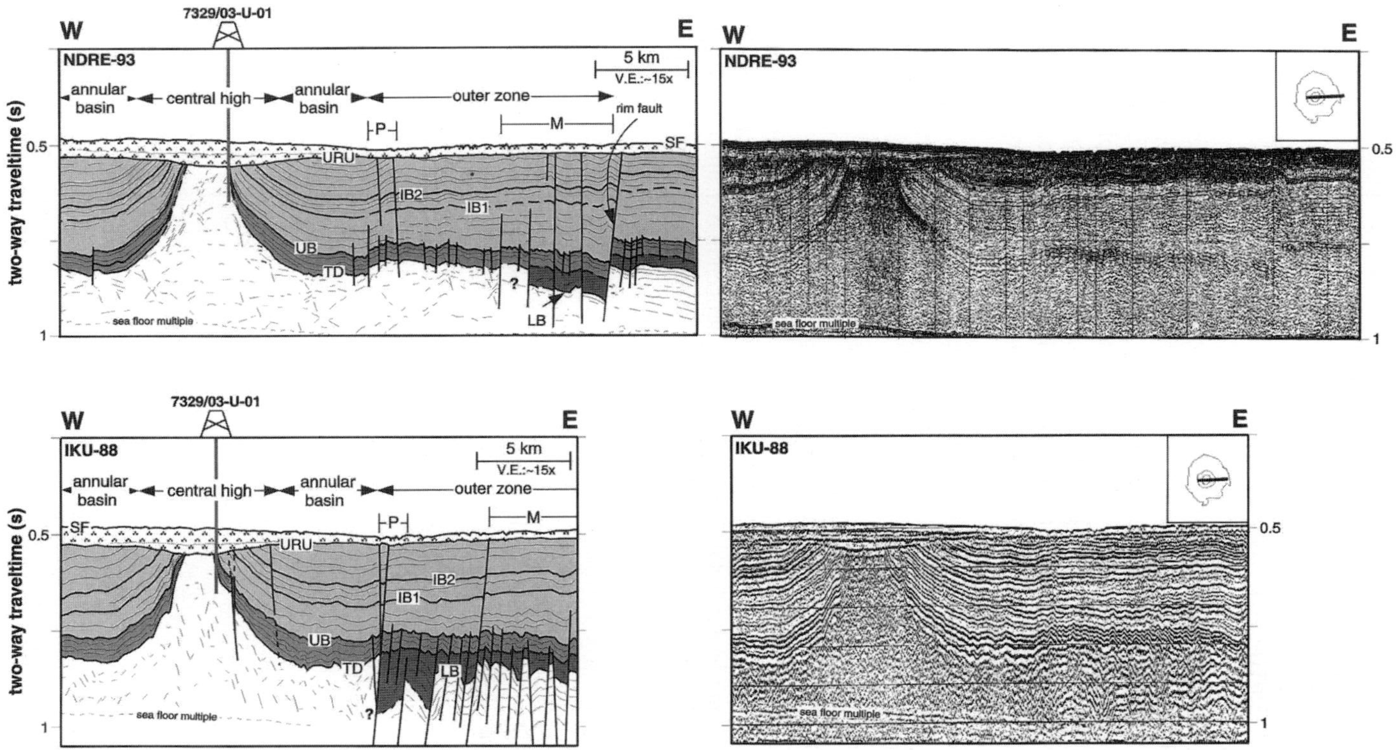

Fig. 4. Interpreted (left) and uninterpreted (right) examples of a single-channel (NDRE-93) and a shallow multichannel (IKU-88) profile crossing the center of the structure and borehole 7329/03-U-01. *P*, peak ring; *M*, marginal fault zone; *IB1/IB2*, intra-Barremian reflectors. Other annotations as in Fig. 3.

3 Seismic Correlation

3.1 Borehole 7430/10-U-01

The shallow borehole 7430/10-U-01 was drilled in 1988, prior to the interpretation of the Mjølnir Structure as an impact crater (Gudlaugsson 1993), and before acquisition of the high-resolution seismic, gravity, and magnetic profiles (Tsikalas et al. 1998b, c). The borehole is ~30 km northeast of the crater periphery (Fig. 2) and was drilled to a total depth of 67.6 m. A 57.1-m-long continuous core, ranging in age from late Kimmeridgian (Late Jurassic) to early Barremian (Early Cretaceous), was recovered from beneath 10.5 m of Quaternary sediments (Dypvik et al. 1996).

We used continuous seismic reflection profiles to correlate the impact deformation event at Mjølnir to the borehole (Fig. 5; Table 3). Information from the borehole, combined with regional sequence correlation (Worsley et al. 1988; Gabrielsen et al. 1990), show that the impact event occurred between a lower Barremian limestone unit (UB, upper boundary, ~130 Ma) and a top Callovian unconformity (LB, lower boundary, ~159 Ma). Reflector UB can be confidently traced throughout the dataset and has been used as a marker horizon to produce seismic images of the crater relief (Tsikalas et al. 1998b). Reflector LB marks the top of a thick unit of parallel reflectors, which are broken into distinct, tilted fault blocks along the crater periphery. The LB is not recognized within the crater,

Table 3. Correlation of seismic reflectors with lithological units in core 7430/10-U-01. Note that interval velocities for the seismic units are based on average stacking velocities around the borehole. TD_{min}/TD_{max} define the seismic tie bracket of impact horizon. Annotations in Fig. 3.

SR	DE (twt, s)	IV (kms^{-1})	DB (m)	CU
SF	0.456		0	
		1.80		glacial deposits
URU	0.469		12	
		2.10		post-impact succession
UB	0.491		34	
		2.45		post-impact succession
TD_{min}	0.501		46	
		2.30		proximal ejecta/platform sediments
TD_{max}	0.511		57	
		2.70		platform sediments
LB	0.540		98	

SR, seismic reflector; *DE*, depth; *twt*, two-way travel time; *IV*, interval velocity; *DB*, depth below sea floor; *CU*, core unit

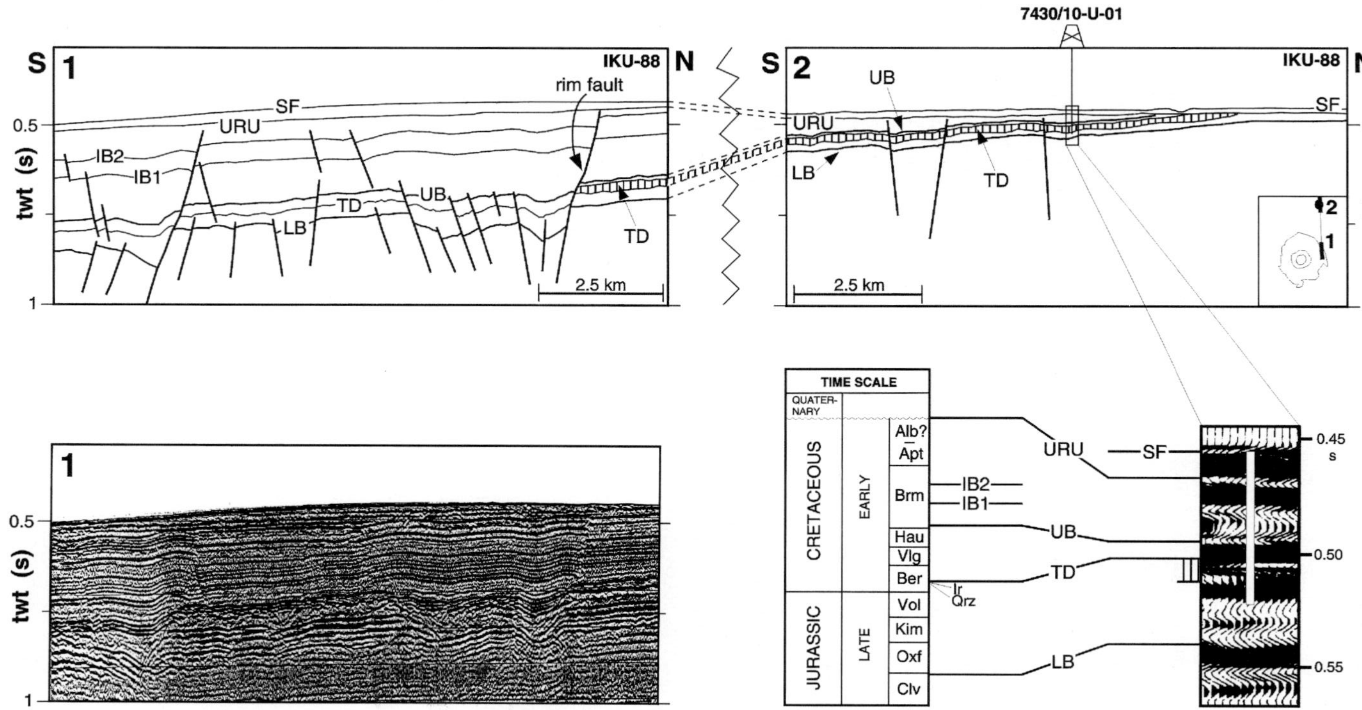

Fig. 5. Stratigraphic relationships between the main seismic units at Mjølnir and borehole 7430/10-U-01. Vertical-bars raster in profiles 1 and 2 denotes the uncertainty in the seismic tie of the impact horizon (reflector TD). Core recovery is indicated by the white bar. *Qrz/Ir*, position of shocked quartz grains and iridium peak, respectively (Dypvik et al. 1996). Other annotations as in Figs. 3-4.

where the strata reflections of the older platform change into a chaotic and diffraction-dominated seismic zone (Figs. 3 and 4).

The fault blocks are overlain by a variably thick and incoherent seismic unit LB-TD (Figs. 3 - 5) ascribed to a polymictic allochthonous breccia, which formed during impact and was covered by post-impact strata (Tsikalas et al. 1998b). Reflector TD (top disturbance) defines the impact horizon and intersects the borehole at a level where a 0.8-m-thick core section contained impact ejecta. Within the ejecta, quartz grains showed planar fractures and recrystallized planar deformational features. An iridium peak of about 1000 ppt, was recorded approximately at the same level (Fig. 5) (Dypvik and Ferrell 1998; Dypvik and Attrep 1999).

The configuration of rim faults at the crater periphery and the regional northward thinning of the Upper Jurassic-Lower Cretaceous strata (LB-UB unit) are the main sources of uncertainty in the seismic tie (Fig. 5). However, the prominent character of reflectors LB and UB together with the smooth and continuous character of both the LB-TD and TD-UB units provide a robust correlation (Fig. 5; Table 3).

3.2
Borehole 7329/03-U-01

The study of the Mjølnir crater led to a new shallow borehole, drilled in 1998, specifically to recover both impact-related and post-impact strata. The 171-m-deep borehole was drilled at the edge of the up-dipping flank of the crater central high. A 121-m-long core was recovered below 50 m of glacial cover (Fig. 6).

Table 4. Correlation of seismic reflectors with lithological units in core 7329/03-U-01. Annotations in Fig. 3.

SR	DE (twt, s)	IV (kms^{-1})	DB (m)	CU
SF	0.488		0	
		1.77		glacial deposits
URU/UB	0.545		50	
		2.51		post-impact succession
R1	0.556		57	
		2.51		post-impact succession
R2	0.562		65	
		2.51		post-impact succession
R3	0.570		75	
		2.59		current reworked material
TD	0.581		89	
		2.38		impact breccia
(borehole bottom)	0.650		171	

SR, seismic reflector; *DE*, depth; *twt*, two-way travel time; *IV*, interval velocity; *DB*, depth below sea floor; *CU*, core unit

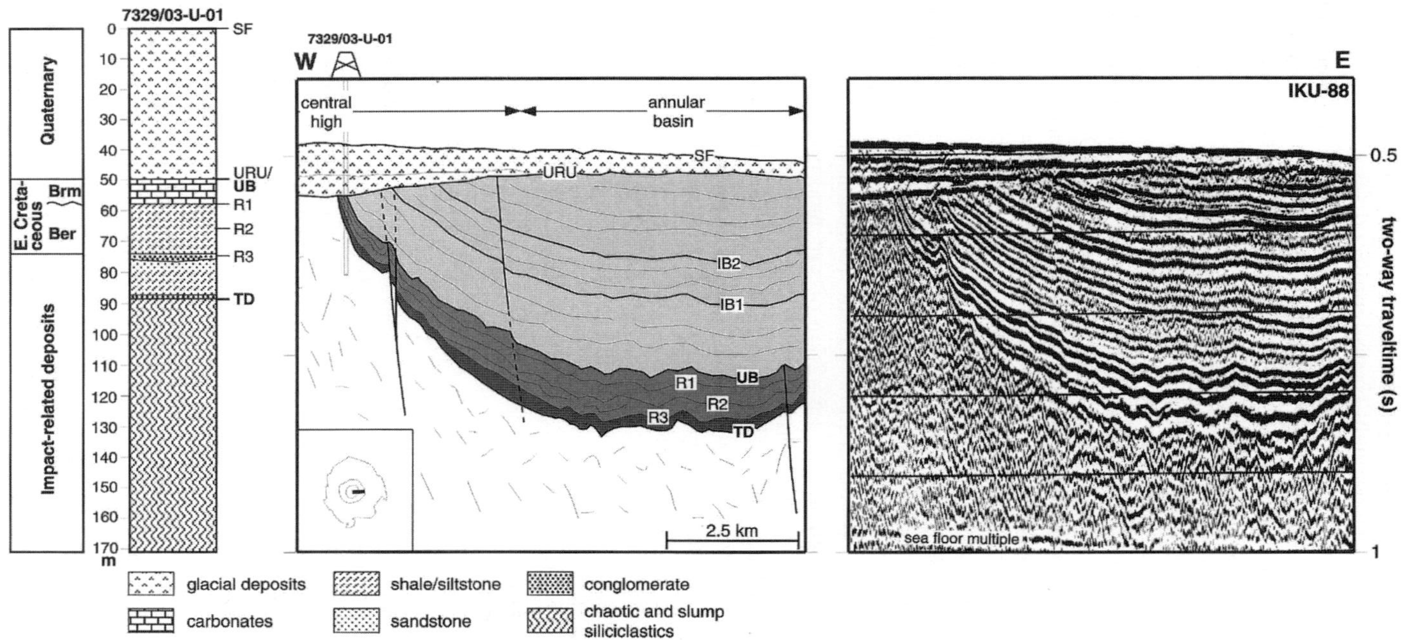

Fig. 6. Generalized lithostratigraphic log of borehole 7329/03-U-01 and interpreted shallow multi-channel seismic profile across the borehole. R1/R2/R3, continuous reflectors within the TD-UB unit at the crater center. Annotations as in Figs. 3-4.

A detailed seismic stratigraphy has been established based on correlation with the high-resolution shallow multi-channel profile crossing the crater center (Fig. 6; Table 4). The interval velocities for the seismic units in Table 4 are based on one stacking velocity analysis at the borehole location. Reflector TD represents the top of a unit of chaotic slump deposits, reflecting impact-induced deformation (Smelror et al. 2001). Away from the crater center, the reflector defines the top of the uppermost impact-related breccia unit, the allochthonous breccia (Table 2) (Tsikalas et al. 1998a). This unit exhibits minor reflectivity in places but is mostly characterized by reflection-free zones. The allochthonous breccia is expected to consist of crushed fallback, collapsed crater-wall, and washback material (e.g., Melosh 1989). A velocity inversion below reflector TD (Table 4) is possibly associated with the brecciated character of the impact strata. The allochthonous breccia is underlain by the *in situ* fractured autochthonous breccia. The autochthonous breccia comprises the tilted and internally stratified fault blocks at the periphery, but changes into disrupted and chaotic reflections, diffractions, and reflection-free zones toward the crater center. The top of the allochthonous breccia corresponds to the apparent crater floor. Theoretically the base of the allochthonous breccia defines the true crater floor (Grieve 1991) but this level is difficult to identify on the shallow seismic reflection profiles. However, MCS profiles provided estimates of 1.5 - 2 km of structural uplift at the center of the structure (Tsikalas et al. 1998a). Furthermore, gravity and seismic traveltime modelling, together with the MCS profiles demonstrate that the allochthonous breccia unit exhibits a lenticular cross-sectional shape beneath the annular basin pinching out towards the central high and towards the periphery (Table 2) (Tsikalas et al. 1998a, c; Tsikalas et al. 2002). At the borehole location, it is, therefore, not obvious whether reflector TD defines the top of the allochthonous breccia wedge or marks a mixture of allochthonous and autochthonous breccia.

Reflector TD is overlain by a unit, TD-R3 (Fig. 6), which contains sediments related to the final current activity imposed by the Mjølnir impact (Smelror et al. 2001). In a broad sense, the unit can be considered the topmost part of the allochthonous breccia because it is related to processes triggered by the impact event. It contains current-reworked, fine-grained, fallout debris. Between reflector R3 and the glacial cover the post-impact succession comprises 16 m of Berriasian shales and 8 m of early Barremian carbonates (Fig. 6). The regionally prominent reflector UB correlates with the top of the carbonate bed, which is truncated at the borehole location (as are all the cited reflectors) by the glacial URU unconformity (Fig. 6). Reflector R1 defines the base of the carbonate bed, whereas reflector R2 is an internal Berriasian reflector (Fig. 6; Table 4).

4
Impact Timing

Stratigraphic control in borehole 7430/10-U-01 indicated that the Mjølnir impact took place in the latest Jurassic to earliest Cretaceous, 141 - 149 Ma (Dypvik et al.

1996). New biostratigraphic evidence from the same borehole and, in particular, from borehole 7329/03-U-01 (Smelror et al. 2001) further constrains the age of impact to the Volgian-Ryazanian boundary, i.e., 142 ± 2.6 Ma according to the geological time scale of Gradstein et al. (1994, 1999). This age is based primarily on macrofauna recovered just at the base of the post-impact succession in the central crater core (R3 - R2 unit, Fig. 6), which represents a return to normal sedimentary conditions after the impact.

We believe that the 142 ± 2.6 Ma age should be considered as a minimum age estimate for the impact; it may have been slightly older, because the shallow reflection profiles provide evidence for crater-influenced sedimentation (Figs. 4 and 6). In particular, the depression around the central high exhibits infilling by progressive accumulation of sediments into the original annular basin. The depression is completely smoothed above reflector IB1 as indicated by the uniform thickness of the IB1-IB2 (intra-Barremian) unit when the crater-relief control on sedimentation was minor (Fig. 6) (Tsikalas et al. 1998b). More importantly, we observe onlap against the high for most of the early, pre-IB1, post-impact deposits, and especially for those immediately after-impact, between reflectors TD and UB, which were penetrated by the borehole (Fig. 6). The increase in reflection amplitude between the strata onlapping the central high also suggests lateral changes in depositional facies (Fig. 4) (Tsikalas et al. 1998b). Therefore, the lateral thickness variations of the immediately post-impact sequences, especially the R3-R2 unit (Fig. 6), suggest that the impact took place somewhat earlier than the borehole biostratigraphic age, but most likely still within the Upper Volgian stage (144.2 ± 2.6 to 142 ± 2.6).

5 Summary

From correlation of the seismic reflection profiles with boreholes 7329/03-U-01 (near the crater center) and 7430/10-U-01 (~30 km from the crater periphery), we infer:

1. A detailed seismic stratigraphy for the impact-related and post-impact strata.
2. Dating of the impact event between a prominent lower Barremian limestone unit (UB) and the base reflector of the Upper Jurassic sequence corresponding to a top Callovian unconformity (LB).
3. The impact horizon (TD) within the LB-UB unit corresponds to the top of the impact-induced disturbance within the crater and to the proximal ejecta layer outside the crater periphery.
4. In view of the considerable thickening of the immediately after-impact deposits in the depression surrounding the central high the biostratigraphic 142 ± 2.6 Ma age (Smelror et al. 2001) is considered the best minimum impact age. Thus, the actual impact event may be slightly older but most likely still within the Upper Volgian stage (144.2 ± 2.6 to 142 ± 2.6).

Acknowledgements

We gratefully acknowledge the Norwegian Defense Research Establishment, the Norwegian Petroleum Directorate, and SINTEF Petroleum Research for providing seismic reflection profiles and information from boreholes 7430/10-U-01 and 7329/03-U-01. We also thank C.W. Poag and B. Milkereit for helpful comments and suggestions as reviewers.

References

Dypvik H, Gudlaugsson ST, Tsikalas F, Attrep M Jr, Ferrell RE Jr, Krinsley DH, Mørk A, Faleide JI, Nagy J (1996) Mjølnir structure: An impact crater in the Barents Sea. Geology 24: 779-782

Dypvik H, Ferrell RE Jr (1998) Clay mineral alterations associated with a submarine meteorite impact. Clay Minerals 33: 51-64

Dypvik H, Attrep M Jr (1999) Geochemical signals of the late Jurassic, marine Mjølnir impact. Meteoritics & Planetary Science 34: 393-406

Faleide JI, Vågnes E, Gudlaugsson ST (1993) Late Mesozoic-Cenozoic evolution of the south-western Barents Sea in a regional rift-shear tectonic setting. Marine and Petroleum Geology 10: 186-214

Gabrielsen RH, Færseth RB, Jensen LN, Kalheim JE, Riis F (1990) Structural elements of the Norwegian continental shelf. Part I: The Barents Sea region. Norwegian Petroleum Directorate Bulletin No. 6, 33 pp

Gradstein FM, Agterberg FP, Ogg JG, Hardenbol J, van Veen P, Thierry J, Huang Z (1994) A Mesozoic time scale. Journal of Geophysical Research 99: 24051-24074

Gradstein FM, Agterberg FP, Ogg JG, Hardenbol J, Bäckström S (1999) On the Cretaceous time scale. Neues Jahrbuch für Geologie und Paläontologie, Abhandlungen 212: 3-14

Grieve RAF (1991) Terrestrial impact: The record in the rocks. Meteoritics 26: 175-194

Grieve RAF, Pesonen LJ (1996) Terrestrial impact craters: their spatial and temporal distribution and impacting bodies. Earth, Moon, and Planets 72: 357-376

Gudlaugsson ST (1993) Large impact crater in the Barents Sea. Geology 21: 291-294

Gudlaugsson ST, Faleide JI, Johansen SE, Breivik AJ (1998) Late Palaeozoic structural development of the south-western Barents Sea. Marine and Petroleum Geology 15: 73-102

Lawver LA, Gahagan LM, Campbell DA, Brozena JM, Childers V (1999) Mid-Jurassic to recent tectonic evolution of the Arctic region (*Powerpoint* animation, using the *PLATES'* animation software) [abs]. In: Lawver LA, Bronzena JM, Kovacs LC, Dhilders V (eds) Compilations in the Canada Basin, Aerogeophysical Anomalies. EOS, Transactions, American Geophysical Union, Fall Meeting 1999, San Francisco, 80 (46): 1000

Melosh HJ (1989) Impact Cratering - A Geologic Process. Oxford University Press, New York, 245 pp

Smelror M, Kelly SRA, Dypvik H, Mørk A, Nagy J, Tsikalas F (2001) Mjølnir (Barents Sea) meteorite impact ejecta offers a Volgian-Ryazanian boundary marker. Newsletter on Stratigraphy 38: 129-140

Tsikalas F, Gudlaugsson ST, Faleide JI (1998a) The anatomy of a buried complex impact structure: the Mjølnir Structure, Barents Sea. Journal of Geophysical Research 103: 30,469-30,484

Tsikalas F, Gudlaugsson ST, Faleide JI (1998b) Collapse, infilling, and post-impact deformation at the Mjølnir impact structure, Barents Sea. Geological Society of America Bulletin 110: 537-552

Tsikalas F, Gudlaugsson ST, Eldholm O, Faleide, JI (1998c) Integrated geophysical analysis supporting the impact origin of the Mjølnir Structure, Barents Sea. Tectonophysics 289: 257-280

Tsikalas F, Gudlaugsson ST, Faleide JI, Eldholm O (1999) Mjølnir Structure, Barents Sea: a marine impact crater laboratory. In: Dressler BO, Sharpton VL (eds) Large Meteorite Impacts and Planetary Evolution II. Geological Society of America Special Paper 339: 193-204

Tsikalas F, Gudlaugsson ST, Faleide JI, Eldholm O (2002) The Mjølnir marine impact crater porosity anomaly. In: Gersonde R, Deutsch A, Kyte F (eds) Oceanic impacts - Mechanisms and Environmental Consequences. Deep Sea Research II, in press

Worsley D, Johansen R, Kristensen SE (1988) The Mesozoic and Cenozoic succession of Tromsøflaket. In: Dalland A, Worsley D, Ofstad K (eds) A lithostratigraphic scheme for the Mesozoic and Cenozoic succession offshore mid- and northern Norway. Norwegian Petroleum Directorate Bulletin No. 4, 42-65

Numerical Modeling of Impacts into Shallow Sea

Valery V. Shuvalov

Institute for Dynamics of Geospheres, Leninsky Prospect 38-6, 117979 Moscow, Russia. (valery@vshuvalov.mccme.ru)

Abstract. The SOVA multi-material hydrocode is used to model impacts of 200 - 300-m-diameter cosmic bodies (both asteroids and comets) into an ocean with depths ranging from 100 to 500 m. The size of the transient cavity (in the crystalline basement, sediment layer, and water), as well as the propagation of tsunami-like waves and the resurge flow, are calculated. The maximum ratio of diameters of the transient cavity in water to that of the central depression in the basement is found to be approximately 2. It is also shown that 200 - 300 m impactors can be disrupted, deformed and even decelerated during the flight through the Earth's atmosphere. These effects and their influence upon cratering are also investigated.

1 Introduction

Most cosmic bodies impacting the Earth hit the sea, but modern geological data reveal an appreciable deficiency of underwater craters. At least two reasons are evident: the water column weakens impact cratering, and dynamic plate-tectonic movements destroy even giant craters. It is a well-known fact that the oceanic crust is appreciably younger than the continental. Due to subduction the oldest oceanic plate material is only 150 - 180 Ma old. Modern geological and geophysical investigations, including deep sea drilling, may reveal impact affected materials buried under the seafloor.

Recently several marine-target craters (hydroblems) in the shelf sea have been found and studied in detail (Ormö and Lindström 2000 and references therein). Most of these craters were formed in shallow water, where the impactor size d probably exceeded the sea depth H. The size of known hydroblems varies from 2 - 3 km for Tvären (Lindström et al. 1994) and Granby (Bruun and Dahlman 1982) to 40 km for Mjølnir (Dypvik et al. 1996), up to 90 km for Chesapeake Bay (Poag 1997). Eltanin (Gersonde et al. 1997) is the only documented deep-ocean (~4 km) impact, although there are indications of other similar events (Simonson et al. 1998).

The process of formation and the characteristics of marine-target craters are still poorly known. It is because craters covered by the sea are less accessible for geological investigation, and our knowledge about impact processes is based mainly on the study of impacts on land.

The first numerical simulations of meteoroid impacts into the ocean have been done for very large (about 10 km) impactors, which are interesting from the viewpoint of terrestrial life extinctions (O'Keefe and Ahrens 1982a; Roddy et al. 1987). Laboratory experiments (Gault and Sonett 1982) and detailed numerical simulations (Artemieva and Shuvalov 1999) have allowed to determine the critical sea depth at which a crater is formed and/or shock modified material can be found to identify the impact site. The influence of shallow water (and water depth) upon the crater morphology is less investigated.

Impacts on the seabed are typical examples of impacts against a laminated (layered) target. At least three target layers can be distinguishcd: low density water, low strength sediments, and high-strength crystalline basement. The qualitative description of impacts against the laminated target was given by experiments of Quaide and Oberbeck (1968). In this study the layer of relatively high-strength material was covered by strengthless sand. Different morphology impact structures formed depending on the impactor diameter to sand depth ratio. Ormö and Lindström (2000) referred to these results in an attempt to explain the concentricity noticed in field data from some of the marine-target craters.

In shallow water relative to the size of the impact the resulting crater is similar to a land-target crater. In relatively deeper water a small crater is formed in the crystalline basement and a second (outer) larger crater arises in sediments (Ormö and Lindström 2000). The size of the outer crater is considered to coincide with the size of the transient water. The ratio of diameters of the outer crater and central depression depends on the relative (to the impactor size) water depth. This qualitative description of the impact cratering at the sea was partially justified by experiments of Gault and Sonett (1982) and McKinnon (1982). However, these results were obtained for impactor velocities of 2 to 7 kms^{-1}, which are considerably smaller than typical planetary impact velocities. Moreover, the scale factor can significantly change the results.

The purpose of the present paper is to study impacts into shallow water using detailed numerical simulations. The SOVA multi-material hydrocode (Shuvalov 1999) is used to model the impacts of cosmic bodies (both asteroids and comets), 200 - 300 m in diameter, into a sea with depth ranging from 100 to 500 m. SOVA is an Eulerian material response code with some Lagrangian features. It allows to model strong hydrodynamic flows with accurate description of the boundaries between different materials (e.g., rock, gas, water, etc.). The code is conceptually similar in to the CTH hydrocode (McGlaun et al. 1990), which is widely used in the USA. ANEOS (Thompson and Lauson 1972) and Tillotson (Tillotson 1962) equations of state are used to describe thermodynamical properties of water and solids.

The simplest assumption is used to determine the size of the transient cavity in the crystalline basement. The cratering flow in the basement is artificially stopped when its velocity decreases to 200 ms^{-1}, which corresponds to a material crushing strength of about 10^5 Pa (Melosh 1989). This assumption slightly underestimates the size of the transient cavity.

2 Impacts into Water Layer Overlaying Crystalline Basement

In the first set of numerical simulations the size of the transient cavity in the crystalline basement and overlaying water is determined for different impactor diameter to sea depth ratios. The computational grid consists of 300 × 500 cells in vertical and horizontal directions respectively. The space step is 10 m (10 - 50 cells across sea depth) for the central 100 × 200 cell region around the point of impact, and increases progressively moving away from the central region. The grid boundaries are approximately 10 km away from the impact point.

Figure 1 represents density distributions 5 s after the vertical impact of a stony body 200 m in diameter. Impact velocity is 15 kms^{-1}, water depth varies from 100 m to 500 m. For comparison the results for deep water are also presented. The flow resulting from the impact strongly depends on the water depth. For the depth of 100 m and 200 m the sea only slightly influences the formation of the

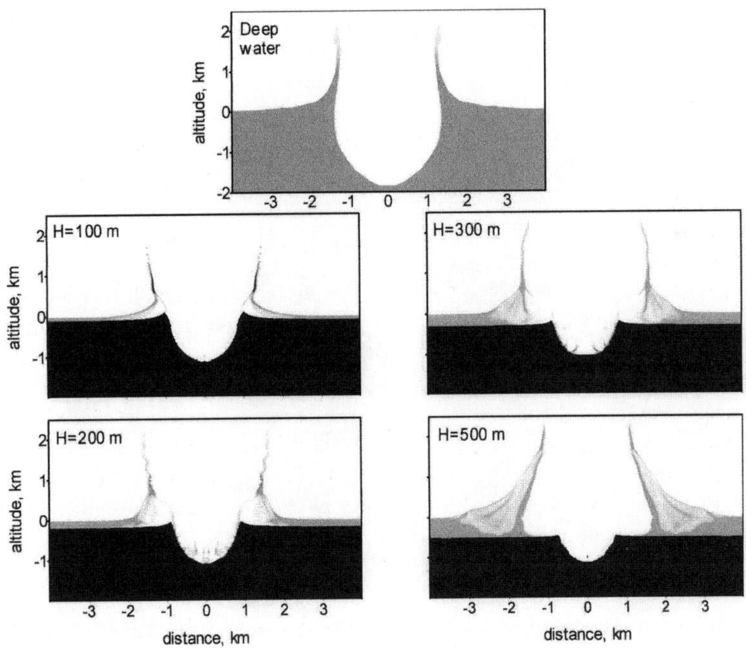

Fig. 1. Density distributions (darker = denser) 5 s after the vertical impact of a stony body 200 m in diameter for various sea depths H. Impact velocity is 15 kms^{-1}.

cavity in the basement. The motion of water is defined by a combination of the initial shock wave (created by the penetration of the meteoroid into the sea), the reflected shock wave (originated from the water-soil boundary) and the seismic waves propagating through the ground. The latter causes perturbations (rarefaction) of the water layer outside the water crater. In the case of water depth of 300 m and 500 m a well-defined water cavity is formed. Moreover its size proves to be greater than the size of the transient cavity in deep water. This increase is explained by the reflection of the shock waves and the flow itself from the water-soil boundary. In deep water the cavity has reached the maximum depth at 5 seconds, but its radius continues to grow.

Figure 2 shows the same distributions at 10 s. A well defined water cavity is clearly seen in all cases, its size increases with the sea depth H. For depth exceeding 100 m, the size of water cavity is larger than that for the case of deep water.

Figure 3 shows the crater shapes 15 s after impact. For water depth of 100 and 200 m a resurge flow starts at this moment, and the water cavity begins to decrease.

Twenty seconds after the impact (Fig. 4) the resurge flow is evident for all cases. At the same time the formation of tsunami-like waves begins. The maximum size of the transient water cavity proves to be comparable to that of the water crater in deep water, although the flow itself is not the same.

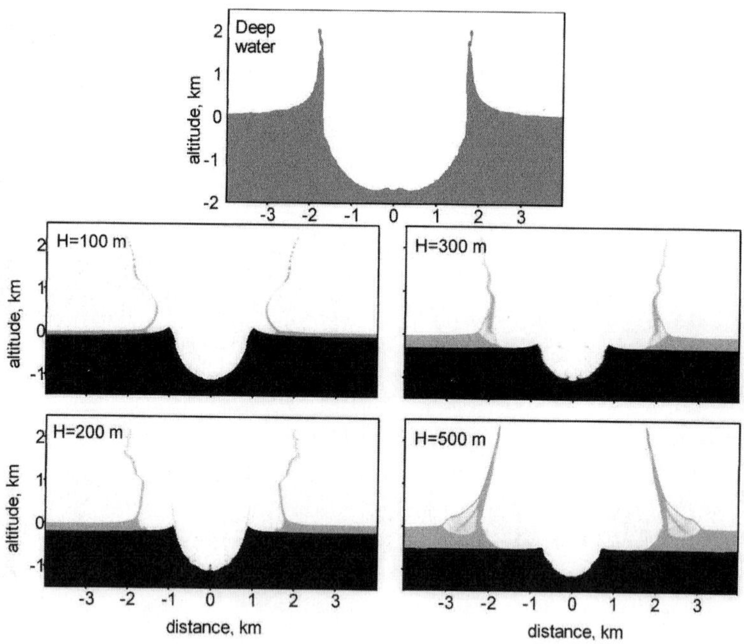

Fig. 2. Density distributions (darker = denser) 10 s after the vertical impact of a stony body 200 m in diameter for various sea depths H. Impact velocity is 15 kms^{-1}.

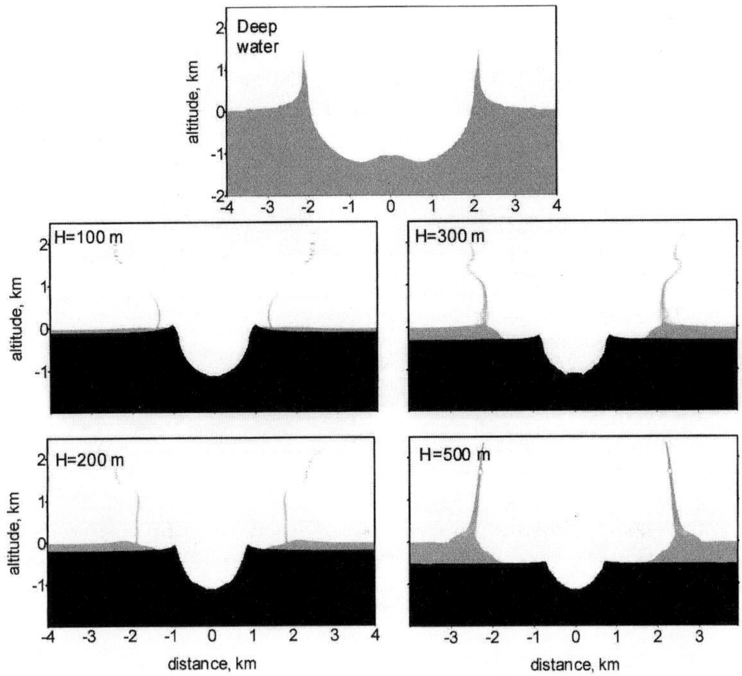

Fig. 3. Density distributions (darker = denser) 15 s after the vertical impact of a stony body 200 m in diameter for various sea depths *H*. Impact velocity is 15 kms^{-1}.

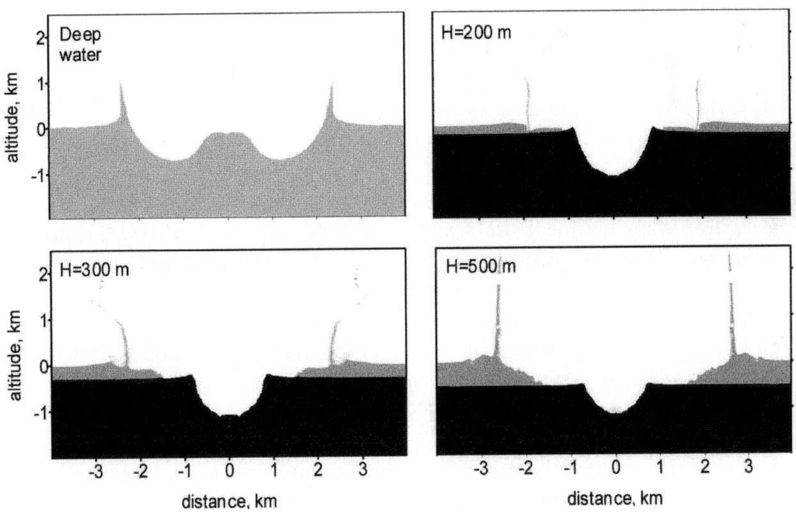

Fig. 4. Density distributions (darker = denser) 20 s after the vertical impact of a stony body 200 m in diameter for various sea depths *H*. Impact velocity is 15 kms^{-1}.

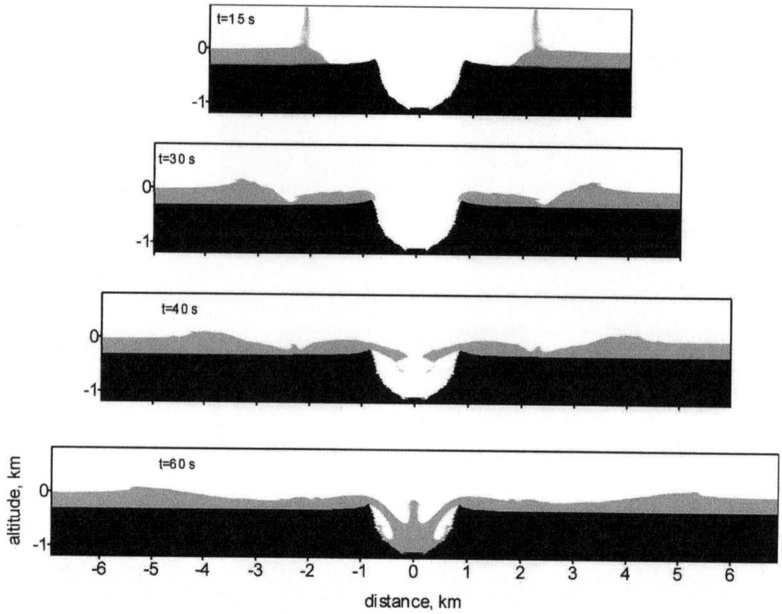

Fig. 5. Tsunami-like waves generated by the impact of a stony body 200 m in diameter into a sea of 300 m depth at different times (t) after the contact.

The numerical simulations suggest that the maximum ratio of diameters of the water transient cavity and central depression in the basement does not exceed a factor of 2.5 if the water depth is less than three impactor diameters. This value seems to be an upper estimate, because the final size of the central depression is slightly bigger due to crater modification, which is not taken into account in the present simulations. In the case of very shallow water (H = 100 m) the transient water cavity proves to be no bigger than the final (after modification) size of the basement depression.

The formation of tsunami-like waves for one of the runs (H = 300 m) is shown in Fig. 5. The wavelength of the waves is about 1 km, with an initial amplitude around 100 m. The time needed to fill the cavity by resurge is about 60 - 70 s, whereas the maximum velocity of the flow is ~65 ms^{-1}. Because the basement crater modification is not included into the models (a natural crater is more sloping), the process of filling by water could have been modelled somewhat inaccurately. A rough estimate of the resurge flow velocity can be made by an approach of collapsing fluid column (Nichols and Hirt 1971), giving a value of 55 ms^{-1} for a 300 m water layer. However, the real height, and, therefore, speed, of the resurge water column is larger than the original pre-impact thickness due to water surge forming near the water crater boundary. Moreover, the axial symmetry also leads to an increase of the velocity (so called cumulation effect) as well.

3
Impact Against a Three-layer Target

In reality it is not obvious that the size of outer crater in sediments coincides with the size of transient water cavity, because these substances have considerably different densities. To clarify this question the direct numerical simulations of an impact against a three-layer target were performed. A 100 m layer of strengthless sediments was added between a 300 m water column and crystalline basement. For simplicity the strength of sediments is assumed to be negligible and the density is assumed to be the same as that of the crystalline material. The results are shown in Fig. 6. For a strengthless sedimentary layer the size of the outer crater proves to be approximately the same as the size of the water transient cavity. Neglecting the density difference between water and sediments one can consider the target as two-layered, consisting of layers with (crystalline basement) and without (water + sediments) strength, whereas the size of the outer crater is defined by the transient cavity in the strengthless layer. This means that the results described in the previous section can be used for a rough (upper) estimate of the size of possible outer crater in sediments.

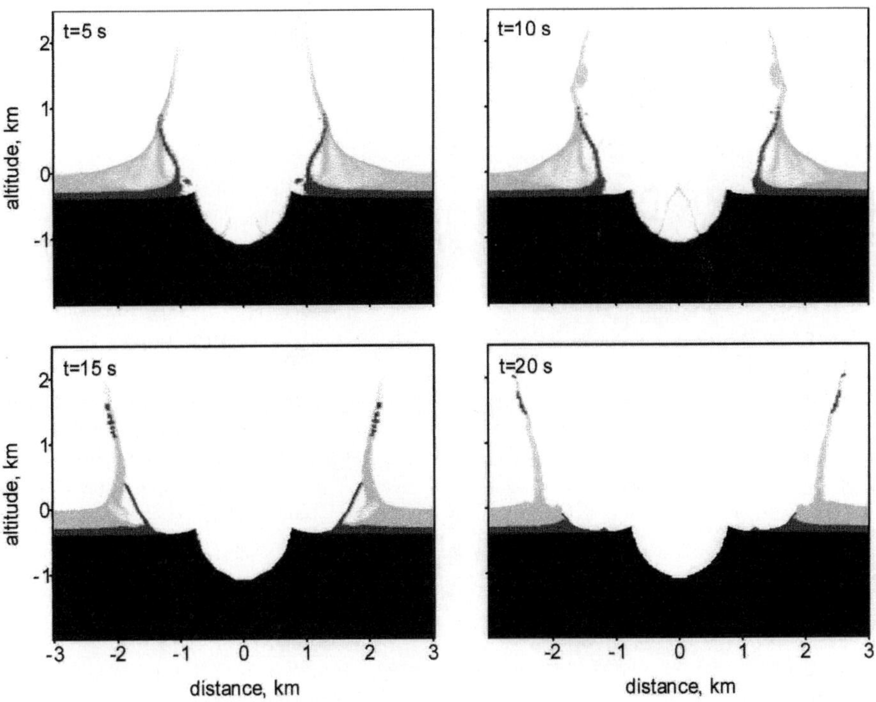

Fig. 6. The evolution of the cavity for an impact of a 200 m asteroid into a three-layered target (300 m water and 100 m sediments overlying crystalline basement) at different times (t) after the contact.

However, to obtain the final crater topography and morphology one should consider the modification stage in detail and include a strength model both for the basement material and the sediments.

4 Impacts of Disrupted Projectiles

As is shown above, only rather small impactors (about 200 - 300 m in diameter) can produce concentric craters in shallow (~100 - 200 m) sea, whereas larger projectiles (with larger impactor size to water depth ratio) form usual (nonconcentric) craters. However, impactors can be deformed, fragmented and even decelerated during the flight through the Earth's atmosphere. This effect, in turn, can influence the cratering process. To study this influence, the passage of small comets and asteroids through the atmosphere has been modeled together with the cratering process. The simulation of the deformation, fragmentation, and deceleration of the meteoroid due to aerodynamic loading is similar to that used

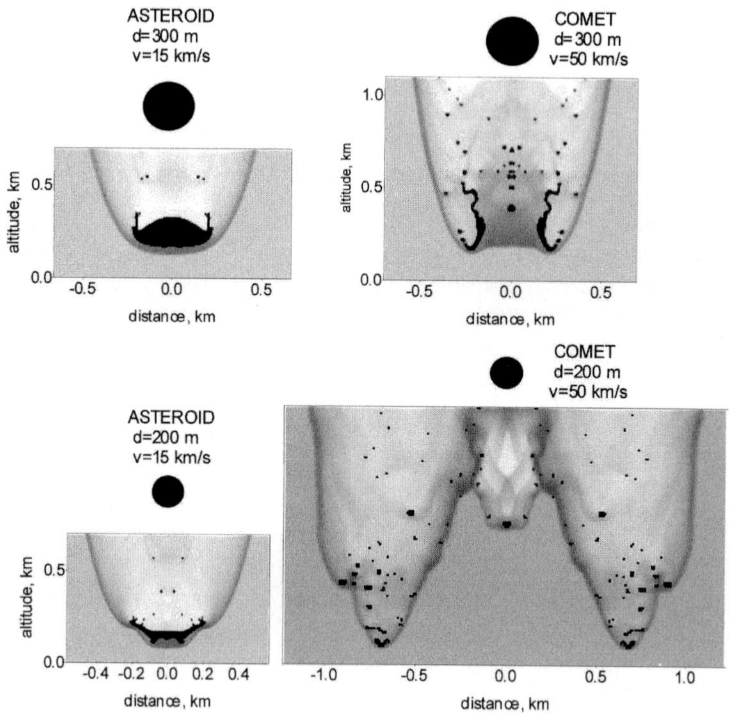

Fig. 7. Density distributions (darker = denser) and projectile shapes just before impact. *d*, diameter; *v*, pre-entry velocity.

for modeling the interaction of comet Shoemaker-Levy 9 with Jovian atmosphere (Shuvalov et al. 1999). Numerical simulations start at an altitude of 40 km.

Figure 7 represents the condition of different meteoroids just before striking the sea. The pre-entry velocity of 200 and 300-m-diameter spherical asteroids is considered to be 15 kms^{-1}, and comets of the same size and shape are assumed to have a pre-entry velocity of 50 kms^{-1}. The 300 m asteroid is slightly deformed. The smaller stony impactor (200 m) is strongly deformed, appearing like a typical pancake (Chyba et al. 1993), as its cross-section exceeds that of the parent body by an order of magnitude. However, both asteroids are practically not decelerated. Comets are more subjected to atmospheric action. Even a 300 m icy body is fragmented, although only slightly decelerated (from 50 to 45 kms^{-1}). The 200 m comet is totally disintegrated and transformed into a debris jet, and its velocity is ~40 % lower of the pre-entry velocity.

It should be noted that all these results concern vertical impacts. An oblique impact, which is more probable, would intensify the deformation and fragmentation.

No analyses about the fragmentation level, critical for the cratering process, exist. A rough estimate can be done assuming that fragmentation is not important if the effective diameter d_{ef} of the disrupted meteoroid or debris jet is smaller than the size (diameter) D_{cr} of the transient cavity produced by the nondisrupted parent meteoroid:

$$d_{ef} < D_{cr} \tag{1}$$

However, this condition would not suffice, at least, for the case of highly fragmented impactors. For example, the fivefold increase of the effective impactor size is equivalent to the decrease of its effective density by two orders of magnitude. However, low density impactors are known to produce small shallow craters, the depth and diameter of which are considerably smaller than those of the crater resulting from the impact of the dense meteoroid with the same energy (O'Keefe and Ahrens 1982b).

The drop in cratering efficiency induced by a highly fragmented meteoroid results from the deceleration of each successive fragment by the expanding ejecta generated by the impacts of the previous fragments. In some way this process is similar to the propagation of a reflected shock wave in the case of impact of a homogeneous low density meteoroid. This gives one more constraint:

$$\frac{d_{ef}}{V} < \tau_f, \tag{2}$$

where V is the meteoroid velocity and τ_f is the effective time for ejecta formation due to the impact of a single fragment. In other words, the crater size is different for a sequence of time-separated small impacts (induced by each fragment) and for a simultaneous large impact. The value τ_f can be estimated as

$$\tau_f = \alpha \frac{d_f}{V}, \tag{3}$$

where d_f is the size of a single fragment. The constant α can be determined, for example, from numerical simulations of the ordinary impact, which give the value $10 < \alpha < 20$. Using this value it follows from Eqs. (2 - 3) that

$$d_f < (0.05 \div 0.1) \cdot d_{ef}. \qquad (4)$$

From Eqs. 1 and 4 one can conclude that only the impact of a 200-m-diameter comet would produce a crater that differs significantly from its counterpart generated by the nondisrupted parent (i.e., original) body. These speculations have been verified by numerical simulations of the formation of the transient cavity resulting from the impact of meteoroids presented in Fig. 7.

Figure 8 shows the evolution of the enlarging cavity after the impact of the 300 m comet both for the fragmented impactor and nondisturbed parent body. There is a great difference between the two runs in the initial stage of the cratering process. Several small craters are formed by individual fragments in the case of the disrupted projectile. However, approximately 0.4 s after the impact both craters have almost the same shape.

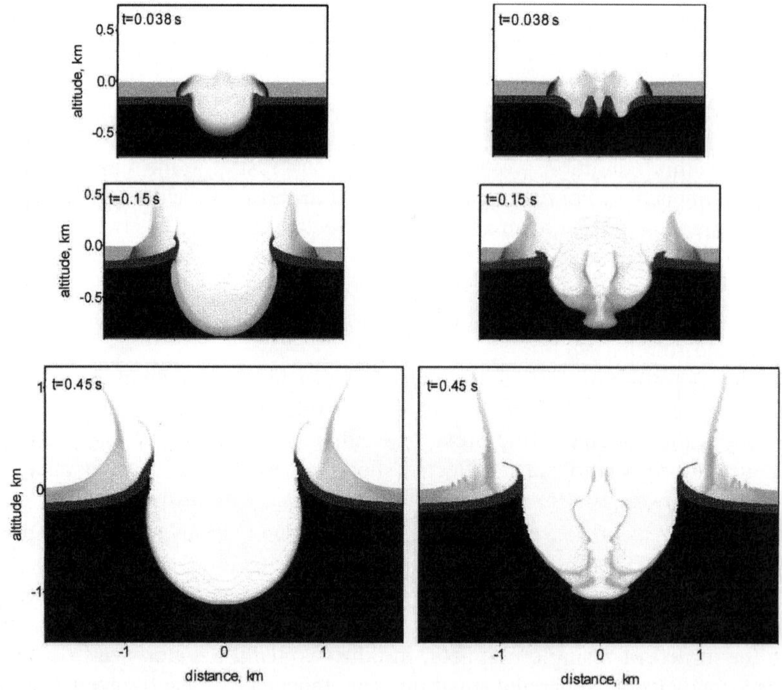

Fig. 8. Water density distributions (darker = denser) and crater shapes after the impact of a comet 300 m in diameter. Impact velocity is 50 kms^{-1}. Left panels correspond to the impact of a parent nonfragmented body, right panels show the cavity evolution for an impactor fragmented while passing through the atmosphere.

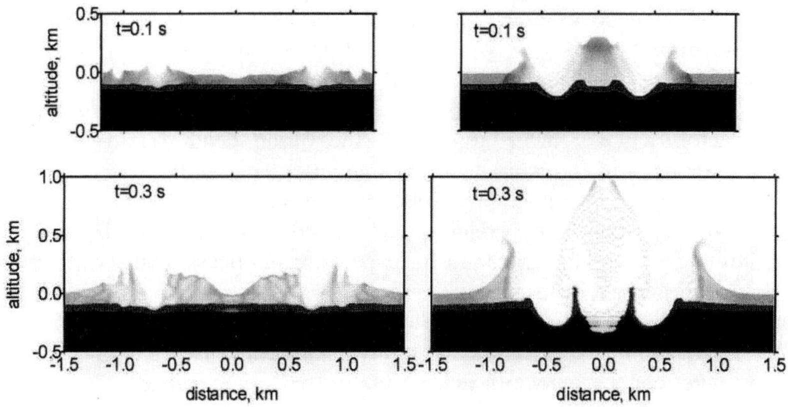

Fig. 9. Water density distributions (darker = denser) and crater shapes at different times (t) after the impact of 200-m-diameter (left) and 240-m-diameter (right) comets.

Similarly, the deformation of 200 m and 300 m asteroids is found to have no influence on the final crater morphology. However, this is not the case for small comets. Figure 9 represents the initial stage of cratering resulting from the impact of originally 200 and 240 m wide fragmented comets. In the first case no well-defined crater is formed, although the nonfragmented parent icy body would produce approximately the same crater as the 200 m asteroid does. The 240 m fragmented comet produces a smaller depression in the crystalline basement and approximately the same (in width and shape) water transient cavity as the nonfragmented parent body.

5 Concluding Remarks

The processes accompanying impacts into the ocean strongly depend on the sea depth. If water depth exceeds two impactor diameters, the projectile is totally disrupted and decelerated as it passes through the water column, and only a small depression in the sea bottom is formed under the action of shock compressed and downwards accelerated water. No crater forms if the sea depth exceeds four impactor diameters. In such a case only a small amount of shock modified target material is produced and ejected from the crater (see Artemieva and Shuvalov 2000). The following transport and redeposition of the ejecta is controlled both by the air-water-vapor-debris plume evolution, water flow and gravity. Craters formed in deep water probably lack melt sheets and rim walls but have deposits and radial gullies formed by the resurge of the sea (Ormö and Lindström 2000). The resurge flow and tsunami-like waves are important mechanisms of sea bottom erosion. Sediment deposits and radial gullies serve as important indicators of marine-target impacts.

Numerical results presented above show that the water column very slightly influences the formation of the cavity in the crystalline basement and/or sediments if the impactor diameter d is larger than the ocean depth H. In the range $d < H < 4d$ a crater still forms in the crystalline basement, but the water layer can considerably influence its topography and morphology.

The formation of concentric craters is not directly related to the depth of the water column. Such craters appear to form because of special underwater target lithology (layered targets). In particular, such craters are created in low strength sediments or regolith layer overlying the crystalline basement. The size of the transient cavity in the strengthless water layer can be considered as an upper estimate of outer crater diameter in sediments.

Numerical simulations show that the size of the water transient cavity depends on the sea depth. In the range of $d < H < 2.5d$ the radius of water transient cavity increases with the sea depth H, and proves to be larger than the counterpart in deep water. The maximum ratio of diameters of the water transient cavity and central depression in the basement is approximately 2 for $H < 2.5d$.

The data presented in Ormö and Lindström (2000) reveal that most discovered "hydroblems" (except Eltanin) formed in very shallow water where $H < d$. In those cases the sea only slightly influenced the formation of transient cavity, controlled mainly by the strength of solid target layers. However, the physical properties of target itself are different on the land and sea bottom. As a result, craters on the seafloor can differ from those on land. Moreover, the presence of the water layer becomes important in the stage of crater modification (collapse). Resurge flow and tsunami-like waves can influence the collapse and cause additional erosion of the seabed, both inside and outside the crater (Sturkell 1998). Perturbed and moving water should also influence the redeposition of ejecta (Sturkell et al. 2000). The process of crater collapse after impacts on land is qualitatively known, however, a sophisticated quantitative model (describing all scales) has not yet been developed (Melosh and Ivanov 1999). The presence of a water layer adds some complexity.

All icy and stony impactors with diameter of <300 m are deformed and/or even fragmented and decelerated while passing through the Earth's atmosphere. Therefore, most simple craters (with a diameter of <3 - 4 km) have been created by deformed and/or fragmented projectiles, except for those formed by iron projectiles. Fragmentation influences the cratering process, especially in the case of cometary and oblique impacts, because the critical projectile size for fragmentation is known to be inversely proportional to the cosine of the trajectory angle. Some specific features of the crater shape can probably be explained by impactor fragmentation.

Acknowledgements

This study was partially supported by the Russian Fund of Basic Research (grant #00-05-81152 Bel2000_a). The author would like to thank Irina Trubetskaya for assistance with preparing this paper; also to thank the reviewers Dr. Elisabetta Pierazzo, Dr. Jens Ormö, and Dr. Jüri Plado for valuable remarks.

References

Artemieva NA, Shuvalov VV (1999) Shock zones on the ocean floor – numerical simulations [abs]. In: Gersonde R, Deutsch A, Milkereit B, Smit J (eds) Abstracts ESF Workshop: Oceanic impacts mechanisms and environmental perturbations. Bremerhaven, Germany: pp 14-17

Artemieva NA, Shuvalov VV (2000) Shock metamorphic effects caused by impacts into water basin [abs]. In: Catastrophic events and mass extinctions: impacts and beyond. Lunar and Planetary Institute Contribution No. 1053: p 27

Bruun A, Dahlman B (1982) Den paleozoiska berggrunden. Sveriges Geologiska Undersokning Af 120: 76-109

Dypvik H, Gudlaugsson ST, Tsikalas F, Attrep M Jr, Ferrell R Jr, Krinsley DH, Mork A, Faleide JI, Nagy J (1996) Mjølnir structure: an impact crater in the Barents Sea. Geology 24: 779-782

Gault DE, Sonett CP (1982) Laboratory simulations of pelagic asteroidal impact: atmospheric ejection, benthic topography, and the surface wave radiation field. In: Silver LT, Schultz PH (eds) Geological Implications of Impacts of Large Asteroids and Comets on the Earth. Geological Society of America Special Paper 190: 69-92

Gersonde R, Kyte FT, Bleil U, Diekmann B, Flores JA, Gohl K, Grahl G, Hagen R, Kuhn K, Sierro FJ, Völker D, Abelmann A, Bostwick JA (1997) Geological record and reconstruction of the late Pliocene impact of the Eltanin asteroid in the Southern Ocean. Nature 390: 357-363

Lindström M, Flodén T, Grahn Y, Kathol B (1994) Post-impact deposits in Tvären, a marine Ordovician crater south of Stockholm, Sweden. Geological Magazine 131: 91-103

McGlaun JM, Thompson SL, Elrick MG (1990) CTH: a three-dimensional shock wave physics code. International Journal of Impact Engineering 10: 351-360

McKinnon WB (1982) Impact into the Earth's ocean floor: preliminary experiments, a planetary model, and possibilities for detection. In: Silver LT, Schultz PH (eds) Geological Implications of Impacts of Large Asteroids and Comets on the Earth. Geological Society of America Special Paper 190: 129-142

Melosh HJ (1989) Impact Cratering: A Geologic Process. Oxford University Press, New York, 245 pp

Melosh HJ, Ivanov BA (1999) Impact crater collapse. Annual Reviews of Earth and Planetary Science 27: 385-415

Nemtchinov IV, Popov SP, Teterev AV (1994) Estimates of the characteristics of waves and tsunami produced by asteroids and comets falling into oceans and seas. Solar System Research 28: 81–99

Nichols BD, Hirt CW (1971) Improved free surface boundary conditions for numerical incompressible-flow calculation. Journal of Computational Physics 8: 434-448

O'Keefe JD, Ahrens TJ (1982a) The interaction of the Cretaceous/Tertiary extinction bolide with atmosphere, ocean, and solid earth. In: Silver LT, Schultz PH (eds) Geological Implications of Impacts of Large Asteroids and Comets on the Earth. Geological Society of America Special Paper 190: 103-119

O'Keefe JD, Ahrens TJ (1982b) Cometary and meteorite swarm impact on planetary surface. Journal of Geophysical Research 87: 6668-6680

Ormö J, Lindström M (2000) When a cosmic impact strikes the sea bed. Geological Magazine 137: 67-80

Poag CW (1997) The Chesapeake Bay bolide impact: a convulsive event in Atlantic coastal plain evolution. Sedimentary Geology 108: 45-90

Quaide WL, Oberbeck VR (1968) Thickness determinations of the lunar surface layer from lunar impact craters. Journal of Geophysical Research 73: 5247-5270

Roddy DJ, Schuster SH, Rosenblatt M, Grant LB, Hassig PJ, Kreyenhagen KN (1987) Computer simulations of large asteroid impacts into oceanic and continental sites - preliminary results on atmospheric, cratering and ejecta dynamics. International Journal of Impact Engineering 5: 525-541

Shuvalov VV (1999) 3D hydrodynamic code SOVA for interfacial flows, application to thermal layer effect. Shock Waves 9: 381-390

Shuvalov VV (2000) Numerical modeling of the impacts into shallow sea [abs]. In: Plado J, Pesonen LJ (eds) Abstracts of ESF Workshop: Meteorite Impacts in Precambrian Shields. Geological Survey of Finland and University of Helsinki, Espoo, p 57

Shuvalov VV, Artemieva NA, Kosarev IB (1999) 3D hydrodynamic code SOVA for multimaterial flows, application to Shoemaker-Levy 9 comet impact problem. International Journal of Impact Engineering 23: 847-858

Simonson BM, Davies D, Wallace M, Reeves S, Haller BM (1998) Iridium anomaly but no shocked quartz from Late Archean microkrystite layer: oceanic impact ejecta? Geology 26: 195-198

Sturkell EFF (1998) Resurge morphology of the marine Lockne impact crater, Jämtland, central Sweden. Geological Magazine 135: 121-127

Sturkell E, Ormö J, Nõlvak J, Wallin Å (2000) Distant ejecta from the Lockne marine-target impact crater, Sweden. Meteoritics & Planetary Science 35: 929-936

Thompson SL, Lauson HS (1972) Improvements in the Chart D radiation-hydrodynamic CODE III: Revised analytic equations of state. Sandia National Laboratory Report SC-RR-71 0714, Albuquerque, 119 pp

Tillotson JH (1962) Metallic equations of state for hypervelocity impact. General Atomic Report GA-3216, 137 pp

75^{00} #115317
Geol. Garage LOFT
"Plado"